Genetically Engineered Organisms in Bioremediation

Genetically Engineered Organisms in Bioremediation provides comprehensive coverage of biotechnological applications of genetically engineered microorganisms for the bioremediation of polluted environments.

Chapters are contributed by international scientists with in-depth knowledge, expertise, vision and commitment in their scientific profession. They detail several genetically engineered microorganisms and their enzymes that could be applied to biologically break down persistent organic pollutants and recombinant DNA technologies which entail development of "suicidal-GEMs" for effective and safe remediation of heavily polluted sites.

FEATURES:

- Highlights genes that encode catabolic enzymes involved in the biodegradation of pollutants.
- Explores combining genetically engineered microorganisms with bioaugmentation, biostimulation and bioattenuation strategies.
- Details the application of genetic engineering of bacteria for managing aromatic organic compounds under hypoxic conditions.
- Discusses tracking techniques and suppression strategies of genetically modified microorganisms.

Written for researchers, engineers and academics working in bioremediation, microbiology and biotechnology, this book is both timely and important.

Genetically Engineered Organisms in Bioremediation

Edited by

Inamuddin
Aligarh Muslim University,
Aligarh, India

Charles Oluwaseun Adetunji
Edo State University Uzairue Iyamho,
Edo State, Nigeria

Mohd Imran Ahamed
Aligarh Muslim University,
Aligarh, India

Tariq Altalhi
Taif University,
Saudi Arabia

CRC Press
Taylor & Francis Group
Boca Raton London New York

CRC Press is an imprint of the
Taylor & Francis Group, an **informa** business

Cover image: Shutterstock ID 1009503226

First edition published 2024
by CRC Press
6000 Broken Sound Parkway NW, Suite 300, Boca Raton, FL 33487–2742

and by CRC Press
4 Park Square, Milton Park, Abingdon, Oxon, OX14 4RN

CRC Press is an imprint of Taylor & Francis Group, LLC

ISBN: 978-1-032-03696-0 (hbk)
ISBN: 978-1-032-03697-7 (pbk)
ISBN: 978-1-003-18856-8 (ebk)

DOI: 10.1201/9781003188568

Typeset in Times
by Apex CoVantage, LLC

Contents

Chapter 5   Application of Genetically Modified Microorganisms as Biosorbents
for Polluted Environments..74

Nagma Parveen, Amrita Kumari Panda,
Rojita Mishra, Aseem Kerketta and Satpal Singh Bisht

Chapter 6   Application of Genetically Modified Microorganisms for Remediation
of Petrol Discharges and Related Polluted Sites 86

Noreen Sajjad, Ayesha Sultan, Gulzar Muhammad, Aiza Azam,
Muhammad Arshad Raza, Muhammad Ajaz Hussain and Liaqat Ali

Chapter 10 Application of Genetically Modified Microorganisms for the
Reduction of the Toxicity of Hazardous Compounds 151

*Bhagwan Toksha, Saurabh Tayde, Ajinkya Satdive, Shyam Tonde
and Aniruddha Chatterjee*

Chapter 11 Bioremediation of Heavy Metals Using Microorganisms 168

*M.S. Nagmote, A.R. Rai, R. Sharma, M.F. Desimone, R.G. Chaudhary
and N.B. Singh*

Editors

Dr. Inamuddin is an assistant professor at the Department of Applied Chemistry, Zakir Husain College of Engineering and Technology, Faculty of Engineering and Technology, Aligarh Muslim University, Aligarh, India. He has extensive research experience in multidisciplinary fields of analytical chemistry, materials chemistry, electrochemistry, renewable energy, and environmental science. He has worked on different research projects funded by various government agencies and universities and is the recipient of awards, including the Department of Science and Technology, India, Fast-Track Young Scientist Award and Young Researcher of the Year Award 2020 from Aligarh Muslim University. He has published about 210 research articles in various international scientific journals, 18 book chapters, and 170 edited books with multiple well-known publishers. His current research interests include ion exchange materials, a sensor for heavy metal ions, biofuel cells, supercapacitors, and bending actuators.

Prof Charles Oluwaseun Adetunji is presently a faculty member at the Microbiology Department, Faculty of Science, Edo State University Uzairue, Edo State, Nigeria. He is currently the Director of Research and Innovation at EDSU. He is a Fellow of the Royal Society of Biology; UK; Fellow of Biotechnology Society of Nigeria; Fellow of Nigerian Young Academy. He is a Visiting Professor and the Executive Director for the Center of Biotechnology, Precious Cornerstone University, Ibadan. He is an affiliate member of the African Academy of Science. He has won several scientific awards and grants from renowned academic bodies. He was recently listed in Stanford University's World Top 2% Scientists Rankings for 2023. He has published many scientific journals, conference proceedings in refereed national and international journals with over 580 manuscripts including over 40 books and many scientific patents with Google Scholar i10-Index 136. He was ranked recently as number 1st among the top 500 prolific authors in Nigeria between 2020 till date by SciVal/SCOPUS and recently he was ranked number 1 in two different fields from Biological Sciences and Agricultural science among 500 prolific authors from 58 different countries in Africa.

Mohd Imran Ahamed, PhD, is working as Research Associate at Department of Chemistry, Aligarh Muslim University, Aligarh, India. He has published several research and review articles in various international scientific journals. He has co-edited books of international repute. His research work includes ion-exchange chromatography, wastewater treatment, and analysis, bending actuator and electrospinning.

Tariq Altalhi, PhD, is working as Associate Professor in the Department of Chemistry at Taif University, Saudi Arabia. He received his doctorate degree from University of Adelaide, Australia in the year 2014 with Dean's Commendation for Doctoral Thesis Excellence. He has worked as head of Chemistry Department at Taif university and Vice Dean of Science College. In 2015, one of his works was nominated for Green Tech awards from Germany, Europe's largest environmental and business prize, amongst top 10 entries. He has co-edited various scientific books. His group is involved in fundamental multidisciplinary research in nanomaterial synthesis and engineering, characterization, and their application in molecular separation, desalination, membrane systems, drug delivery, and biosensing. In addition, he has established key contacts with major industries in Kingdom of Saudi Arabia.

Contributors

Sofía Abad-Sojos
Institute of Biology
ELTE Eötvös Loránd University
Budapest, Hungary

Charles Oluwaseun Adetunji
Applied Microbiology, Biotechnology
    and Nanotechnology Laboratory,
    Department of Microbiology, Edo
    State University Uzairue, Iyamho,
    Edo State, Nigeria

Mohd Imran Ahamed
Department of Chemistry
Faculty of Science, Aligarh Muslim University
Aligarh, India

Liaqat Ali
Department of Chemistry
Government College University
Lahore, Pakistan

Tariq Altalhi
Department of Chemistry,
College of Science, Taif University, Taif
Saudi Arabia

Aharon Azagury
Department of Chemical Engineering
    and Biotechnology
Ariel University
Ariel, Israel

Aiza Azam
Department of Chemistry
Government College University
Lahore, Pakistan

Satpal Singh Bisht
Department of Biotechnology
Sant Gahira Guru University
Ambikapur, India

Emilio Bucio
Department of Radiation Chemistry
    and Radiochemistry
Institute of Nuclear Sciences
National Autonomous University
    of Mexico
Ciudad de México, Mexico

Moises Bustamante-Torres
Biomedical Engineering Department,
    School of Biological
    and Engineering
Yachay Tech University
San Miguel de Urcuquí, Ecuador
and
Department of Radiation Chemistry and
    Radiochemistry
Institute of Nuclear Sciences, National
    Autonomous University of Mexico
Ciudad de México, Mexico

Aniruddha Chatterjee
Centre for Advanced Materials Research and
    Technology
Plastic and Polymer Engineering Department,
    Maharashtra Institute of Technology
Aurangabad, India

R.G. Chaudhary
Post Graduate Department of Chemistry
S.K. Porwal College
Kamptee, India

Ajit Debnath
Department of Physics
National Institute of Technology Agartala
Agartala, India

M.F. Desimone
Universidad de Buenos Aires
Facultad de Farmacia y Bioquímica
Buenos Aires, Argentina

Sumeyra Gurkok
Department of Biology
Science Faculty, Ataturk
    University
Erzurum, Turkey

Muhammad Ajaz Hussain
Institute of Chemistry
University of Sargodha
Sargodha, Pakistan

Inamuddin
Department of Applied Chemistry
Zakir Husain College of Engineering
    and Technology, Faculty of
    Engineering and Technology
Aligarh Muslim University
Aligarh, India

Sahidul Islam
Department of Chemistry
Krishnath College
Murshidabad, India
and
Department of Chemistry
The University of Burdwan
Bardhaman, India

Pallavi Jain
Department of Chemistry,
    SRM Institute of Science
    and Technology,
    Delhi-NCR Campus
New Delhi, India

Shamim Ahmed Khan
Department of Chemistry
National Institute of Technology
    Agartala
Agartala, India

Carlos Llerena-Bustamante
Biology Department
Faculty of Biology
Universidad Central del Ecuador
Quito, Ecuador

Ujjwal Mandal
Department of Chemistry
The University of Burdwan
Bardhaman, India

Rojita Mishra
Department of Botany
Polasara Science College
Polasara, India

Gulzar Muhammad
Department of Chemistry
Government College University Lahore
Lahore, Pakistan

M.S. Nagmote
Post Graduate Department of Chemistry
S.K. Porwal College
Kamptee, India

Pinku Chandra Nath
Department of Bio Engineering
National Institute of Technology Agartala
Agartala, India

El Asri Ouahid
Laboratory of Microbial Biotechnology
    and Vegetal Protection
Ibn Zohr University
Agadir, Morocco

Amrita Kumari Panda
Department of Biotechnology
Sant Gahira Guru University
Ambikapur, India

Samantha Pardo
Environmental Engineering Faculty
Salesian Polytechnic University
Cuenca, Ecuador

Nagma Parveen
Department of Zoology
Kumaun University
Nainital, India

Dhaval Patel
Assistant Professor
Post Graduate Department of Biosciences
Sardar Patel University
Gujarat, India
and
Department of Chemical Engineering
    and Biotechnology
Ariel University
Ariel, Israel

A.R. Rai
Post Graduate Department of
  Microbiology
S.K. Porwal College
Kamptee, India

Anirudh Pratap Singh Raman
Department of Chemistry
SRM Institute of science and
  Technology
Delhi-NCR Campus, New Delhi, India

Anchal Rana
Dr. Yashwant Singh Parmar University of
  Horticulture and Forestry
Solan, India

Muhammad Arshad Raza
Department of Chemistry
Government College University
Lahore, Pakistan

David Romero-Fierro
Chemical Engineering Department
School of Biological and Engineering,
  Yachay Tech University
San Miguel de Urcuquí, Ecuador
and
Department of Radiation Chemistry
  and Radiochemistry
Institute of Nuclear Sciences, National
  Autonomous University of Mexico
Ciudad de México, Mexico

Biplab Roy
Department of Chemical Engineering
National Institute of Technology
  Agartala
Agartala, India

Noreen Sajjad
Department of Chemistry
University of Lahore
Lahore, Pakistan

Ajinkya Satdive
Centre for Advanced Materials Research
  and Technology,
Plastic and Polymer Engineering
  Department,
Maharashtra Institute of Technology
Aurangabad, India

Ankita Sharma
Dr. Yashwant Singh Parmar University of
  Horticulture and Forestry
Solan, India

Nandita Sharma
CSIR Institute of Microbial Technology
Chandigarh, India

R. Sharma
Department of Rasa Shastra and
  Bhaishajya Kalpana
Institute of Medical Science, India

Samriti Sharma
Dr. Yashwant Singh Parmar University of
  Horticulture and Forestry
Solan, India

N.B. Singh
Department of Chemistry and
  Biochemistry, and RDC
Sharda University
Greater Noida, India

Prashant Singh
Department of Chemistry
Atma Ram Sanatan Dharma College
New Delhi, India

Jyoti Solanki
Post Graduate Department of Biosciences
Sardar Patel University
Gujarat, India

R.B. Subramanian
Post Graduate Department of Biosciences
Sardar Patel University
Gujarat, India

Ayesha Sultan
Department of Chemistry
University of Education
Lahore, Pakistan

Saurabh Tayde
Centre for Advanced Materials Research and
  Technology
Plastic and Polymer Engineering
  Department
Maharashtra Institute of Technology
Aurangabad, India

Ajita Tiwari
Department of Agricultural
    Engineering
Assam University
Silchar, India

Bhagwan Toksha
Basic Sciences and Humanity
    Department
Maharashtra Institute of Technology
Aurangabad, India

Shyam Tonde
Centre for Advanced Materials Research and
    Technology
Plastic and Polymer Engineering Department
Maharashtra Institute of Technology
Aurangabad, India

Odalys Torres
Institute of Biology
ELTE Eötvös Loránd University
Budapest, Hungary

1 Applications of Genetically Modified Microorganisms for the Removal of Hydrocarbons

Moises Bustamante-Torres, David Romero-Fierro,
Carlos Llerena-Bustamante and Emilio Bucio

INTRODUCTION

Hydrocarbons are mostly derived from crude oil. The degradation of these compounds is more complicated than for other compounds derived from oils. Hydrocarbons supply 96% of the energy generally consumed worldwide: 40% comes from crude oil, 34% comes from coal and 22% from natural gas. The remaining 4% comes from hydraulic power (Obodovskiy 2019). However, its extraction or use may generate contamination in water and soils due to constant accidental spills.

Several chemical and physical techniques have been developed to treat hydrocarbon contamination. However, these processes present low efficiency and high costs of production. Therefore, bioremediation has become an attractive and promising alternative to traditional physicochemical techniques for the remediation of compounds that pollute the environment. Bioremediation is understood as the application of microorganisms, fungi, plants, or enzymes derived from them for the restoration of the environment (Rani et al. 2019). This technology acts through biological interventions to achieve mitigation of the negative effects caused by the environmental contamination to which a specific place was subjected.

Microorganisms have always existed, affecting living beings directly or indirectly (Bustamante-Torres et al. 2021a). They play an important role in nature, and are found throughout soil, water and air. Many microorganisms perform the positive function of cleaning up and mitigating the effects of environmental pollutants. Several microorganisms can obtain their energy from the degradation of organic compounds (OCs) such as hydrocarbons (energetically favorable) (Abbasian et al. 2015). However, these OCs contain hydrogen (H_2) and carbon (C). Moreover, various indigenous microorganisms present enzymes that carry out hydrocarbon degradation. This process depends on nature in hydrocarbons and enzymes. However, indigenous microorganisms sometimes present a low rate and low efficiency. Additionally, during bioaugmentation *in-situ*, some microorganisms get the benefits of surrounding nutrients, while other microorganisms lose them.

Bioremediation is the use of microbial species to clean up soil and groundwater that has been contaminated by discharged chemicals (Speight 2018). Despite being an organic substance, hydrocarbons are not usually biodegradable. However, research has been developed to try to solve this problem. Genetically engineered microorganisms (GEMs) are of great application for environmental bioremediation, produced through human actions and environmental and meteorological factors. GEMs have been reported to have a higher degradation capacity for some pollutants (Singh et al. 2020). For this, a bacterial combination manages to break down the hydrocarbon molecule to make it susceptible to another bacterium (Cerniglia 1992). Research has been conducted on the degradation of environmental pollutants by various GEMs such as *E. coli AtzA, Pseudomonas fluorescens HK44, Burkholderia cepacia VM1468* and *Pseudomonas putida PaW85* (Strong et al. 2000; Sayler and Ripp 2000; Taghavi et al. 2005; Jussila et al. 2007).

DOI: 10.1201/9781003188568-1

HYDROCARBONS

Hydrocarbons are compounds comprised exclusively of carbon and hydrogen and they are by far the dominant components of crude oil and other derivates (Abrajano et al. 2007). These compounds are usually altered to get some beneficial uses. Hydrocarbons can be gaseous, liquid, or solid. They are found in nature due to the contamination of biomass for millions of years, or are produced synthetically. The smallest hydrocarbon is methane, which contains one carbon and four hydrogen atoms. The molecular weight of known hydrocarbons ranges from 16.04 in methane to 9,000 in synthesized paraffin of high molecular weight. Additionally, hydrocarbons are also known as a significant source of energy.

PHYSICAL AND CHEMICAL PROPERTIES

An important characteristic is the reaction of hydrocarbons with oxygen, a process known as combustion. Optimal combustion of hydrocarbons produces water and carbon dioxide, while improper combustion can also produce carbon or carbon monoxide (soot) and water. Improper combustion releases less energy, so optimal hydrocarbon combustion is important. If a hydrocarbon burns with a sooting flame, this may be an indication of higher carbon content in the compound (longer chain length) (Speight 2011a).

Various isomers of hydrocarbons present a similar empirical formula but different structural formulas. They are found in the alkanes of butane and most other hydrocarbons. Special isomerism is the *cis-trans* isomerism that usually occurs in double bonds and is a spatial arrangement of the groups that make it up (Speight 2011c). Additionally, both unsaturated hydrocarbons and saturated hydrocarbons present similar physical properties.

Hydrocarbon is a non-polar compound. This implies its insolubility in water, though it is easily soluble in most organic solvents. Their intermolecular bonds are weak van der Waals bonds. The melting and boiling points of small molecules are lower than for larger and heavier molecules, which present higher values for both temperatures.

The reactivity of alkanes depends on the length of their chain. Long-chain alkanes are relatively inert. In addition to redox reactions in its combustion, substitution reactions can occur, in which hydrogen atoms are exchanged for other atoms and groups of atoms—mainly halogens. Alkenes and alkynes, on the other hand, are quite reactive due to the electron delocalization of the multiple bonds of these compounds. Among the physical properties to be highlighted are the boiling point and the density of these compounds, which increase as the size of the hydrocarbon increases. This is because the intermolecular forces are greater when the molecule is larger (Speight 2011a).

HYDROCARBON CLASSIFICATION

Saturated Hydrocarbons

In all saturated hydrocarbons, carbon exhibits tetragonal hybridization (Obodovskiy 2019). Figure 1.1 shows some examples of saturated hydrocarbons. Each carbon atom binds to other carbon atoms and hydrogen atoms, so it is said to be saturated, which means that its valence capacity is complete. Furthermore, in each member of this series, the molecules are arranged in straight or

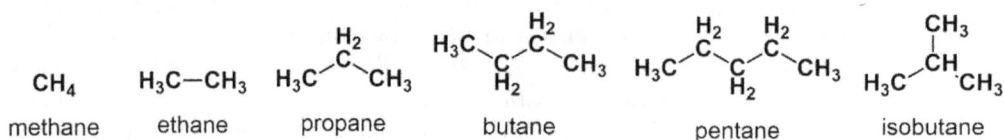

FIGURE 1.1   Some examples of saturated hydrocarbons.

branched (aliphatic) chains and not in closed rings. Additionally, saturated hydrocarbons are usually degraded by different physical, chemical, or biological processes, just like polar OCs.

The first members of the alkanes (methane to butane) are employed in domestic life and industry. They are obtained primarily from fossil fuels found in underground natural gas deposits or gases associated with oil deposits. In some areas—for instance, under the North Sea—natural gas is almost pure methane. In the United States, it may contain a significant percentage of other hydrocarbons such as ethane, propane and butane. Long-chain alkanes such as gasoline, kerosene, paraffin, mineral oils, fats, waxes and bitumen are obtained by petroleum distillation.

Methane is used mainly as fuel. It is also an essential raw product in the production of other chemical compounds. Methane can be converted, for example, to methanol and other long-chain alcohols, as well as long-chain alkanes such as gasoline.

Ethane is mainly used as a raw material to obtain ethene, obtaining significant substances such as polyethylene (PE). Propane and butane are easily converted into liquids that can be stored under pressure in containers, which allow the transportation of the fuel. Likewise, they are used to obtain other compounds such as ethene and propene.

Unsaturated Hydrocarbons

Unsaturated hydrocarbons have fewer hydrogens than the corresponding bonds of paraffin. Hence, the molecules contain at least one carbon-carbon multiple bond (double or triple bond). The four valences of carbon are not bound to other atoms; hence it is said that its valence capacity is not complete. Double bond compounds are also commonly called alkenes or olefins. This last name is due to the ease with which they give oily or oily liquids when reacting with chlorine. Its general formula is C_nH_{2n}. Figure 1.3 show some examples of unsaturated hydrocarbons.

Ethene is a colorless, odorless, tasteless hydrocarbon with a faint ethereal odor. It burns brighter than methane. It is obtained as a byproduct of high scale petroleum cracking. Ethene is the most important alkene, followed by propene and diolefin. The compounds called alkynes contain one or more carbon-carbon triple bonds in their molecules. Their general formula is C_nH_{2n-2}. The first and main member of this series is ethyne.

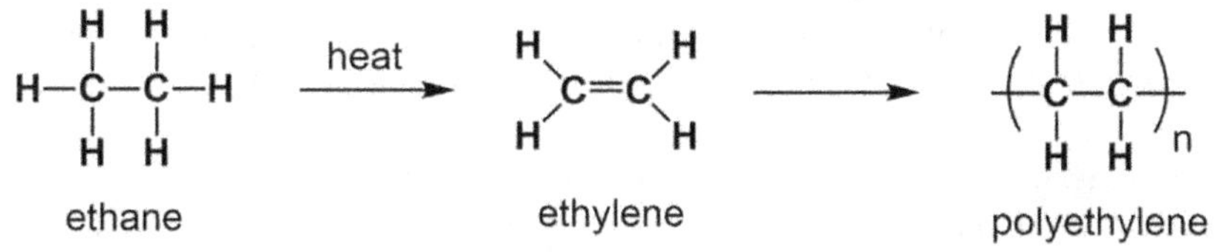

FIGURE 1.2 Preparation of PE from ethane.

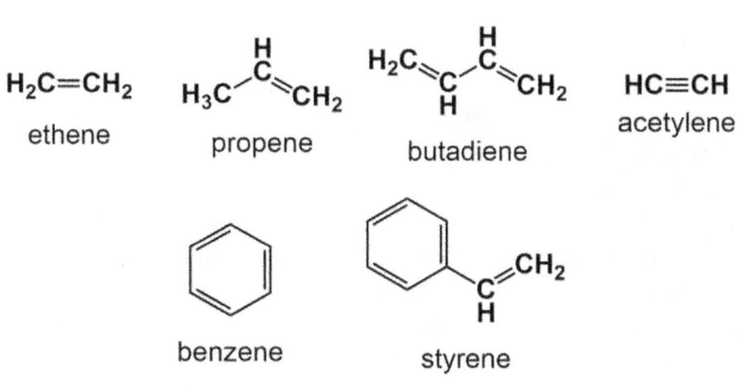

FIGURE 1.3 Some examples of unsaturated hydrocarbons.

a)

b)

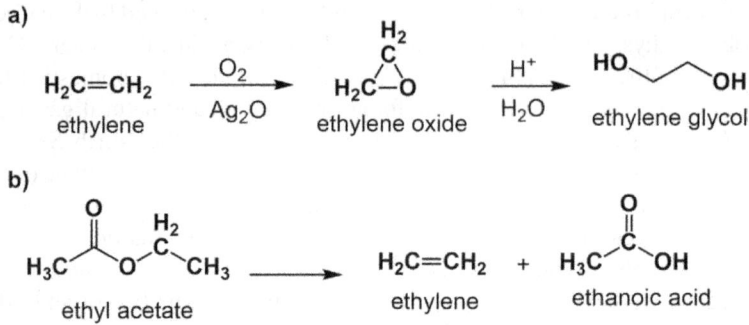

FIGURE 1.4   Production of a) ethylene glycol and b) ethanoic acid.

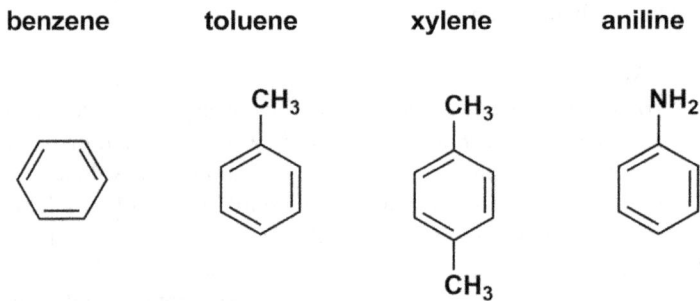

benzene   **toluene**   **xylene**   **aniline**

FIGURE 1.5   Chemical structure of some aromatic compounds.

Simple alkenes are produced by industry by cracking naphtha, a straw-colored liquid obtained from the refining of crude oil or some types of natural gas. Naphtha, a mixture of saturated hydrocarbons, composed mainly of molecules that can have 4 to 12 carbon atoms, undergoes a thermal cracking of 540–650 °C in the absence of air, which decomposes it into products such as ethene, propene and butadiene. Although natural gas is primarily methane, some reservoirs contain significant amounts of ethane, which can be converted to nearly pure ethene.

Currently, ethylene is not only the most important organic compound in the world but also an important natural product that can interfere with the ripening of fruits. It is a colorless gas, highly flammable, and has a sweet smell and taste. Figure 1.4 describes the synthesis of ethylene glycol and ethanoic acid.

Aromatics

Aromatic compounds have the form of a cyclic hydrocarbon containing different alkyl groups. They are important hydrocarbons, such as toluene, xylene, benzene (Dincer and Zamfirescu 2014) and aniline (Sadeghbeig 2012). Figure 1.5 illustrates the chemical structure of some aromatic compounds. Additionally, they can produce polymers and have a planar molecular structure with at least one carbon ring and 4_{n+2} electrons to form a covalent bond (Dincer and Zamfirescu 2014). Because polycyclic aromatic hydrocarbons (PAHs) have an impact on human health and the environment, PAHs are considered important pollutants (Alegbeleye et al. 2017).

Benzene has been employed for a long time in a wide range of applications. It is known as a carcinogenic compound (Abel and DiGiovanni 2008). Benzene is rapidly absorbed through the lungs; as such, benzene readily crosses the alveolar membranes and is taken up by circulating blood in pulmonary vessels (Barton 2014).

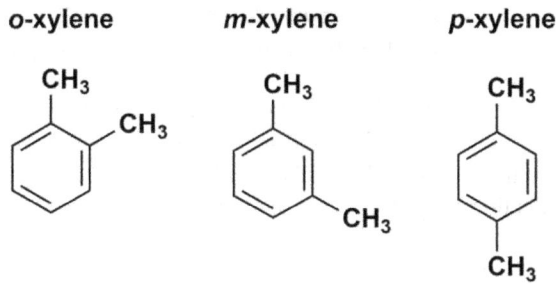

FIGURE 1.6 Chemical structure ortho-, meta- and para- xylene.

Toluene is a colorless, flammable liquid that is about 15% lighter than water. It has a sweet, pungent, benzene-like odor (Clough 2014). Toluene is a seven-carbon aromatic hydrocarbon compound that is a minor component (~5%) of all gasoline. Toluene is a flammable hydrocarbon distillation product of petroleum with uses as both a fuel additive in motor vehicles and an anthelmintic in dogs and cats (Page 2008).

Xylene is a derived from dimethyl benzene. It is classified as the ortho-, meta- and para- forms of the molecule. o-Xylene has been used for the production of o-dicarboxylic anhydride. m-Xylene is known to produce isophthalic acid. And finally, p-Xylene is an organic chemical crucial for the synthesis of terephthalate (to produce resins and films) and dimethyl terephthalate (Zhou et al. 2012). Figure 1.6 shows the chemical structure of three isomers of xylene.

Resins

Hydrocarbon resins are defined as low molecular weight (usually below 2,000) thermoplastic polymers whose appearance ranges from liquids to amorphous viscosity (Gandini 1989). Hydrocarbon resins tend to have high glass transition temperatures. Therefore, they melt upon processing, which can improve the fluidity of complex viscosity molds. However, they harden at room temperature, thereby maintaining the hardness and modulus of the compound (Rodgers and Waddell 2013).

Asphaltenes

Asphaltenes are found in animals and plants. In geological periods, the effects of temperature and pressure only partially decompose them. They contain most of the inorganic components of crude oil, including sulphur, and nitrogen, as well as metals such as nickel and vanadium. Asphaltenes form a micelle, which is a solid structure represented by the polynuclear aromatic layers with folded alkane chains (Bai and Bai 2019).

THE FATE OF HYDROCARBONS IN THE ENVIRONMENT

The hydrocarbon exploration and exploitation sector has shown exponential growth in recent years, becoming a fundamental element in the growth of the world economy (Alegbeleye et al. 2017). Petroleum and its derivatives produced as a result of industrial development pose significant risks to the environment.

Due to their chemical composition, hydrocarbons prevent gaseous exchange with the atmosphere, initiating a series of simultaneous physical-chemical processes, such as evaporation and penetration. This process depends on the type of hydrocarbon and the environmental conditions, which can constitute relatively slow mechanisms. Significant toxicity generates serious environmental consequences for both flora and fauna.

In humans, petroleum distillates cause fulminant and sometimes fatal pneumonitis when aspirated. Central nervous system (CNS), gastrointestinal (GI), hepatic, renal, cardiovascular and hematologic toxicity may also occur (Lewander and Aleguas 2007). Therefore, crude oil spills cause adverse effects on the health of first responders and the general public, which is of utmost importance (Wnek et al. 2018).

Several ecotoxicity tests on crude oil hydrocarbons in soils have been conducted in recent years. These tests were done on invertebrate plants and animals in contaminated soils (Efroymson et al. 2004). Testing is usually done by physical and chemical measurements, but chemical measurement presents a problem which is that it cannot give a biological measurement.

IMPACT ON WATER

The contamination of water by hydrocarbons in storage systems, in underground and surface supply sources, as well as in other bodies of water, occurs with relative frequency. This type of contamination produces a change in the organoleptic characteristics of the water that may cause ill effects to those who consume it; its ingestion represents a health risk. Major sources of hydrocarbon contamination in bodies of water include industrial wastewater, through the atmosphere, urban fluvial and oil spills. The effects on the environment depend on the type of hydrocarbon and the source components (Abdel-Shafy and Mansour 2016). According to the National Academy of Sciences, about 9 million tons of oil are spilled into the water every year, 70% of which are caused by human activities (Shankar and Srivastava 2017).

IMPACT ON THE AIR

Air pollution is due to changes in air quality and purity levels caused by natural emissions or chemical and biological substances. At present, pollution generated by the combustion of hydrocarbons (gasoline, diesel gas) from automobiles is the first cause of air pollution in industrialized environments, while inefficient industrial plants are the major sources of air pollution for non-industrialized areas (Abdel-Shafy and Mansour 2016).

Respiratory diseases, asthma and allergies are associated with air pollution. The relationship between air pollution and health is becoming increasingly well known. Asthma and allergies have increased over the past decades across Europe (Dejmek et al. 2000). Approximately 10% of the child population suffers from one of these diseases. The environmental agents involved are nitrogen and sulfur oxides, suspended particles, ozone, metals, volatile organic compounds (VOCs) and hydrocarbons (Hall et al. 2003; Wu et al. 2006).

IMPACT ON SOILS

The soil is a mixture of water, air, inorganic and organic matter. Soil organic matter (SOM) is commonly defined as the organic fraction of soil residues. SOM encompasses the totality of organic material (living and nonliving) present in soils, including living microorganisms and undecayed residues (Bernoux and Cerri 2005). Besides, during coal combustion, a large proportion of inorganic matter is collected as ash (Pudasainee et al. 2020), and then it is released into the environment. The mixture of these compounds in the soil makes it difficult to remove hydrocarbons (Barathi and Vasudevan 2001).

Soil pollutants have caused loss of soil fertility, poor crop yields and possible harmful consequences for humans and the entire ecosystem. The determination of these units is very complex and the grouping criterion is made up of the geomorphological landscape and the climate; therefore, each spill situation is unique. Each affected place has its particularity of temperature, pH, humidity, type of soil, for which there is no generalized formula for a route of actions to mitigate the different spills or environmental effects that may occur. The biological activity of the soil, such as the microbial biomass of the soil and its enzymatic activity, can be affected by a series of parameters and physical, chemical and environmental disturbances. The mechanisms of hydrocarbon degradation consist in either aerobic and anaerobic conditions (Das and Chandran 2011). Aerobic or anaerobic degradations have several reactions such as oxidation, reduction, hydroxylation and dehydrogenation (Abbasian et al. 2015; Wilkes et al. 2016). Besides, each mechanism employs a great number of specific enzymes.

BIOREMEDIATION OF HYDROCARBONS

Environmental remediation is based on the treatment of pollutants in water, air, or soil (Bustamante-Torres et al. 2021b). In bioremediation, microorganisms with biological activity, including algae, bacteria, fungi and yeast, can be used in their naturally occurring forms (Coelho et al. 2015), in order to transform contaminants to less toxic or nontoxic forms (Fennel et al. 2011). However, under certain circumstances, hydrocarbons have a certain opposition to microbial degradation, which is given by the type, molecular weight and the number of rings, the latter corresponding to PAHs (Varjani 2017).

BIODEGRADATION

OCs break down into fragments or inorganic molecules, in both aerobic and anaerobic conditions. A huge amount of living beings contribute to this process. Traditional pollutant degradation methods involve the separation of one or more organisms that can degrade target pollutants in the environment (Pushpanathan et al. 2014). For instance, some plants associated with bacteria as endophytic and rhizospheric bacteria improve phytoremediation, enhancing biodegradation. Currently, many reports indicate that bioaugmentation and biostimulation enhance hydrocarbon biodegradation in soil contaminated by oil spillage. Bioaugmentation involves the inoculation of exogenous degrading microorganisms into the soil. Biostimulation avoids metabolic limitation by adding nutrients, thereby stimulating the degrading capacity of indigenous communities. Since the bioremediation process is carried out by various microorganisms in the soil, bioaugmentation and biostimulation affect the population of hydrocarbon degradants (Mariano et al. 2009). In general, these techniques are very promising methods for remediating oil-contaminated soil.

Bioaugmentation has been employed to eliminate several compounds—such as naphthalene, phenol, pyridine, quinoline and carbazole—present in coking wastewater with *Paracoccus denitrificans* and five *Pseudomonas* sp. strains. Similarly, one study used a highly effective naphthalene-degrading bacterial strain that was isolated from acclimated activated sludge from a coal gasification wastewater plant and identified as a *Streptomyces* sp. The naphthalene degradations occur under specific conditions (pH 7.0 and at 35 °C). Additionally, degradations can be enhanced through the addition of glucose and methanol (Xu et al. 2014).

MINERALIZATION

It consists of the biodegradation of OCs into inorganic compounds (such as CO_2 or H_2O_4). Mineralization occurs when biodegradation is completed.

PHYTOREMEDIATION

Phytoremediation is an emerging technology that uses plants to deal with various environmental pollution problems (Das and Chandran 2011). It is depicted as a green and eco-friendly approach. This bioremediation process has great advantages because it is 10-fold cheaper than conventional technologies for the remediation of contaminated soils and wastes (Pandey et al. 2020).

SEPARATION OF HYDROCARBONS

Separation is a physical process through which the different phases that make up a hydrocarbon stream can be obtained at ambient pressure and temperature conditions. Before 1925, the separation of hydrocarbons was considered an unimportant operation, since it was frequent to see that the crude was passed directly from the mouth of the wells to the storage tanks discovered in the atmosphere. However, it has been observed that when passing the oil through an attachment for the separation, a higher yield was obtained, having oil with a higher percentage of gasoline. The main reason that led to the separation of the hydrocarbons was the problems that arose from having two fluids with different characteristics and behavior in the same collection, transport, and storage systems, for which a mechanical device called a separator was created (Haussard et al. 2003). The separation can be carried out in different ways, depending on the phase of the bodies to be separated and the type of hydrocarbons involved.

To establish the most appropriate separation conditions, according to the characteristics of the fluids produced, the following control variables must be considered: the type, size and internal devices of the separator, the oil resistance time, the stages of separation, the pressures, and operating temperatures and the place of installation of the separators. There will be a combination of all these variables that allows us to obtain the required separation at a minimum cost (Khan et al. 2007). The selection of the separation conditions depends mainly on the established production objectives. These objectives are aimed at obtaining:

High Efficiency in the Separation of Oil and Gas

This efficiency in a separator depends fundamentally on its design. The characteristics of the fluids and the costs determine the type and dimensions of the separator for each particular case.

Higher Production Rates

When the exploitation conditions of the producing fields are favorable, the production rate of their wells can be increased by reducing their counter-pressure on the surface. The lowest back pressure and, consequently, the highest expense is obtained by placing the spaced as closely as possible to the wells, simultaneously adjusting their operating pressure to the minimum value that the production conditions allow.

Greater Recovery of Liquid Hydrocarbons

Since the hydrocarbons with the highest commercial value are liquids, the efficiency of the separation process is frequently related to the number of liquefiable hydrocarbons contained in the gas phase that leaves the separators.

Stabilized Oil and Gas

So that the oil does not experience substantial losses due to evaporation during its storage, when it is handled under surface conditions in refineries, or when loading ships for export, it is necessary to stabilize it previously. The oil is stabilized by adjusting its vapor pressure in such a way that it is less than the atmospheric one at the maximum temperature expected in the environment.

Liquid-liquid extraction is the most widespread technique and the first that was used industrially, being able to simultaneously extract aromatics from a mixture in very different concentrations. The solvents used are selective for aromatics such as sulfolane, dimethylformamide, which reduces the relative volatility of aromatics (Figure 1.7) (Sander et al. 2015). This extraction has been replaced

sulfolane   dimethylformamide

FIGURE 1.7 Solvents used in liquid-liquid extraction

in some applications by extractive distillation, obtaining energy, and investment savings, but it continues to be the most widely used technique for the separation of aromatics from reformed gasoline.

In extractive distillation, practically the same solvents are used for liquid-liquid extraction. It is used mainly when the aromatic content is of the order of 65–90% as in pyrolysis gasoline and it is more restricted in application, although some solvents allow the joint separation of, for example, benzene and toluene.

Temperature

At high temperatures (30–40 °C), the metabolic rate of hydrocarbons reaches its highest level (Bossert and Bartha 1984). For example, hydrocarbon degradation has a wide range of temperatures. This degradation depends on the environment affected, in either soil, water, or air. In soil, marine, and freshwater environments, the degradation rate is highest in the temperature range of 30–40, 20–30, and 15–20 °C, respectively (Bossert and Bartha 1984; Cooney 1984). It is also affected at low temperatures.

ADVANCES IN BIOREMEDIATION

Bioremediations as a biological process involve microorganisms to remove a significant amount of environmental contaminants. The microbial activity presents high promising results to remove several pollutants. However, despite bioremediation a great advance in the use of microorganisms, these processes are very complicated to be carried out, leading to adverse accumulation and expensive. GEMs are highly used to remove hydrocarbons found in the soil. The genetic modifications in specific microorganisms enhanced their abilities to survive in extreme conditions.

A wide amount of strains is studied due to their hydrocarbon degradation activity. For instance, Kafilzadeh et al. (2011), reported that among 80 bacterial strains, which belonged to 10 genus as follows: *Bacillus, Corynebacterium, Staphylococcus, Streptococcus, Shigella, Alcaligenes, Acinetobacter, Escherichia, Klebsiella,* and *Enterobacter, Bacillus* was the best hydrocarbon-degrading bacteria (Kafilzadeh et al. 2011). Aromatic hydrocarbons are mainly degrading from isolated strains from the gram-negative strains as genus *Pseudomonas*. However, the performance of these microorganisms can be improved through synergic biological processes. For example, within a mixed culture, the bacteria *Alcaligenes* and *Sphingobacterium* are not affected by being in crude oil, while the bacteria *Citrobacter* had an increase in its population as more crude oil is found (Maddela et al. 2016). In contrast to the previous bacteria, *Achromobacter, Comamonas, Pseudomonas* and *Variovorax* had a suppression by crude oil, because the more oil there is, the greater the suppression that the bacteria will have (Lisiecki et al. 2014).

In almost the majority of bacteria, they required nutrients and suitable conditions for their growth. Therefore, the environment is poorly suitable for the growth of bacteria. The physical and chemical factors of the contaminated part may determine the microbial population that can survive and reproduce in this harsh environment that uses pollutants. Bacteria and fungi that use hydrocarbons can be easily isolated and can be increased by the presence of oils or fatty wastes in the soil. The most important bacteria that can degrade hydrocarbons in soil are Achromobacter, *Acinetobacter, Alcaligenes, Arthrobacter, Bacillus, Flavobacterium, Nocardia,* and *Pseudomonas*. The main fungi causing hydrocarbons biodegradation are *Trichoderma* and *Mortierella* (Chandra et al. 2012).

One of the main causes of seawater pollution is the release of petroleum and petroleum byproducts into the marine environment. Therefore, it is necessary to mitigate contamination with bioremediation, which use specialized microorganism to biodegrade contaminants (Maier and Gentry 2015). Some microorganisms have a potential for biodegradation and transformation contaminants into less toxic components (Cerqueira et al. 2011). However, these microorganisms are usually low in abundance in marine environments, hence the isolation and selection process must be extremely important (Nikolopoulou et al. 2013). The principal mechanism to remove hydrocarbons from polluted seawater is the use of microorganisms because they are available to overcome the insoluble and persistent nature of polycyclic aromatic hydrocarbons. Nevertheless, the catabolism process has several drawbacks over time

ENZYMATIC MECHANISMS

Hydrocarbon degradation is carried out by many processes. The biological process usually employs microorganisms containing enzymatic activity to degrade hydrocarbons. As increased the enzymatic activity from microorganisms, the degradation of these OCs as well.

An interesting point is that these enzymes present a significant capacity and a versatile genetic material, which can be modified. The degradation of each or hydrocarbon depends on its nature and the nature of the enzyme. For instance, methane monooxygenases (MMO), alkane hydroxylases, and Cytochrome P450 carried out some degradation processes. These enzymes work better in suitable conditions (temperature, pH, and substrate available) for them. Normally, they operate at mesophilic temperatures (20 to 40 °C) for optimal performance.

The MMO reaction mechanism include the splitting of the dioxygen bond, the interaction of spin unpaired (triplet) oxygen with spin paired (singlet) substrates, and the large thermodynamic driving force (Messerschmidt 2010). MMO is a metalloenzyme (multicopper enzyme) that catalyzes the conversion of methane to methanol in methanotrophic bacteria (Chan et al. 2011). These enzymes have a relatively wide substrate specificity and can catalyze the oxidation of a range of substrates including ammonia, methane, halogenated hydrocarbons, and aromatic molecules (Arp et al. 2002). Aerobic methanotrophs are a unique group of gram-negative bacteria capable of utilizing methane as the sole carbon and energy source. Methanotrophs are present in a wide variety of environments and play an important role in the oxidation of methane in the natural world (Jiang et al. 2011).

Cytochrome P450 is an enzymes groups that function as MMO and affect the oxidation of OCs by transfer of one oxygen atom through several steps (Saha 2018). Cytochrome P450 enzymes have been isolated from Candida species, including *Candida, Candida maltosa, and Candida tropicalis.* (Scheller et al. 1998). Naphthalene, vanfluorine, acenaphthalene, acenaphthylene, and 9-methylanthracene are some of the most harmful pollutants which are significantly degraded by the cytochrome P450 and MMO (Kumar et al. 2013). Table 1.1 summarizes some of the most important enzymes and the microorganisms from which they are obtained to degrade some compounds.

Compounds containing more than 17 Carbons are degraded by complex enzymes that are still studied. The modification of these enzymes takes great effort from scientists. *Rhodococcus* isolates grow well on purified alkanes up to C_{32} by unknown enzyme systems, whereas an uncharacterized alkane oxygenase allows *P. fluorescens* to grow on C_{18}–C_{28} alkanes (Van Beilen et al. 2002; Smits et al. 2002).

GENETICALLY ENGINEERED MICROORGANISMS (GEMs)

The degradation activity of hydrocarbons can be enhanced through genetic engineering. Molecular biology is an interesting alternative to modify and optimize the microorganism's activity. The essential strategy of using genetically engineered microbes in bioremediation is to design such novel strains which can degrade high molecular weight compounds as hydrocarbons. Therefore, the genes responsible to eliminate or degrade different have been identified and studied. The genes that carried out the coding for hydrocarbon degradation are located on large conjugative plasmids (Boronin and Kosheleva 2014). Table 1.2 describes some plasmids found in various microorganisms responsible to degrade hydrocarbon pollutants. In contrast, a serious issue is that GEMs for bioremediation do not always utilize the target substrate as well as expected, even if the pathway seems well designed biochemically (Janssen and Stucki 2020).

In wastewater, the choice of plasmids and donor bacteria is controlled by the operator (Nzila et al. 2016). Therefore, the feasibility to use GEMs to clean up the polluted water has been widely

TABLE 1.1

Enzymes obtained from some microorganisms involved in hydrocarbon and its derivate degradation.

Enzyme	Microorganism	Compounds degraded	References
MMO	Methanotrophs: *Methylosinus trichosporium* OB3b *Methylomonas*, *Methylobacter*, *Methylococcus* (type I), *Methylosinus* *Methylocystis (type II)* *Methylosphaera* *Methylocaldum, Methylothermus* *Methylohalobius* *Methylocella* *Methylocapsa* *Methylococcus capsulatus*	Ammonia Methane halogenated Hydrocarbons Aromatic molecules C_1-C_4 (methane to butane)	(Miyaji 2011; Messerschmidt 2010; Arp et al. 2002; Van Beilen and Funhoff 2007).
Alkane hydroxylases	*Pseudomonas (P. putida* GPo1*)* *Acinetobacter* sp. ADP1 *Mycobacterium tuberculosis* H37Rv	Alkanes Alkyl benzenes Cycloalkanes Fatty acids C_5-C_{12} n-alkanes.	(Van Beilen and Funhoff 2007)
Cytochrome P450	*Candida* *Candida maltose* *Candida tropicalis*	Alkanes Fatty acids C_5-C_{16} (pentane to hexadecane	(Scheller et al. 1998; Van Beilen and Funhoff 2007)

TABLE 1.2

Plasmids that coded for hydrocarbon degradation

Microorganism	Plasmid	Pollutants degraded	Reference
Pseudomonas oleovorans	OCT	n-octane 5 to 12 carbon linear alkanes	(Van Beilen et al. 1994)
Pseudomonas	TOL	BTEX (benzene, toluene, ethylbenzene, xylene) compound.	
Pseudomonas putida G7 *Pseudomonas putida pDTG1* *Pseudomonas putida NCIB 9816*	NAH 7	Naphthalene	(Boronin and Kosheleva 2014)
Pseudomonas, Shingomonas and Burkholderia		Polycyclic aromatic hydrocarbons	

discussed. The petroleum industry generates more than one billion tons of waste sludge globally every year.

BIOREMEDIATIONS BASED ON GEMS REPORTED

GEMs have showed a high degrade capacity. Four strains of *Pseudomonas* contain the genes in the extrachromosomal part on the plasmid for pollutant degrading. Genetics tools have approached to get those plasmids and then insert into a single strain of *Pseudomona* (Kumar et al. 2013).

These gram-negative *Pseudomonas* bacteria have a great ability to degrade some hydrocarbons pollutants. *Pseudomonas putida* contain the *NAH* and *XYL* plasmid hybrid can grow in oil crude and can metabolize the hydrocarbons. Based on that, Anand Mohan Chakrabarty, an Indian-born American Microbiologist, is a pioneer for the development of oil-eating "superbug" getting an awarded US patent. He isolated plasmids from four *Pseudomonas*, and then he inserted these plasmids into the one strain of *Pseudomonas putida* which can degrade octane, camphor, xylene, and naphthalene.

Pseudomonas fluorescens HK44 represents the firsts GEM approved for field testing in the United States for bioremediation purposes. Strain *HK44* has an introduced *lux* gene as a reporter gene fused within a naphthalene degradative pathway, showing a bioluminescence as it degrades the naphthalene (Ripp et al. 2000) *Pseudomonas fluorescens HK44* is a whole-cell (since 1990) endowed with a bioluminescent (*luxCDABE*) phenotype for naphthalene degradative pathway (Trögl et al. 2012).

Three *pseudomonas* strains were studied and compared: *Pseudomonas fluorescens* HK44, *Pseudomonas putida* RB1351 (containing *nah-lux*), and *Pseudomonas putida* RB1401 (*xyl-lux* for detection of toluene degradation). The response of *nah-lux* strain RB1351 was comparable to HK44 and both assays got similar results (Burlage et al. 1994).

Suyame et al. (1996) reported GEMs based on *Pseudomonas* strains in which the *bphA1* gene is substituted with the *todC1* gene (coding for toluene dioxygenase of *Pseudomonas putida* F1) within chromosomal biphenyl-catabolic *bph* gene clusters (Suyama et al 1996). These GEMs showed great activity against aromatic hydrocarbons. Besides, De Lorenzo et al. (1993) describes four recombinant *mini-Tn5* transposons containing the three systems of promoters from *Pseudomonas putida* their cognate wild-type regulatory genes (*xylS, xylR, nahR*) or mutant varieties (*xylS2*). These were: *Pu* promoter of the TOL plasmid (pWWO); while, the *Pm* promoter of the TOL plasmid and the *Psal* promoter of the NAH7 plasmid. Transcription of those promoters leads to activation in the host bacteria encounters aromatic hydrocarbons (alkyl-and halobenzoates, alkyl and halotoluenes or salicytes). pWWO is consider as the first plasmid assigned for TOL degradation.

A few years later, Panke et al. (1998) describe the upper TOL operon of plasmid pWW0 of *Pseudomonas putida that* was fully reassembled as a single gene cassette along with its cognate regulatory gene, *xylR*. This microorganism need a carbon source from toluene to survive, resulting in the conversion of toluene into benzoate, mediated by the upper TOL enzymes (Panke et al. 1998).

On another hand, Layton et al. (1999) reported the recombination of a gram-negative and bioluminescent bacteria known as *Vibrio fischeri*. Surfactant-resistant bioluminescent bacterial strains were constructed by transferring a broad host range plasmid containing the bioluminescent genes. The *Stenotrophomonas 3664* and *Alcaligenes eutrophus 2050* strains were useful for toxicity reduction evaluations of remediation processes that use surfactants for solubilization of hydrophobic pollutants (Layton et al. 1999).

Deinococcus radiodurans strains were genetically modified for toluene degradation. This genetic strain is consider as radiation-resistant organism. These GEMs present presents plasmid used to increase resistance to some radioactive elements like toluene (Ezezika and Singer 2010).

Fujita and Ike (1993), reported a technique employ for water bioremediations using GEMs. *Pseudomonas Putida* was the host, while *E. coli C600* recombined was added to enhance the remediation potential into activated sludge (Fujita and Ike 1993). Likewise, Fujita et al. (1991) describe the activity of some *E. coli* and *Pseudomonas Putida* strains to harbor the recombinant plasmids containing salicylate oxidase gene or catechol 2,3 oxygenase gene (*nahG* or *pheB*). Moreover, these researches reported the plasmid *NAH*, that was transferred to a plaque-forming bacterium, *Pseudomonas lemoignei 551* by conjugation, and plaque-forming and salicylate-degrading bacteria were reared. This flocculant growing genetically modified bacterium appears to be stable activated sludge or in the wastewater treatment process (Fujita et al. 1991). Table 1.3 summarizes some GEMs used to degrade some hydrocarbon compounds.

TABLE 1.3

GEMs reported to degrade some types of hydrocarbons.

GEMs	Compounds degraded	Reference
Pseudomonas	BTEX (benzene, toluene, ethylbenzene, xylene) compounds	(Boronin and Kosheleva 2014)
Pseudomonas putida.	Octane Camphor Xylene Naphthalene	(Renneberg et al. 2000)
	Toluene Naphthalene Aromatic hydrocarbons: alkyl-and halobenzoates, alkyl and halotoluenes or salicytes.	(De Lorenzo et al. 1993)
	Toluene	(Panke et al. 1998)
Pseudomonas spp	Naphthalene Salicylate Camphor Octane Xylene Toluene	(Ozcan et al. 2012).
Pseudomonas putida F1	Aromatic hydrocarbons	(Suyama et al. 1996)
Pseudomona.putida BH-1 and E. coli.	Phenol	(Fujita and Ike 1993).
Pseudomonas putida RB1351	Napthalene Toluene	(Burlage et al. 1994)
Pseudomonas fluorescens HK44.	Aromatic hydrocarbons	
Vibrio fischeri (Stenotrophomonas 3664 and Alcaligenes eutrophus 2050)	Hydrophobic pollutants	(Layton et al. 1999).
Deinococcus radiodurans.	Radioactive elements: toluene	(Ezezika and Singer 2010).

CONCLUSION

The oil industry remains one of the main and powerful worldwide. The constant production of hydrocarbons tends to have severe adverse effects on nature. Therefore, several techniques have been developed to eliminate these contaminants but present inefficient results.

Thus, bioremediation is a novel solution that can reduce or remove potentially hazardous waste present in the environment, being effective in cleaning both land and contaminated water. Bioremediation has better approach than traditional physicochemical techniques for its profitability and because it is functional without the need to damage the native flora and fauna of the places where it is used. The molecular structure of the pollutant is the a significant chemical factor in bioremediation since by affecting the physical and chemical properties it affects its ability to be biodegraded. In addition, it must be taken into account that the biodegradability of a hydrocarbon depends on its chemical structure, therefore, analyzing its molecular composition is one of the most important points to determine the biodegradation capacity of a hydrocarbon (Speight 2011b).

Bioremediation using GEMs present better properties than native or indigenous microorganisms. The genes responsible to remove different pollutants are already identified, studied. For instance, the decomposition of certain oil-derived pollutants are carried out by the plasmids present in the biological structure of *Pseudomonas spp*. Therefore, the researcher can take advantage of those identified genes. Despite the great advance in science, still there is a huge gap between academic research and technology applied. In general, the problems must be solved before GEMs can provide an effective cleanup process at low cost (Das and Chandran 2011).

REFERENCES

Abbasian, Firouz et al. 2015. "A comprehensive review of aliphatic hydrocarbon biodegradation by bacteria". *Applied Biochemistry and Biotechnology* 176 (3): 670–699. Springer Science and Business Media LLC. doi:10.1007/s12010-015-1603-5.

Abdel-Shafy, Hussein I. and Mansour, Mona S.M. 2016. "A review on polycyclic aromatic hydrocarbons: Source, environmental impact, effect on human health and remediation". *Egyptian Journal of Petroleum* 25 (1): 107–123. Elsevier BV. doi:10.1016/j.ejpe.2015.03.011.

Abel, Erika L. and DiGiovanni, John. 2008. "Environmental carcinogenesis". *The Molecular Basis of Cancer*: 91–113. Elsevier. doi:10.1016/b978-141603703-3.10007-x.

Abrajano, T.A. et al. 2007. "High-molecular-weight petrogenic and pyrogenic hydrocarbons in aquatic environments". *Treatise on Geochemistry*: 1–50. Elsevier. doi:10.1016/b0-08-043751-6/09055-1.

Alegbeleye, Oluwadara Oluwaseun, Opeolu, Beatrice Oluwatoyin and Jackson, Vanessa Angela. 2017. "Polycyclic aromatic hydrocarbons: A critical review of environmental occurrence and bioremediation". *Environmental Management* 60 (4): 758–783. Springer Science and Business Media LLC. doi:10.1007/s00267-017-0896-2.

Arp, Daniel, Sayavedra-Soto, Luis and Hommes, Norman. 2002. "Molecular biology and biochemistry of ammonia oxidation by Nitrosomonas europaea". *Archives of Microbiology* 178 (4): 250–255. Springer Science and Business Media LLC. doi:10.1007/s00203-002-0452-0.

Bai, Yong and Bai, Qiang. 2019. "Wax and asphaltenes". *Subsea Engineering Handbook*: 435–453. Elsevier. doi:10.1016/b978-0-12-812622-6.00016-6.

Barathi, S. and Vasudevan, N. 2001. "Utilization of petroleum hydrocarbons by pseudomonas fluorescens isolated from a petroleum-contaminated soil". *Environment International* 26 (5–6): 413–416. Elsevier BV. doi:10.1016/s0160-4120(01)00021-6.

Barton, C. 2014. "Benzene". *Encyclopedia of Toxicology*: 415–418. Elsevier. doi:10.1016/b978-0-12-386454-3.00364-x.

Bernoux, M. and Cerri, C.E.P. 2005. "Geochemistry | soil, organic components". *Encyclopedia of Analytical Science*: 203–208. Elsevier. doi:10.1016/b0-12-369397-7/00245-4.

Boronin, Alexander M. and Kosheleva, Irina A. 2014. "The role of catabolic plasmids in biodegradation of petroleum hydrocarbons". *Current Environmental Issues and Challenges*: 159–168. Springer Netherlands. doi:10.1007/978-94-017-8777-2_9.

Bossert, I. and Bartha, R. 1984. "The fate of petroleum in soil ecosystems". *Petroleum microbiology*, ed. Atlas, Ronald M. New York: Macmillan, pp. 435–473.

Burlage, Robert S. et al. 1994. "Bioluminescent reporter bacteria detect contaminants in soil samples". *Applied Biochemistry and Biotechnology* 45–46 (1): 731–740. Springer Science and Business Media LLC. doi:10.1007/bf02941845.

Bustamante-Torres, Moisés et al. 2021a. "Natural antimicrobial materials". *Environmental and Microbial Biotechnology*: 149–169. Springer Singapore. doi:10.1007/978-981-15-7098-8_6.

Bustamante-Torres, Moisés et al. 2021b. "Basics and green solvent parameter for environmental remediation". *Green Sustainable Process for Chemical and Environmental Engineering and Science*: 219–237. Elsevier. doi:10.1016/b978-0-12-821884-6.00007-3.

Cerniglia, Carl E. 1992. "Biodegradation of polycyclic aromatic hydrocarbons". *Biodegraation* 3 (2–3): 351–368. Springer Science and Business Media LLC. doi:10.1007/bf00129093.

Cerqueira, Vanessa S. et al. 2011. "Biodegradation potential of oily sludge by pure and mixed bacterial cultures". *Bioresource Technology* 102 (23): 11003–11010. Elsevier BV. doi:10.1016/j.biortech.2011.09.074.

Chan, Sunney I. et al. 2011. "Overexpression and purification of the particulate methane monooxygenase from methylococcus capsulatus (bath)". *Methods in Methane Metabolism, Part B: Methanotrophy*: 177–193. Elsevier. doi:10.1016/b978-0-12-386905-0.00012-7.

Chandra, Subhash et al. 2012. "Application of bioremediation technology in the environment contaminated with petroleum hydrocarbon". *Annals of Microbiology* 63 (2): 417–431. Springer Science and Business Media LLC. doi:10.1007/s13213-012-0543-3.

Clough, S.R. 2014. "Toluene". *Encyclopedia of Toxicology*: 595–598. Elsevier. doi:10.1016/b978-0-12-386454-3.00438-3.

Coelho, Luciene M. et al. 2015. "Bioremediation of polluted waters using microorganisms". *Advances in Bioremediation of Wastewater and Polluted Soil*. InTech. doi:10.5772/60770.

Cooney, J.J.1984. "The fate of petroleum pollutants in fresh water ecosystems". *Petroleum microbiology*, ed. Atlas, Ronald M. New York: Macmillan, pp. 399–434.

Das, Nilanjana and Chandran, Preethy. 2011. "Microbial degradation of petroleum hydrocarbon contaminants: An overview". *Biotechnology Research International* 2011: 1–13. Hindawi Limited. doi:10.4061/2011/941810.

De Lorenzo, Victor et al. 1993. "Engineering of alkyl- and haloaromatic-responsive gene expression with mini-transposons containing regulated promoters of biodegradative pathways of Pseudomonas". *Gene* 130 (1): 41–46. Elsevier BV. doi:10.1016/0378-1119(93)90344-3.

Dejmek, J. et al. 2000. "The impact of polycyclic aromatic hydrocarbons and fine particles on pregnancy outcome". *Environmental Health Perspectives* 108 (12): 1159–1164. Environmental Health Perspectives. doi:10.1289/ehp.001081159.

Dincer, Ibrahim and Zamfirescu, Calin. 2014. "Fossil fuels and alternatives". *Advanced Power Generation Systems*: 95–141. Elsevier. doi:10.1016/b978-0-12-383860-5.00003-1.

Efroymson, Rebecca A., Sample, Bradley E. and Peterson, Mark J. 2004. "Ecotoxicity test data for total petroleum hydrocarbons in soil: Plants and soil-dwelling invertebrates". *Human and Ecological Risk Assessment: An International Journal* 10 (2): 207–231. Informa UK Limited. doi:10.1080/10807030490438175.

Ezezika, Obidimma C. and Singer, Peter A. 2010. "Genetically engineered oil-eating microbes for bioremediation: Prospects and regulatory challenges". *Technology in Society* 32 (4): 331–335. Elsevier BV. doi:10.1016/j.techsoc.2010.10.010.

Fennell, D. et al. 2011. "Dehalogenation of polychlorinated dibenzo-*p*-dioxins and dibenzofurans, polychlorinated biphenyls, and brominated flame retardants, and potential as a bioremediation strategy". *Comprehensive Biotechnology* 6: 143–157. Elsevier. doi:10.1016/B978-0-444-64046-8.00487-0.

Fujita, M. and Ike, M. 1993. "Application of genetically engineered microorganisms to bioremediation and wastewater treatment". Idenshi sosa biseibutsu no kankyo joka mizushori eno tekiyo. Japan.

Fujita, M., Ike, M. and Hashimoto, S. 1991. "Feasibility of wastewater treatment using genetically engineered microorganisms". *Water Research* 25 (8): 979–984. Elsevier BV. doi:10.1016/0043-1354(91)90147-i.

Gandini, Alessandro. 1989. "Polymers from renewable resources". *Comprehensive Polymer Science and Supplements*: 527–573. Elsevier. doi:10.1016/b978-0-08-096701-1.00237-8.

Hall, Charles et al. 2003. "Hydrocarbons and the evolution of human culture". *Nature* 426 (6964): 318–322. Springer Science and Business Media LLC. doi:10.1038/nature02130.

Haussard, M. et al. 2003. "Separation of hydrocarbons and lipid from water using treated bark". *Water Research* 37 (2): 362–374. Elsevier BV. doi:10.1016/s0043-1354(02)00269-5.

Janssen, Dick B. and Stucki, Gerhard. 2020. "Perspectives of genetically engineered microbes for groundwater bioremediation". *Environmental Science: Processes & Impacts* 22 (3): 487–499. Royal Society of Chemistry (RSC). doi:10.1039/c9em00601j.

Jiang, H. et al. 2011. "Methanotrophs". *Comprehensive Biotechnology*: 249–262. Elsevier. doi:10.1016/b978-0-08-088504-9.00374-3.

Jussila, Minna M. et al. 2007. "TOL plasmid transfer during bacterial conjugation in vitro and rhizoremediation of oil compounds in vivo". *Environmental Pollution* 146 (2): 510–524. Elsevier BV. doi:10.1016/j.envpol.2006.07.012.

Kafilzadeh, Farshid et al. 2011. "Isolation and identification of hydrocarbons degrading bacteria in soil around Shiraz Refinery". *African Journal of Microbiology Research* 5 (19): 3084–3089. Academic Journals. doi:10.5897/ajmr11.195.

Khan, A. et al. 2007. "Novel modified alumina: Synthesis, characterization and application for separation of hydrocarbons". *Separation and Purification Technology* 55 (3): 396–399. Elsevier BV. doi:10.1016/j.seppur.2007.03.010.

Kumar, Sandeep et al. 2013. "Genetically modified microorganisms (GMOs) for bioremediation". *Biotechnology for Environmental Management and Resource Recovery*: 191–218. Springer India. doi:10.1007/978-81-322-0876-1_11.

Layton, A.C. et al. 1999. "Validation of genetically engineered bioluminescent surfactant resistant bacteria as toxicity assessment tools". *Ecotoxicology and Environmental Safety* 43 (2): 222–228. Elsevier BV. doi:10.1006/eesa.1999.1792.

Lewander, Willian J. and Aleguas, Alfred. 2007. "Petroleum distillates and plant hydrocarbons". *Haddad and Winchester's Clinical Management of Poisoning and Drug Overdose*: 1343–1346. Elsevier. doi:10.1016/b978-0-7216-0693-4.50097-9.

Lisiecki, Piotr et al. 2014. "Biodegradation of diesel/biodiesel blends in saturated sand microcosms". *Fuel* 116: 321–327. Elsevier BV. doi:10.1016/j.fuel.2013.08.009.

Maddela, Naga Raju et al. 2016. "Removal of petroleum hydrocarbons from crude oil in solid and slurry phase by mixed soil microorganisms isolated from Ecuadorian oil fields". *International Biodeterioration & Biodegradation* 108: 85–90. Elsevier BV. doi:10.1016/j.ibiod.2015.12.015.

Maier, Raina M. and Gentry, Terry J. 2015. "Microorganisms and organic pollutants". *Environmental Microbiology*: 377–413. Elsevier. doi:10.1016/b978-0-12-394626-3.00017-x.

Mariano, Adriano Pinto et al. 2009. "Investigation about the efficiency of the bioaugmentation technique when applied to diesel oil contaminated soils". *Brazilian Archives of Biology and Technology* 52 (5): 1297–1312. FapUNIFESP (SciELO). doi:10.1590/s1516-89132009000500030.

Messerschmidt, Albrecht. 2010. "Copper metalloenzymes". *Comprehensive Natural Products II*: 489–545. Elsevier. doi:10.1016/b978-008045382-8.00180-5.

Miyaji, Akimitsu. 2011. "Particulate methane monooxygenase from methylosinus trichosporium OB3b". *Methods in Methane Metabolism, Part B: Methanotrophy*: 211–225. Elsevier. doi:10.1016/b978-0-12-386905-0.00014-0.

Nikolopoulou, M., Pasadakis, N. and Kalogerakis, N. 2013. "Evaluation of autochthonous bioaugmentation and biostimulation during microcosm-simulated oil spills". *Marine Pollution Bulletin* 72 (1): 165–173. Elsevier BV. doi:10.1016/j.marpolbul.2013.04.007.

Nzila, Alexis, Razzak, Shaikh and Zhu, Jesse. 2016. "Bioaugmentation: An emerging strategy of industrial wastewater treatment for reuse and discharge". *International Journal of Environmental Research and Public Health* 13 (9): 846. MDPI AG. doi:10.3390/ijerph13090846.

Obodovskiy, Ilya. 2019. "Basics of Biochemistry". *Radiation*: 399–427. Elsevier. doi:10.1016/b978-0-444-63979-0.00033-1.

Page, Stephen W. 2008. "Antiparasitic drugs". *Small Animal Clinical Pharmacology*: 198–260. Elsevier. doi:10.1016/b978-070202858-8.50012-9.

Pandey, Vimal Chandra et al. 2020. "Case studies of perennial grasses—phytoremediation (holistic approach)". *Phytoremediation Potential of Perennial Grasses*: 337–347. Elsevier. doi:10.1016/b978-0-12-817732-7.00016-x.

Panke, Sven, Sánchez-Romero, Juan M. and de Lorenzo, Víctor. 1998. "Engineering of quasi-natural pseudomonas putida strains for toluene metabolism through an ortho-cleavage degradation pathway". *Applied and Environmental Microbiology* 64 (2): 748–751. American Society for Microbiology. doi:10.1128/aem.64.2.748-751.1998.

Pudasainee, Deepak, Kurian, Vinoj and Gupta, Rajender. 2020. "Coal". *Future Energy*: 21–48. Elsevier. doi:10.1016/b978-0-08-102886-5.00002-5.

Pushpanathan, Muthuirulan et al. 2014. "Microbial bioremediation". *Microbial Biodegradation and Bioremediation*: 407–419. Elsevier. doi:10.1016/b978-0-12-800021-2.00017-0.

Rani, Nisha et al. 2019. "Microbes". *Microbial Wastewater Treatment*: 83–102. Elsevier. doi:10.1016/b978-0-12-816809-7.00005-1.

Ripp, Steven et al. 2000. "Controlled field release of a bioluminescent genetically engineered microorganism for bioremediation process monitoring and control". *Environmental Science & Technology* 34 (5): 846–853. American Chemical Society (ACS). doi:10.1021/es9908319.

Rodgers, Brendan and Waddell, Walter. 2013. "The science of rubber compounding". *The Science and Technology of Rubber*: 417–471. Elsevier. doi:10.1016/b978-0-12-394584-6.00009-1.

Sadeghbeigi, Reza. 2012. "FCC feed characterization". *Fluid Catalytic Cracking Handbook*: 51–86. Elsevier. doi:10.1016/b978-0-12-386965-4.00003-3.

Saha, Nilanjan. 2018. "Clinical pharmacokinetics and drug interactions". *Pharmaceutical Medicine and Translational Clinical Research*: 81–106. Elsevier. doi:10.1016/b978-0-12-802103-3.00006-7.

Sander, Aleksandra et al. 2015. "Separation of hydrocarbons by means of liquid-liquid extraction with deep eutectic solvents". *Solvent Extraction and Ion Exchange* 34 (1): 86–98. Informa UK Limited. doi:10.1080/07366299.2015.1132060.

Sayler, Gary S. and Ripp, Steven. 2000. "Field applications of genetically engineered microorganisms for bioremediation processes". *Current Opinion in Biotechnology* 11 (3): 286–289. Elsevier BV. doi:10.1016/s0958-1669(00)00097-5.

Scheller, Ulrich et al. 1998. "Oxygenation cascade in conversion of n-alkanes to α,ω-Dioic acids catalyzed by cytochrome P450 52A3". *Journal of Biological Chemistry* 273 (49): 32528–32534. Elsevier BV. doi:10.1074/jbc.273.49.32528.

Shankar, M.K. and Srivastava, S.K. 2017. *Genetically modified microbes for bio-remediation of oil spills in marine environment*. Bioremediation: Current Research and Applications. www.researchgate.net/publication/324112035_Genetically_Modified_Microbes_for_Bio-remediation_of_Oil_Spills_in_Marine_Environment.

Singh, Tripti et al. 2020. "An effective approach for the degradation of phenolic waste". *Abatement of Environmental Pollutants*: 203–243. Elsevier. doi:10.1016/b978-0-12-818095-2.00011-4.

Smits, Theo H.M. et al. 2002. "Functional analysis of alkane hydroxylases from gram-negative and gram-positive bacteria". *Journal of Bacteriology* 184 (6): 1733–1742. American Society for Microbiology. doi:10.1128/jb.184.6.1733-1742.2002.

Speight, James G. 2011a. "Chemical and physical properties of hydrocarbons". *Handbook of Industrial Hydrocarbon Processes*: 325–353. Elsevier. doi:10.1016/b978-0-7506-8632-7.10009-x.

Speight, James G. 2011b. "Environmental effects of hydrocarbons". *Handbook of Industrial Hydrocarbon Processes*: 539–576. Elsevier. doi:10.1016/b978-0-7506-8632-7.10015-5.

Speight, James G. 2011c. "Fuels for fuel cells". *Fuel Cells: Technologies for Fuel Processing*: 29–48. Elsevier. doi:10.1016/b978-0-444-53563-4.10003-3.

Speight, James G. 2018. "Mechanisms of transformation". *Reaction Mechanisms in Environmental Engineering*: 337–384. Elsevier. doi:10.1016/b978-0-12-804422-3.00010-9.

Strong, Lisa C. et al. 2000. "Field-scale remediation of atrazine-contaminated soil using recombinant Escherichia coli expressing atrazine chlorohydrolase". *Environmental Microbiology* 2 (1): 91–98. Wiley. doi:10.1046/j.1462-2920.2000.00079.x.

Suyama, A. et al. 1996. "Engineering hybrid pseudomonads capable of utilizing a wide range of aromatic hydrocarbons and of efficient degradation of trichloroethylene". *Journal of Bacteriology* 178 (14): 4039–4046. American Society for Microbiology. doi:10.1128/jb.178.14.4039-4046.1996.

Taghavi, Safiyh et al. 2005. "Horizontal gene transfer to endogenous endophytic bacteria from poplar improves phytoremediation of toluene". *Applied and Environmental Microbiology* 71 (12): 8500–8505. American Society for Microbiology. doi:10.1128/aem.71.12.8500-8505.2005.

Trögl, Josef et al. 2012. "Pseudomonas fluorescens HK44: Lessons learned from a model whole-cell bioreporter with a broad application history". *Sensors* 12 (2): 1544–1571. MDPI AG. doi:10.3390/s120201544.

Van Beilen, Jan B. et al. 2002. "Alkane hydroxylase homologues in Gram-positive strains". *Environmental Microbiology* 4 (11): 676–682. Wiley. doi:10.1046/j.1462-2920.2002.00355.x.

Van Beilen, Jan B. and Funhoff, Enrico G. 2007. "Alkane hydroxylases involved in microbial alkane degradation". *Applied Microbiology and Biotechnology* 74 (1): 13–21. Springer Science and Business Media LLC. doi:10.1007/s00253-006-0748-0.

Van Beilen, Jan B., Wubbolts, Marcel G. and Witholt, Bernard. 1994. "Genetics of alkane oxidation by Pseudomonas oleovorans". *Biodegradation* 5 (3–4): 161–174. Springer Science and Business Media LLC. doi:10.1007/bf00696457.

Varjani, S.J. 2017. "Microbial degradation of petroleum hydrocarbons". *Bioresource Technology* 223: 277–286. Elsevier BV. doi:10.1016/j.biortech.2016.10.037.

Wilkes, Heinz et al. 2016. "Metabolism of hydrocarbons in n-alkane-utilizing anaerobic bacteria". *Journal of Molecular Microbiology and Biotechnology* 26 (1–3): 138–151. S. Karger AG. doi:10.1159/000442160.

Wnek, Shawn M. et al. 2018. "Forensic aspects of airborne constituents following releases of crude oil into the environment". *Oil Spill Environmental Forensics Case Studies*: 87–115. Elsevier. doi:10.1016/b978-0-12-804434-6.00005-7.

Wu, Ben-Zen et al. 2006. "Measurement of non-methane hydrocarbons in Taipei city and their impact on ozone formation in relation to air quality". *Analytica Chimica Acta* 576 (1): 91–99. Elsevier BV. doi:10.1016/j.aca.2006.03.009.

Xu, Peng et al. 2014. "Isolation of a naphthalene-degrading strain from activated sludge and bioaugmentation with it in a MBR treating coal gasification wastewater". *Bulletin of Environmental Contamination and Toxicology* 94 (3): 358–364. Springer Science and Business Media LLC. doi:10.1007/s00128-014-1366-7.

Zhou, Yong, Wu, Jiangtao and Lemmon, Eric W. 2012. "Thermodynamic properties of o-Xylene, m-Xylene, p-Xylene, and ethylbenzene". *Journal of Physical and Chemical Reference Data* 41 (2): 023103–023103–26. AIP Publishing. doi:10.1063/1.3703506.

2 Application of Genetically Modified Microorganisms for Bioremediation of Polluted Environments

Moises Bustamante-Torres, Odalys Torres,
Sofía Abad-Sojos, Samantha Pardo and Emilio Bucio

INTRODUCTION

A vast amount of pollutants is produced worldwide, causing significant damage to ecosystems. Conventional technologies have been established to remove contaminants from the environment. However, these methods are expensive and come up with high processing costs, toxic and expensive reagent necessities, and involve many chemicals that produce secondary pollution, making their applications very restricted (Dasgupta et al. 2015).

The decades of contamination generated by industrialization has culminated in one of the most severe problems of our times (Sayler and Ripp 2000; Saxena et al. 2020). Several contaminants generated by anthropogenic activities pollute significant parts of the environment, causing considerable damage to ecosystems and human health (Samaksaman et al. 2016). Chemical, petrochemical, agricultural, and pharmaceutical industries produce a considerable quantity of pollutants. Table 2.1 describes some types of contaminants containing toxic, mutagenic, carcinogenic, and teratogenic effects. For instance, excessive mining, fossil fuel burning, and agricultural waste can lead to the liberation of high amounts of toxic heavy metals with acute secondary effects on human health (Wijnhoven et al. 2007). The prior physical and chemical techniques used to remediate these problems are not making progress. They are inadequate at eliminating existing recalcitrant pollutants that are challenging to degrade (Saxena et al. 2020).

Soil, water and air are essential resources on our planet. Therefore, bioremediation arises as an effective and low-cost solution to rescue those resources. Bioremediation is the application of biological means to the degradation of pollutants. These pollutants can be materials or substances that are hazardous due to their high toxicity, accumulation, and difficult removal. Bioremediation is being developed to treat air, water, and soil against such pollutants.

BIOREMEDIATION

Bioremediation is an emerging technology which can be simultaneously used with other physical and chemical treatment methods for complete management of a variety of environmental pollutants (Singh et al. 2020). A wide range of microorganisms are highly used for this bioremediation. Bioremediation using microorganisms is cost-effective and environmentally friendly. Microorganisms can degrade and transform pollutants without causing secondary contamination and can handle the most hostile environments. Increasing environmental pollution problems have led to new methods to improve bioremediation (Saxena et al. 2020).

DOI: 10.1201/9781003188568-2

TABLE 2.1

Examples of Hazardous Pollutants and Their Effects

Source of contaminants	Type of contaminants	Effects	Reference
Excessive mining, agricultural waste, burning fossil fuels	Toxic heavy metals: U, Cd, Cr, Ni, Pb, Zn, Ag, Co, Cu, and metalloids (As)	Mutagenic and carcinogenic effects	(Saxena et al. 2020)
Chemical industries	Benzenes, toluene, polychlorinated biphenyls (PCBs), polyaromatic hydrocarbons (PAHs), dioxins, nitro-aromatics, dyes, polymers, pesticides, explosives, chlorinated organic, and pharmaceuticals	Hazardous compounds, high toxicity, alteration of the ecological balance	

Microbe-assisted bioremediation typically involves the secretion of enzymes that participate in metabolic pathways that allow the degradation of a toxic substance into harmless compounds (Rucká, Nešvera, and Pátek 2017). These enzymes help the degradation of dangerous and recalcitrant pollutants. Bacteria of the genera *Sphingomonas, Pseudomonas, Burkholderia, Dehalococcoides, Rhodococcus, Comamonas, Alcaligenes*, and *Ralstonia* have been used for the conversion of contaminating compounds into non-toxic compounds. These microorganisms were able to reduce or even eliminate the pollutants (Liu et al. 2019).

Humans provide the suitable conditions to the microbes to regulate the bioremediation process. Rate, pH, temperature, food, and nutrient availability are essential for optimal microbial biodegradation (Chen, Wang, and Yang 2016). Microbial degradation is an extensive process that occurs all the time in nature. It takes place everywhere, and is commonly seen in the degradation of metals. However, microbes have limitations, such as low degradability and accumulation. Managing pollutants spends a lot of time and energy, and in some cases, pollutants are impossible to degrade completely (Saxena et al. 2020). Given the levels of contamination that humanity has reached today, the properties of microbial degradation are not sufficient. There are limits to the methods currently used in bioremediation. Consequently, genetically modified microbes (GEMs) have entered the discussion (Pant et al. 2020).

GENETICALLY ENGINEERED MICROORGANISMS (GEMS)

Scientists have endeavored to take full advantage of the properties of these microorganisms. Many enzymes, genes, and metabolic pathways are studied today for applications against environmental pollution. GEMs have potential that shows superior results compared to previously established conventional remediation methods. Figure 2.1 depicts the comparison of both standard and engineered bacteria against pollutants. The development of efficiency and optimization strategies is essential to increase these microorganisms' benefits (Liu et al. 2019). For instance, bacteria, fungi, and algae grind to bioremediate important organic pollutants and even reduce oil spills (Pant et al. 2020).

Genetic engineering techniques can be effectively be used to degrade heavy metals and other contaminants. Microbiologists and molecular biologists have observed the effective use of these GEMs. Heavy metal pollution is widespread, disturbing the natural composition of the environment. Further investigation into potential microbial strains is needed. Genetically modified organisms can be successfully employed against different pollutants under laboratory conditions. The scaling of this technology to optimal *in-situ* bioremediation is imperative for boosting bioremediation processes.

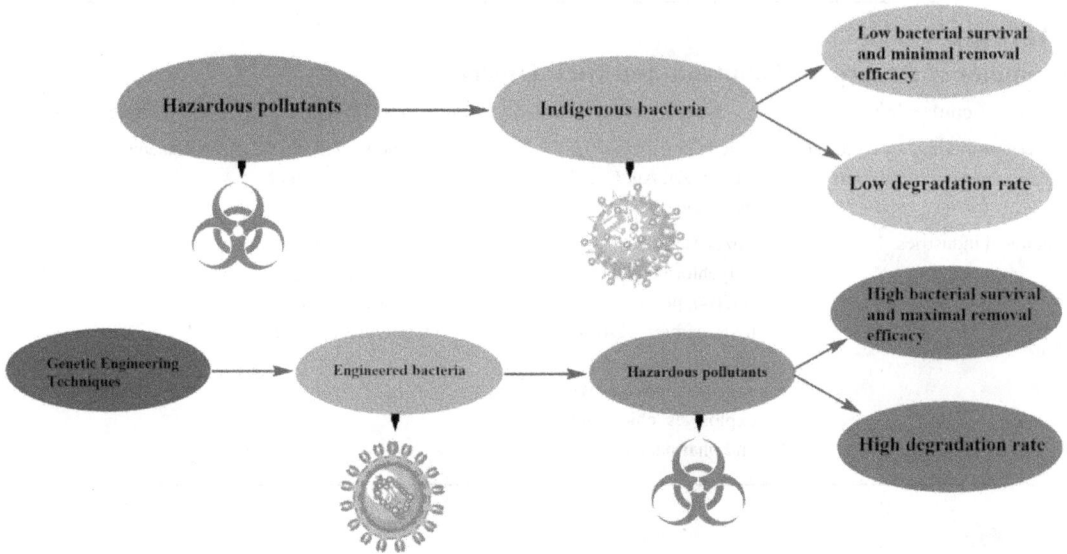

FIGURE 2.1 Comparison of a standard and engineered bacteria against pollutants.

GEMs are an effective alternative to counter contamination. GEM-based bioremediation uses living organisms, especially plants and microorganisms, to reduce, eliminate, transform, and detoxify the adverse products present in soils, water, and air. According to Kumar et al. 2020, there are four ways that bioremediation with GEMS could be employed:

1) Enzyme structure and function
2) Degradative metabolic pathways
3) Regulation, appraisal, and generation of metabolic pathways
4) Bioreporter biosensor development using bio affinity.

Additionally, the gene technology process in microorganisms can be monitored by molecular techniques such as ribosomal intergenic spacer analysis (RISA), amplified ribosomal DNA restriction enzymes (ARDRA), and denaturing gradient gel electrophoresis (DGGE) (Kumar et al. 2020).

DNA Recombinant Technology

The microorganisms must be adapted to express their genetic systems properly. The insertion of specific genes can improve that system known as recombinant DNA. With this method, we modify the bacteria, fungi, and other organisms according to the trait we intend them to express (Liu et al. 2019; Singh et al. 2011). A vector that can be a phage, plasmid, or virus incorporates our gene of interest, and this gene is later expressed in the appropriated host of choice (Gupta and Walther 2014). With microbial engineering, the process gets optimized and improves the bioremediation method. Table 2.2 shows the several tools that are required for the process of bioremediation.

Rapid development in this field and advanced molecular biology techniques have enabled:

- Increase in copy number, deletion, and/or changes in desired genes synthesizing particular enzymes of a metabolic routes,
- Reduction in limitations associated with metabolic routes,
- Improvement in energy generation processes, and
- The incorporation of desired genes with novel characteristics (Kumar et al. 2020; Liu et al. 2006; Shimizu 2002; Timmis and Pieper 1999).

TABLE 2.2

Genetic Tools Essential for the GEM Process

GEM technology materials	References
DNA ligase	(Strauss and Sax 2016)
Vector	
Reverse transcriptase	
Alkaline phosphatase	
Exonuclease	
Linker	
Terminal deoxynucleotidyl transferase	
Adapter molecules	

TABLE 2.3

Model Traits of Microbial Strain Characteristics for Heavy Metal Bioremediation

Essential traits for biodegradation	References
• Hold genes of metal homeostasis	(Kamthan et al. 2016)
• Biodegradative enzymes	
• Metal uptake	
• Synthesis of metal chelators	
• Genes to survive biotic and abiotic stress conditions	

GEMs for Ecological Release Alongside Contaminants

Engineered bacteria for bioremediation, biocatalysis, or biosensing *in-situ* necessarily require hosts, genetic tools, and even methodological approaches that vary widely from those used in laboratory conditions (de Lorenzo 2009). The application of these microorganisms for successful *in-situ* bioremediation using GMEs requires an expanded knowledge of biotechnology, ecology, engineering, and biochemical processes (Liu et al. 2019). Cost-effective bioremediation of contaminated sites requires a microbial strain that researchers chose for the function of its potential, growth, nutrition response—then, the engineering process is possible. Many other points must be considered for an optimal engineering design. Traits vary in the necessity of the polluted site and the microbial strain. Table 2.3 describes the specific microbial characteristics needed in the heavy metal bioremediation process. The formation of bacterial consortia for an area contaminated by multiple pollutants is also an alternative.

REGULATIONS REGARDING GEMS FOR BIOREMEDIATION

Since the beginning of civilization, humans have taken advantage of microorganisms for food and agricultural purposes (Wozniak and McHughen 2012). The first GEMs were developed in the 1970s. Known as "superbugs", their development was based on plasmid transfer to degrade petroleum-derived chemicals such as hexane, toluene, xylene, naphthalene, and camphor, developed principally for the potential application of bioremediation of polluted sites (Wozniak and McHughen 2012). These modified organisms can degrade environmental pollutants and detoxify a contaminated matrix to protect the environment and public well-being (Saxena et al. 2020). In 1986, the Coordinated Framework of Regulation of Biotechnology from the Office of Science and

TABLE 2.4

Genetic Modifications Traits of Microorganisms Applied to Crops

Crop GEM benefits	References
• Tolerance to drought and water-related stress • Biofortified crops: more efficient nutrient delivery and provision of micronutrients • Biopharming: High-value-added molecules	(Conko et al. 2016)

Technology Policy appeared in the United States (Wozniak and McHughen 2012). In 1989, the National Research Council (NRC) outlined a rich-based approach, which is currently being reevaluated after the two decades of experience surrounding genetically engineered organisms in order to diminish the unnecessary regulatory obstacles that GEMs face (Conko et al. 2016).

The molecular engineering techniques and tools resulting from this technology were tagged as hazardous for the natural ecosystem. Despite the lack of scientific support for new risks related to these technologies, regulatory strictures are still growing around them. The biological techniques applied to GEMs are still developing. Genome editing, grafting onto transgenic rootstock, precision breeding, agroinfiltration, synthetic biology, oligo-directed mutagenesis, and reverse breeding are techniques still restricted by the regulatory impediments surrounding GEMs (Conko et al. 2016).

Table 2.4 indicates some benefits of genetically modified crops. However, there are still restrictions for these new technologies. Most of the regulatory scrutiny around the commercialization of GE organisms are inversely proportional to the risks of GEMs, are generally are not justifiable, and do not possess any defensible scientific evidence supporting the restrictions.

Consider that the corn that we know and eat today has undergone a gradual amount of modification over the years, due to the crop selectivity we have employed. The plant we eat today does not resemble the original grass-like plant teosinte from which corn originates. The application of GEMs is limited by the fear of the collective towards modified organisms. Genetic modification appears as a concept beyond the understanding of society.

Regulatory Reform

The reform of regulations surrounding GEMs has been discussed for decades, and it is necessary for the following reasons:

- The technologies applied to the genetic engineering of microorganisms do not present unique or different risks in comparison to other breeding or genetic alteration approaches (National Research Council (NRC) 1989; National Research Council (NRC) 1987);
- Authoritative reports in the UK and Canada have accepted the previously mentioned facts (United Kingdom Advisory Committee on Releases to the Environment (ACRE) 2013);
- The commercial production of GEMs started in 1996. After 40 years of experimentation with these organisms, results have shown no contradictions to the initial conclusions. Adverse effects on ecosystems or human health have not been reported. There are no incremental risks associated with the usage of new crop varieties (Nicolia et al. 2013; Van Eenennaam and Young 2014);
- The recombinant DNA technology techniques have been improved. Genome editing approaches are more predictable and precise. Techniques such as Zinc-finger nucleases (ZFNs), Transcription Activator-like effector nucleases (TALENs), and Clustered Regularly Interspersed Short Palindromic Repeats (CRISPR-Cas9) are the future of GEMs.

The Scientific Foundation for the Regulation of GEMs

Many scientists emphasize that the risks of GEMS relate to insufficient control of their unintended ramifications (Then, Kawall, and Valenzuela 2020). The US NRC in 1978 concluded that the aspects to be considered for the modification and selection of a microorganism in field testing must include: 1) complete knowledge of the microorganisms' essential traits of interest, and testing under environmental conditions; 2) the viability of confining and controlling the microorganism, 3) the possibility of harmful effects in case of the breakage of confinement and control (National Research Council (NRC) 1989).

In theory, the fundamental scientific principles are considered by US policymakers. However, the products' risk-related characteristic has been taken as an approach to limit the regulation process of genetic modification (OSTP 1986). The fact that the organism is genetically modified and the focus of the policies in modification by a particular technique or process exposes the central problem in GEMs regulations. The biological techniques for genetic modifications keep improving but the regulations still limit the use of the final products. Canada is the only country with a risk-based product-focused regulation framework that leads to the proper regulation and development of genetically modified organisms.

The US Environmental Protection Agency (EPA) manages pesticide degradation using GEMs. The Toxic Substances Control Act was expedited in 1977 to review, regulate, and recommend experimental investigation related to the application of GEMs at scale. The European Union (EU) possesses the European Economic Community (EEC) directives 90/220 and 90/219 to settle the place of GEMs in the natural ecosystem. The Canadian Government possess a Federal Regulatory framework for Biotechnology expedited in 1993 specifically for the removal and treatment of pollutants using GEMs. In India, the regulatory Environmental Protection Act regulates the application and utilization of GEMs and their derivatives since 1986. Finally, the United Nations Environmental Program has regulations regarding GEMs to control them and the products of biotechnological approaches as laid out in the Convention of Biological Diversity (CBD) (Kumar et al. 2020).

Given the knowledge that we already have about regulatory regimes worldwide, including the EU, US EPA, and USDA regulations, none are scientifically based or justifiable. There is a massive waste of resources associated with limiting these innovative technologies that can improve public and environmental safety instead of compromising it. The insertion of DNA into a genome via DNA technologies does not increase or generate unique risks for the environment or society compared with previous technologies. The excessive regulation of genetic modification technologies affects small companies, public research, and universities dealing with costly regulatory requirements. Grant programs are also limited by the restrictions putting the research groups at competitive disadvantages, and students cannot learn appropriately about these technologies (Conko et al. 2016).

Regulations of the targeted risks of genetically engineered microbes for environmental bioremediation is essential. Through novel technologies, we can improve the risk assessment processes and open the door to applying GE technologies (Wozniak and McHughen 2012). The addition of a transgene in a bacterial genome altering its phenotypical characteristics is guided by the same biochemical and genetic pathways naturally occurring in microbes (National Research Council (NRC) 1989; Wozniak and McHughen 2012). New biological techniques are being developed to mitigate the concerns about the implications of these genetic modification technologies and mitigate the impact of the regulatory limitations. Scientists are implementing these technologies under the regulatory procedures, opening opportunities for these technologies to be applied worldwide.

Moreover, people's scrutiny about putting GEMs on the market is evident. There is polarization in society regarding genetically modified organisms, highlighting the importance of public education in GE field techniques. The fear of new technologies needs to be placated for advancement and development.

ADVANCES IN MONITORING AND CONTROL STRATEGIES FOR BIOREMEDIATION USING GEMs

Monitoring of GEMs by Reporter Genes

The use of a sensing element (promoter) and a reporter gene facilitates the monitoring of GEMs *in situ*. In this case, the plasmid contains a reporter gene fused with a promoter. The promoter acts in the presence of specific environmental pollutants while the sensing element regulates the expression of a gene under the influence of external, physical environmental signals. Consequently, the use of reporter genes allows the measurement of the activity of the promoter, giving a reading of the concentration of the pollutant present in the environment. Microorganisms that carry this function are denominated biosensors (Paitan et al. 2004). When the environmental pollutant activates the sensing element, it promotes the transcription of a protein that performs the bioremediation process. Therefore, the promoter drives the transcription of the reporter gene (Köhler, Belkin, and Schmid 2000). Some of the most used and studied reporter proteins are β-lactamase, β-galactosidase, β-glucuronidase, alkaline phosphatase, and luciferase. The easy detectability and measurement of activity are the significant advantages of these reporter proteins (Kumar et al. 2013).

Pseudomonas fluorescens HK44 was the first GEM approved in the United States for bioremediation of soils. The naphthalene catabolic plasmid (pUTK21) of this GEM carries a *lux* gene transcribed explicitly in the presence of polyaromatic hydrocarbons. Consequently, the transcript of the *lux* gene luciferase produces a detectable bioluminescent signal induced by the presence of the pollutant in the soil during the bioremediation process. Through the measurement of this bioluminescence, it is possible to perform real-time tracking of the GEMs and simultaneously obtain a measurement of polyaromatic hydrocarbon levels in the soil. *P. fluorescens* HK44 can survive approximately two years in the soil, performing bioremediation and bioluminescence functions continuously, activated by the presence of the pollutant (Figure 2.2) (Garbisu and Alkorta 1999; Ripp et al. 2000).

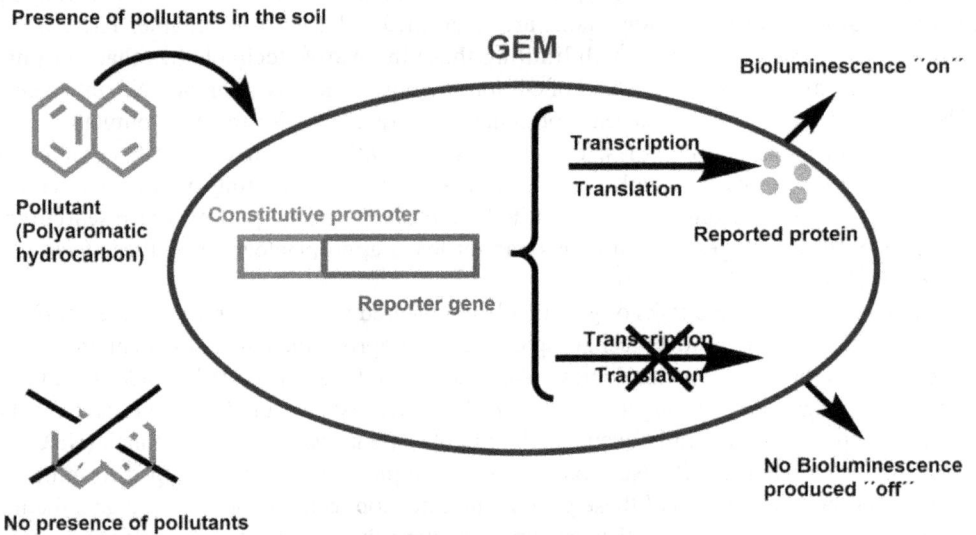

FIGURE 2.2 Bioreported GEM monitored by luminescence. The plasmid of the GEM contains a constitutive promoter fused with a reported gene. Upon exposure to the soil pollutant, the reporter genes transcribed and translated into reporter proteins (luciferase, in the case of *Pseudomonas fluorescent* HK44) provokes a detectable bioluminescent signal. If the pollutant is not present in the soil, the bioluminescent and bioremediation functions are not produced.

Control of GEMs by Mini-Transposon and *gef* Genes

Mini-transposons and the expression of the *gef* genes are two strategies used in *Pseudomonas putida* to control the risks related to horizontal gene transfer. Mini-transposons are functional segments of DNA that could be organized artificially through genetic engineering. These short elements are easy to move or position in other chromosomes or plasmids. Mini-transposons allow stable integration of recombinant genes into the chromosomes of the host. The transfer of genetically modified genes to the environment is avoided by synthesizing mini-transposons with non-antibiotic resistance systems. For the degradation of toluene, mini-transposons were introduced into the genome of *P. putida*. The mini-transposons integrated antibiotic resistance markers are eliminated after the target gene sequence in the host chromosome (De Lorenzo et al. 1998; Herrero, De Lorenzo, and Timmis 1990). The *gef* gene is denominated as a killing gene due to its encoded porin-inducing protein. This porin protein is translated and transcribed depending on the presence or absence of a specific environmental signal. The strain of *P. putida* is capable of degrading the pollutant 3-methylbenzoate (3-MB). The *gef* gene into *P. putida* is induced when there is an absence of the pollutant 3-MB. In this way, after the bioremediation process of 3-MB, when this pollutant is depleted, the *gef* gene is transcribed and translated into porin proteins that kill *P. putida* (Kumar et al. 2017).

Control GEMs Using "Suicidal" Mechanisms

GEMs may be monitored and controlled by killer-antidote mechanisms. Such mechanisms are helpful for the eradication of the GEM population after the bioremediation process. Two plasmid genes are required for this mechanism. One gene codes for toxin protein (killer), and the other codes for the antidote, which inhibits the transcription of the toxic mRNA (Figure 2.3). The

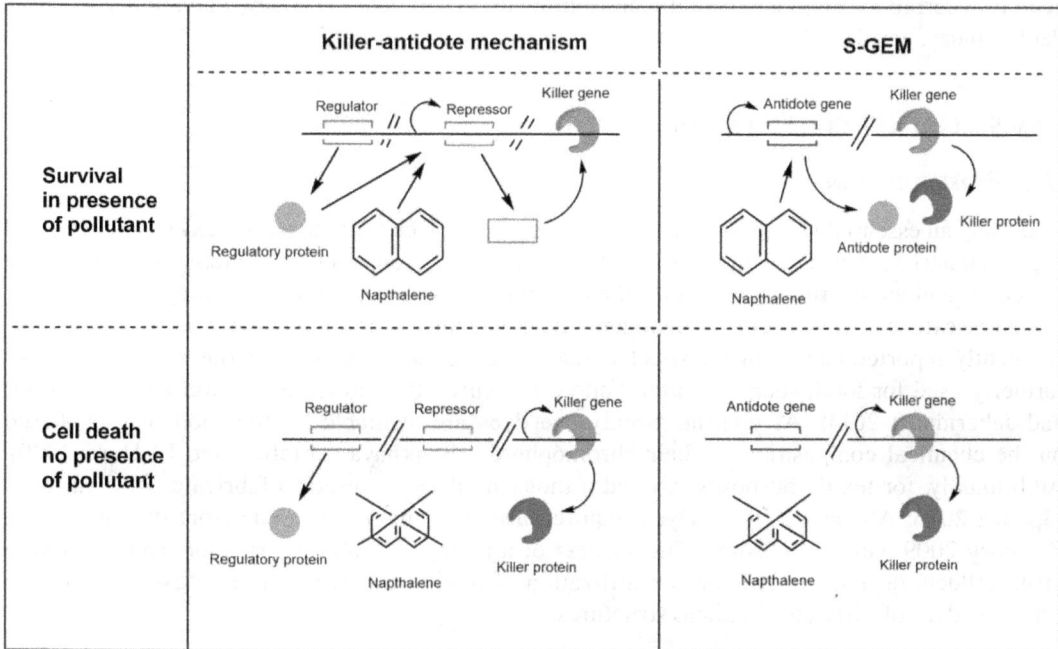

FIGURE 2.3 GEMs controlled by suicidal mechanisms. In the killer-antidote mechanism, the pollutant provokes the transcription of a repressor protein avoiding the transcription of the killer protein to allow GEM survival. Lacking the pollutant in the soil, the killer gene transcribes the killer protein leading to cell death. In the S-GEM mechanism, the pollutant present in the soil allows the transcription of the antidote protein which binds to the killer protein leading to GEM survival. In the absence of the pollutant in the soil, the S-GEM continuously produces the killer protein, leading to cell death.

antidote in this kind of mechanism could be a protein or an antisense RNA; both of them have the function of inactivating the action of the killer toxin and thus preventing cell death. The killer gene is under the control of the pollutant. In the absence of the pollutant, the killer gene is switched on, leading to the death of the microorganism. In the presence of the pollutant, the repression of the antidote avoids the transcription of the killer protein, leading to the survival of the GEM. The antidote protein has a higher decay rate than the killer protein produced in excess with more stability. The differential decay rate of the proteins allows the rapid (within minutes) activation of the killer system after depleting the pollutant. The potential risks of uncontrolled proliferation and survival of GEMs in the environment are significantly reduced using this system (Keasling and Bang 1998; Kumar et al. 2013; Pandey, Paul, and Jain 2005; Paul, Pandey, and Jain 2005).

A novel strategy of Suicidal Genetically Engineered Microorganisms (S-GEMs) was designed to limit their survivability in the environment after performing the bioremediation purpose, reducing potential risks for the environment. S-GEMs are constructed to express a killer gene, which is negatively regulated by the antidote. The killer gene would be continuously expressed while the antidote is explicitly expressed under the presence of the pollutant in the environment. In other words, the pollutant induces the constant production of the antidote, avoiding the killer gene's action. Consequently, the depletion of the pollutant in the environment (bioremediation) drives the "suicidal" action of the microorganism (Figure 2.3). This strategy could be especially useful in the bioremediation of heterogeneous mixtures of pollutants. S-GEMs have a low probability of random mutations in the antidote mechanism because the killer mechanism is always activated, leading to cell death with or without any mutation into the antidote mechanism. The possibility of horizontal gene transfer using this suicidal mechanism is rare because the killer gene transferred to recipients would lead to immediate cell death (Pandey, Paul, and Jain 2005; Paul, Pandey, and Jain 2005).

GEMS AGAINST POLLUTANTS

Dyes Bioremediation

Dyes play an essential role in large industries such as food, cosmetics, paint, textiles, leather, and paper industries. A dye is a molecule with two chemical groups: chromophores, which provide the color, and auxochromes, which fix the dye molecules onto the tissue (Exbrayat 2016). Dyes should be safe, with no toxicity, carcinogenicity, mutagenicity, or allergenicity; however, the most frequently reported causes of unexpected side effects of garments are textile dyes. Some dyes formerly used for food, such as Butter Yellow, are currently known to be carcinogenic (Nikfar and Jaberidoost 2014). At present, twenty-five dyes are available on the food market, based on the chemical composition of their chromophores (Benkhaya, M'rabet, and El Harfi 2020). Additionally, for textile purposes, around a thousand dyes are used to fabricate color varieties (Sponza 2006; Abe et al. 2019). Dyes require various chemical reactions from dye precursors (Gregory 2009; Guo et al. 2018). The success of a biological process for color removal from a given effluent depends in part on the utilization of microorganisms that effectively decolorize synthetic dyes of different chemical structures.

Dye Types

An enormous number of dyes are used in industry. Leather, paper, textile, food coloring, cosmetics, and color fabrics work with thousands of classified dyes. More than three thousand azo dyes have been reported, the most common ones being Sandolan Yellow, Maxilon Blue GRL, and Astrazon Red GTLN (S. Varjani et al. 2020a). As we mentioned before, dyes have at least one chromophore in their structure and can absorb visible light in a spectrum between 400–700nm. Table 2.5 describes dyes classified based on their structure or application.

TABLE 2.5

Dye Classification According to the Function of Their Structure or Application

Classification of dyes

Based on its structure	Based on its application	Reference
Azodye	Acid dye	(S. Varjani et al. 2020b)
Nitro dye	Basic dye	
Phthalein dye	Direct dye	
Triphenyl methane dye	Ingrain dye	
Indigoid dye	Sparse dye	
Anthraquinone dye	Moderate dye	
	Vat dye	
	Reactive dyes	

TABLE 2.6

Required Processes for Dye Intermediates Generation

Reactions for production of dyes intermediates			References
Electrophilic substitution		Monosubstituted products reached by the attack of the Ortho, Meta, or Para position of an unsubstituted benzene ring during an electrophilic attack to the tetrahedral carbon atom; a proton is lost	(Gregory 2009)
Nucleophilic substitution		Nucleophilic reagent with an electron pair (Neutral or charged particle). Replacement by substitutions within the aromatic nucleus	(Yu et al. 2019)
Unit process	Oxidation	Oxygen used for the removal of an H atom from a molecule, replaced by particles with less oxidative responses	(Gregory 2009)
	Reduction	Compound conversion into arylene diamine or arylamine from an aromatic nitro or dinitro takes place	(Gregory 2009)
	Nitration	Introduction of one or more nitro groups into an aromatic ring, and meta-directing groups	(Freeman and Mock 2012)

Dyes Intermediates

Intermediates of dyes are the raw material necessary for producing and synthesizing organic dyes and dye derivates. Table 2.6 shows the use of simple chemical reactions prepared from coal tar elements giving rise to these intermediates (Yu et al. 2019). In marine ecosystems, these dyes have a substantial ecotoxicological impact. Studies using *Hydra attenuatta* as a model found impacts on the health and appearance of marine life, reduced asexual reproduction, and inhibition of feeding behavior (Freeman and Mock 2012).

Dye Degradation

Industrial disposal in water effluents represents a source of water pollution generating direct and indirect risks for human health (Kunz, Mansilla, and Durán 2002; S. J. Varjani and Upasani 2017b; Bencheqroun et al. 2019). Dyes possess a good solubilizing capability. It is not easy to extract dyes from water using conventional methods (Dong et al. 2019; Lellis et al. 2019). Dyes prevent the diffusion of light through water, which generates a reduction in the level of oxygen dissolved in water.

TABLE 2.7

List of the Bacterial Strains Previously Tested in Dye Biodegradation

Bacterial strains	Dye	Reference
Proteus sp., *Pseudomonas* sp. and *Enterococcus* sp	Direct violet 51 and Tatrazine	(Chaube, Indurkar, and Moghe 2010)
Shewanella decolorationis	Amaranth	(Hong et al. 2007)
Aspergillus niger	Congo Red	(Fu and Viraraghavan 2002)
	Basic blue 9	(Fu and Viraraghavan 2000)
Rhizopus arrhizus	Reactive orange	(O'Mahony, Guibal, and Tobin 2002)
	Remazol Black	(Aksu and Teser 2000)
Phanerochaete chrysosporium	Direct Blue Methylene Blue	(Glenn, Akileswaran, and Gold 1986; Kinnunen et al. 2017)

Many physicochemical, chemical, and physical methods present solutions for dye removal (Ajaz, Shakeel, and Rehman 2019).

Lime, Ferric chloride ($FeCl_3$), Ferrous sulfate ($FeSO_4 \cdot 7H_2O$), and Alum (($Al_2SO_4)_3 \cdot 18H_2O$) are commonly employed to modify the physical condition of dye particles (Ayed et al. 2020). Other technologies such as flocculation, membranes for separation, wet oxidation precipitation, and adsorption are widely used physicochemical methods (Wang et al. 2020; Kumar et al. 2020). However, these methods present high costs, are difficult to maintain, and produce large amounts of sludge.

Biological treatment is easy, cheap, and eco-friendly, requires low preparation, and microorganisms are easy to maintain (Crini and Lichtfouse 2018). Certain microorganisms such as yeast, bacteria, and fungi are capable of mineralizing and decolorizing various dyes. Chaube et al. have used the mixed consortia of bacteria consisting of *Proteus* sp., *Pseudomonas* sp. and *Enterococcus* sp. in biodegradation and decolorization of dye (Chaube, Indurkar, and Moghe 2010). However, several researchers have identified single bacterial strains that have very high efficacy for removal of azo dyes, such as *Shewanella decolorationis* (Hong et al. 2007). Table 2.7 lists the microorganisms that come in pure or mixed microbial cultures for dye wastewater treatment. Mixed microbial cultures are very effective due to metabolic pathways that work synergistically in the dye degradation process (Roy et al. 2018; Mandal, Dasgupta, and Datta 2010). Physicochemical factors such as pH, temperature, dye structure, nutrients, heavy metals, and salt solubility of the medium are essential for the proper removal of dyes in environmental conditions. Many technologies are applied to dye degradation, such as amplified polymeric DNA, 16S rDNA sequencing, and PCR, together with new sequencing technologies (Mishra et al. 2021).

In biological degradation, the process is more eco-friendly. It allows the total mineralization of the organic compounds, avoiding the production of considerable amounts of sludge, can be conducted under aerobic or anaerobic conditions, and various microorganisms can be involved in the dye degradation and discoloration processes (S. Varjani et al. 2020a). The different metabolic pathways of the microorganisms allow for different degradation of dyes.

For example, the organisms *E. gallinarum* and *Streptomyces* S27 were reported to degrade azo dyes using the azoreductase enzyme. Other enzymes such as laccase, peroxidase, and exo-enzymes were also reported for dye degradation. Laccase presents a remarkable degradation potential for aromatic compounds (Dong et al. 2019; Bhatia et al. 2017). Currently, the dye removal plant treatment is performed using autochthonous microorganisms, through a process with high output and outstanding efficiency. The dye degradation process using GEMs can improve the process, maintaining the best possible competence due to the design of the treatment plants approaching the maximum population density (Kumar et al. 2020).

In-situ Dye Degradation

Knowledge of the physicochemical factors of wastewater is essential for developing the bacterial strain with the desired metabolic pathway for dye degradation. pH, nutrients, organic matter, dissolved oxygen, organic pollutants, and metals must be considered to design a correct process. Various contaminants (i.e., 2-naphthol, Benzene, P-aminobenzoic acid, Pyrene, Chloroaniline, Ethyldibromide) can affect the growth of the microorganisms and, therefore, the biodegradation process (Al-Amrani et al. 2014; Awad et al. 2019).

The principal factors affecting the process can be classified into two groups:

1) **Environmental Factors:**
   - **pH:** Adjusting the pH to the degrading bacteria or selecting the strain according to the function of the pH (Al-Amrani et al. 2014).
   - **Temperature:** Affects chemical processes and changes pH; extreme temperatures can kill bacteria. For faster degradation, a temperature between 30–40 °C is recommended for most bacterial strains (Das and Mishra 2017).
   - **Oxygen availability and agitation**: These are required conditions for the optimal metabolism of the aerobic and semi-aerobic bacteria. Shaking can be an excellent option for maintaining oxygenation. The presence of oxygen improves the metabolic response of the bacteria to the degradation process (S. J. Varjani and Upasani 2017a).
2) **Nutritional factors**
   - **Soluble salts:** Industry uses salts such as $NaCl$, $NaNO_3$, and Na_2CO_4 to increase ionic strength for dye fixation. These salts are released together with the dye's wastewater and the high concentration of salt affects the physical movement and biodegradation processes (R. Basutkar and T. Shivannavar 2019).
   - **Dye concentration and structure:** As with other contaminants, the dye's low concentration limits the bacterial identification process and the metabolic pathways involved in the production of the enzymes for the degradation of the pollutant. Dyes with low molecular weight cannot be degraded. Also, complex structures do not allow a correct degradation. A low weighted molecule in high concentration is optimally degraded in the ideal case (S. Varjani et al. 2020a). Dye industry wastewater disposal is a dire issue for the environment and human health. Soil and water are highly affected, and improvements in the degradation of these components are imperative.

HEAVY METALS

Heavy Metals Bioremediation

Heavy metals are significant pollutants with toxic and carcinogenic effects (Shivananju et al. 2019). Extensive industrialization and the mismanagement of industrial effluents and their discharge generate severe adverse effects on natural and human health. In China, 20,000,000 acres of farmland remain polluted by heavy metals. Sn, Pb, Cr, and Zn are present in nearly one-fifth of the overall arable zones in the countryside. Most of the crops undergo loss (around 10,000,000 tons) due to the deterioration of the soil generated by heavy metal contamination (Wu et al. 2010). The difference between heavy metals and other pollutants is that heavy metals are indestructible. These metals scale over the food chain and increase their concentration until it reaches the top—humans—and the accumulation of heavy metals in human bodies is inevitable (Wenzel et al. 2003; Crowley et al. 1991).

Physicochemical Available Methods

Heavy metal ion contamination has continuously increased worldwide with increasing urbanization, industrialization, and population. At present, a significant number of methods can be used for the extraction and removal of heavy metals, such as ion exchange, reverse osmosis, solvent extraction, precipitation, and dialysis. However, these techniques can change the original composition of the soil, and most of them are costly and have low efficiency (Xu et al. 2016; Azimi et al. 2017; Pratush, Kumar, and Hu 2018). Bioremediation techniques always appear as the eco-friendliest solution, and in this case, bacteria, fungi, and even plants are included in the degradation of heavy metals.

Bioremediation Processes

The technique applied chiefly for heavy metals bioremediation is the symbiotic system. In this case, we combine the power of phytoremediation with microbial remediation, putting the plant to work together with the microorganisms. Plants are usually great accumulators of Pb, Ni, Cr, Cu, Cr (1,000 ppm), and Zn and Ni (10,000 ppm), so plants are called hyperaccumulators. On the other hand, microbes are good at activating and removing heavy metals due to their large specific areas (Wu et al. 2010).

Heavy metals cannot be degraded but can be effectively neutralized or transformed into a less toxic form, inhibiting their harmful effects using the microbial enzymatic apparatus. There is a significant potential around genetically engineered microorganisms. The microorganisms used for heavy metal remediation can be aerobic and anaerobic. The aerobic are primarily employed. The transformation percentage of heavy metals differs according to the microorganism and the metal itself (Pratush, Kumar, and Hu 2018). Table 2.8 describes the microorganisms involved in heavy metal degradation.

Some strains have evolved and been used in bioremediation, as is shown in Table 2.8. It is important to consider that heavy metals are used daily. Some of the main pollutants derived from heavy metals present in ecosystems are known as radionuclides. These radionuclides are compounds that produce significant problems to biological life.

TABLE 2.8

List of Microorganisms Involved in Heavy Metal Degradation and Differential Transformation Percentages of Various Heavy Metals

Differential transformation percentages		Reference
Cr	27%	(Pratush, Kumar, and Hu 2018)
Co	20%	
Cd	31%	
Pb	22%	
Ni	0.7%	
Zn	0.5%	
Hg & As	18%	

Microorganisms involved in heavy metal remediation

Archaea (phylum)	*Penicillium chrysogenum*
Arthrobacter sp.	*Phanerochaete chrysosporium*
Aspergillus tereus	*Phylum Cyanobacteria sp.*
Aspergillus versicolor	*Pseudomonas aeruginosa*
Bacillus cereus	*Pseudomonas putida*
Bacillus cereus strain XMCr-6	*Pseudomonas veronii*
Bacillus subtilis	*Rhizopus oryzae (MPRO)*
Citrobacter sp.	*Rhodotorula mucilaginosa*
Crenarchaeota sp.	*Rhodotorula rubra GVa5*
Cupriavidus metallidurans	*Saccharomyces cerevisiae*
Enterobacter cloacae	*Sporosarcina ginsengisoli*
Enterobacter cloacae B2-DHA	*Streptomyces sp.*
Fungi Aspergillus fumigatus	*Yeast Candida utilis*
Gloeophyllum sepiarium	*Zoogloea ramigera*
Hansenula anomala	
Kocuria flava	

Phytoremediation

Plants have great natural potential for biodegradation of contaminants, and working together with their associated microbes, can assimilate, sequester, extract, transform, and detoxify organic and inorganic components. This technology is eco-friendly, low cost, and socially accepted but cannot be applied in vast contaminated places. In highly polluted places, plants have a limited region of effectiveness, and bioavailability is a restriction. Phytoremediation can be used in soil and water bodies for pesticides, explosives, radionucleotides, chlorinated solvents, petroleum hydrocarbons, and heavy metals. This technology is currently under-utilized in the public and private sectors (Hussain et al. 2018)

According to Hussain et al. 2018, phytoremediation methods include:

- **Phytoextraction (Phytoaccumulation and phytosequestration):** Uptake of contaminants through the roots, primarily applied in heavy metals
- **Rhizofiltration:** Removal from surface water by roots
- **Blastomycosis**: Use of seedlings
- **Caulofitration**: Use of plant shoots
- **Phytostabilization or phytoimmobilization:** Reduces bioavailability and stabilizes contaminants (sorption, precipitation, complexation, and metal valence reduction)
- **Phytovolatization:** Takes the pollutant and converts it into volatile organic pollutants, heavy metals, and metalloids
- **Phytodesalination:** Halophytic plants for salt removal, applied when there are both salt and heavy metals in the contaminated water
- **Phytotransformation (Phytodegradation):** Transformation of organic pollutants through plant enzymes in aerial parts of the plants; used for petroleum hydrocarbons, chlorinated solvents, synthetic herbicides, and pesticides
- **Rhizodegradation:** The organic components break down due to bacterial and fungal activity.

Mechanisms of Microorganism Remediation

Microorganisms can mineralize some organic pollutants into CO_2 and H_2O or metabolic intermediates because of bioremediation. These intermediates can be utilized as the principal substrate of growth for the bacterial cell. The microorganism responds primarily with the production of degradative enzymes for the selected contaminant and opposition concerning relevant heavy metals. The mechanisms generally used by the bacteria are immobilization of the pollutant, oxidation, transformation, binding, and volatilization. The most well-known methods of bioremediation for heavy metals are bioleaching, biosorption, biomineralization, biotransformation, and metal-microbe interaction. Cell membranes can be disrupted due to the contamination caused by organic solvents. The bacteria can use solvent efflux pumps as a defense mechanism for the outer cell membrane, together with other systems such as energy-dependent and plasmid-encoded metal efflux systems (ion/proton pumps, ATPases) when exposed to As, Cd, and Cr (Verma and Kuila 2019).

According to Ojuederie and Babalola (2017), there are four mechanisms used for microbial bioremediation of heavy metals:

1) Sequestration of toxic metals by cell wall components or by intracellular metal-binding proteins and peptides such as metallothioneins (MT) and phytochelatins, along with compounds such as bacterial siderophores, which are mostly catecholate, compared to fungi that produce hydroxamate siderophore;
2) Biological pathway modification to produce a blockage in metal uptake;
3) Metals are converted to innocuous forms by enzymes;
4) Metal intracellular concentration reduction through the use of precise efflux systems.

Microbes can use a significant number of strategies for heavy metals bioremediation, as summarized previously following the structure described by Pratush, Kumar, and Hu (2018).

Mobilization: The toxic metal dissolves through a redox reaction, and the resultant radionucleotides are converted into mineral organic acids. Additionally, the pH of the pollutant decreases. This process is composed of 4 steps:

1) *Enzymatic oxidation:* Heavy metals lose electrons and are converted into a less toxic state;
2) *Enzymatic reduction:* Remotion of elements from the solution;
3) *Complexation:* Metal complex formation by adding a ligand, easy mobilization, and removal. Suitable for low molecular weight organic acids and high molecular weight ligands like siderophores and toxic metals;
4) *Siderophores:* Iron chelators with specific bindings to catecholate, phenolate, or hydroxamate.

Immobilization: *Ex-situ* bioremediation removes the contaminants from the current place, and microbes immobilize the metal ions in the polluted sample. With *in-situ* biodegradation, the same process occurs in the original polluted place where nitrate-nitrate is converted into organic nitrogen, reversing the mineralization process. Immobilization of heavy metals in wastewater can happen using the following processes:

1) *Precipitation or solidification*: Sulfate-reducing bacteria form metal sulfide precipitates and counters the formation of toxic metal phosphates, thus improving the metal precipitation process;
2) *Biosorption:* Algae, bacteria, and fungi absorb metal ions. Cellulosic materials, extracellular polymeric substances, and fungal biomass are also used;
3) *Bioaccumulation:* Microorganisms accumulate the pollutants into ion pumps, ion channels, endocytosis, and lipid permeation. They make an inactive complex with metal ions.

Xenobiotics

Xenobiotics are compounds synthesized by humans that can reach different ecosystems to produce hazardous effects. The agricultural and military practices of the Word War II era are responsible for the release of organic contaminants around the world. The excessive use of agricultural products has unfavorable implications for human health. Chlorinated solvents, persistent organic pollutants (POPs), and explosives have neurotoxic, mutagenic, and carcinogenic effects. Chlorinated solvents (trichloroethylene, perchloroethylene, carbon tetrachloride, and chloroform) are the most available in the environment and come from plastics, adhesive materials, gasoline, and paints. POPs can resist photochemical degradation and are accumulated persistently in the environment in the long term (Hussain et al. 2018). POPs are fundamentally organochlorine pesticides, industrial chemicals, and their byproducts (Arslan et al. 2015). There is evidence about the bioaccumulation and biomagnification of these components in the fatty acid cells of humans and animals; the most common POP contaminants are called the "dirty dozen". Their persistence and toxicity are affecting human health. Pesticides have also shown persistence and toxicity (Martin et al. 2016).

Industrial-scale and production of explosives in military zones, firing ranges, and production plants strongly contaminate the surrounding zones. Prevalent contaminants are 2,4,6-trinitrotoluene (TNT), hexahydro-1,3,5-trinitro-1,3,5-triazine, hexogen (RDX) and octahydro-1,3,5,7 tetranitro-1,3,5,7-tetrazocine, octogen (HMX) (Ali et al. 2014; Douglas et al. 2011). Disorders of the liver, skin, and immune system were reported due to exposure to total petroleum hydrocarbons. The effects are not only detrimental, but can be lethal (Gerhardt et al. 2017). Xenobiotics are incredibly harmful to human health, causing cancer, mutation, and immune disorders, as well as reproductive

TABLE 2.9

Microorganism Strains Used for Xenobiotics Degradation

Xenobiotics degradation	Reference
Bacteria:	(Mishra et al. 2021)
Pseudomonas, Alcaligenes, Cellulosimicrobium,	
Microbacterium, Micrococcus, Methanospirillum, Aeromonas,	
Bacillus, Sphingobium, Flavobacterium, and Rhodococcus	
Fungi:	
Aspergillus, Penicillium, Trichoderma, and Fusarium	
Yeasts:	
Pichia, Rhodotorula, Candida, Aureobasidium, and Exophiala	

and congenital disabilities (Gilden, Huffling, and Sattler 2010; Kang 2014). Using the catabolic effectiveness of microorganisms, we can bioremediate these recalcitrant compounds effectively. Bioremediation and phytoremediation are usually utilized for xenobiotic degradation. The exact mechanisms for the phytoremediation of heavy metals can also be applied to xenobiotics. Synergistic interactions are also used for the xenobiotic biodegradation.

Microbial Degradation of Xenobiotics

The fundamental bioremediation approach is the removal of xenobiotics from the soil by reducing toxic recalcitrant hazardous components into harmless products, producing CO_2 and water. Table 2.9 describes a group of microorganisms already studied for xenobiotics degradation. All the microbial degradation processes are highly influenced by environmental factors such as soil, salinity, temperature, carbon source, pH, nitrogen sources, moisture, and the concentration of the inoculums (Mishra et al. 2021). Microorganisms use their horizontal gene transfer to pass the biodegradable genes and enzymes via successful application of genome editing and biochemical techniques of strain modification. The microorganism obtained can degrade several xenobiotics (Janssen and Stucki 2020).

Xenobiotics Targeted for GEM Degradation

- **Organochlorine pesticides (OCPs)** of a synthetic origin are xenobiotics highly used worldwide to control insects. Table 2.10 shows some examples of xenobiotics applied in bioremediation. These xenobiotic compounds are toxic, mutagenic, carcinogenic, endocrine disruptors, and immune suppressors. Thus, they affect human and animal health. *Bacillus, Burkholderia, Pseudomonas, Kocuria, Archromobacter, Sphingomonas,* and *Chromohalobacter* have been reported to help with this type of xenobiotics (Zhang et al. 2020).
- **Pyrethroids** such as cyhalothrin, cyfluthrin, bifenthrin, cypermethrin, and deltamethrin are broad-spectrum pesticides used for domestic pests. Pyrethroids can cause molecular toxicity as well as neurological and reproductive toxicity. They can cross the blood-brain barrier, producing motor deficits (Gammon et al. 2019; Kumar Singh et al. 2012). *Acinetobacter, Trichoderma, Roultella, Pseudomonas, Cunninghamella,* and *Bacillus* have helped in the degradation of compounds through pyrethroid hydrolases. Microbial strains can work by co-metabolism. Strains like *Flavobacterium, Sphingomonas, Arthrobacter, Azotobacter, Achromobacter, Microbacterium, Brevibacterium, Rhodococcus, Trichoderma,* and *Aspergillus* described contaminant degradation through the usage of the xenobiotic as an energy source.
- **Polychlorinated bis-phenyls (PCBs)** also possess high toxicity, which can generate a loss of pulmonary function, damage the immune system, cause bronchitis, and interfere with hormonal pathways, leading to cancer development. Researchers use processes

TABLE 2.10

List of the Most Common Xenobiotic Compounds Targeted for Bioremediation

Xenobiotics	Reference
Hexachlorobenzene	(Mishra et al. 2021)
Dichlorodiphenyltrichloroethane (DDT)	
Endosulfan	
Cypermethrin	
Cyhalothrin	
Naphthalene	
Anthracene	
Benzopyrene	
Dioxins	
Carbaryl	
Carbofuran	
Penthachlorobiphenyl (PCB)	
Di(e-Ethylhexyl)phthalate (DEHP)	

like anaerobic dehalogenation and aerobic degradation via *Pseudomonas, Rhosocossus, Comamonas, Burkholderia*, and *Bacillus* to degrade this component. Chlorobenzene acid is produced as a byproduct (Pathiraja et al. 2019; Jing, Fusi, and Kjellerup 2018).

- **Polycyclic aromatic carbons (PAHs)** come from the partial combustion of organic matter and are incredibly hazardous for animal wildlife and aquatic life; birds, especially, can be highly affected (Mishra et al. 2021). In humans, they can produce genotoxicity, mutagenesis, and cancer (Lin et al. 2020). *Sphingomonas, Sphingobium, Novosphingobium, Rhodococcus, Cunninghamella, Pleurotus ostreatus, Oscillatoria, Agmenellum quadriplicatum, Brevibacterium*, and *Nocardiodes* have been reported as helpful for the degradation of PAHs (Auti et al. 2019).
- **Phthalates or phthalic acids** are used as plasticizers to improve the flexibility and hardness of polyvinyl chloride. They possess an endocrine disruptive behaviour and can cause infertility as well as reproductive and developmental toxicity in humans and animals (Mishra). *Microbacterium, Rhodococcus, Arthrobacter, Providencia, Acinetobacter, Gordonia*, and *Pseudomonas* are effective DEHP-degrading microorganisms (Nahurira et al. 2017; Yang et al. 2018).

Finally, a consortium of microbes can be formed instead of the usage of monocultures, presenting an excellent source of bioremediation of xenobiotics. The positive mutual relationship of microorganisms and their increased biodegradation capacities makes this technology worthy of more attention. Microbial consortiums diminish the limitations that a single bacterial strain can present, increasing the degradation potential.

Synergistic Interactions for Phytoremediation of Xenobiotics

The accumulation of xenobiotics in the soil generates detrimental effects for plants. There is a deterioration in soil quality, the nutrients suffer an imbalance, and the hydrophilic conditions become extreme. Consequently, there is reduced seed germination, lower chlorophyll pigment production, reduced root turnover, disturbed root and shoot growth, and a disruption in the architecture of the plant. Thus, plants play an essential role in the bioremediation process, and phytoremediation strategies are focused on them. Legumes are the most eligible actors for the process, followed by some ornamentals, combining ecological and beautification functions. Phytoremediation is a site-specific technology. The plant is associated with microbes that promote plant growth and are also pollutant degraders for yield improvement. Synergistic interactions are a suitable and promising technology (Hussain et al. 2018).

Bacterial association adds a significant number of catabolic diversities to the degradation process. These microorganisms can degrade, transform, and accumulate many organic compounds such as PAHs, PCBs, pesticides, petroleum hydrocarbons, and halogenated hydrocarbons (Xun et al. 2015). Mainly, the process occurs outside the plant. The pollutants that cannot enter the plant because of their hydrophobic nature are taken up by microorganisms (Gerhardt et al. 2017). The rhizosphere becomes more stable than without microbial support due to the production of fewer harmful metabolites. The mineralization of organic pollutants into CO_2 and water can occur (Dzantor 2007).

Bacterial proliferation occurs in the rhizosphere area and the internal tissues. Plants support bacteria giving them an excellent place to live and all the nutrients they need. In exchange, the bacteria metabolize the hazardous components, improving both the degradation process and promoting an excellent mutualistic relationship (Feng et al. 2017). Various bacterial strains have been isolated for targeted toxic compounds, and a wide range of polluted environments can be remediated under aerobic and anaerobic conditions (Ghattas et al. 2017). Plants are capable of synthesizing biosurfactants that improve the bioavailability of organic components. The pollutants are released from the soil particles, and the microorganisms can degrade them quickly (Whitfield Åslund et al. 2010). Rhizosphere engineering is gaining increasing attention because of the potential of nutrient adjustment, insertion of transgenic strains, and exudate regulation for microenvironment modification (Hussain et al. 2018).

GEMs Against Pesticides

Pesticides are defined as mixtures or substances developed to repel, prevent, destroy, or mitigate a pest. The depletion of pests is essential in the sectors of healthcare and agriculture. The main advantages of pesticides include killing vectors of diseases, such as mosquitoes, and preventing loss of crops. Fungicides, herbicides, insecticides, and rodenticides are the types of pesticides most widely used. Certain pesticides in the soil can be degraded by native microorganisms like fungi and bacteria that take advantage of these pesticides as food sources (Vargas 1975). Disadvantageously, the use of pesticides leads to certain unwanted chemicals in soil that can be toxic for a vast number of living organisms, including human beings. High amounts of these chemicals compromise soil quality and contribute to eutrophication. The continuous use of pesticides leads to accumulation of undesirable chemicals in which concentrations can be magnified and enter into the food chain (Bernardes et al. 2015; Bhattacharjee et al. 2020; Lushchak et al. 2018).

Synthetic organophosphates (OPs) represent approximately 38% of total pesticides. The accumulation of OP pesticides in different environments represents a risk for the food chain and all the living organisms involved in it. Soil microorganisms such as *Pseudomonas diminuta* MG and *Flavobacterium* sp. strain ATCC 27551 demonstrate an ability to hydrolyze OP pesticides. These microorganisms can produce the bacterial enzyme denominated organophosphorus hydrolase (OPH) to catalyze the hydrolysis of OP pesticides and other toxic agents. Unfortunately, the rate of hydrolysis of OPs and the permeability barrier of the cell membranes of these microorganisms prevent them from being effective for bioremediation of OP pesticides (Serdar and Gibson 1985; Shimazu, Mulchandani, and Chen 2001; Singh and Walker 2006; Yang et al. 2010).

Stenotrophomonas sp. strain YC-1 is a native soil bacterium capable of surviving in environments with pesticides. This strain can produce the enzyme methyl parathion hydrolase (MPH), which can hydrolyze a wide range of OPs (Purg et al. 2016). *Stenotrophomonas* sp. strain was converted into a GEM containing a truncated ice nucleation protein (INPNC) from *Pseudomonas syringae*. INP is an outer membrane protein that acts as a template for ice nucleation, while INPNC is the truncated version of INP, which contains only the N- and C-terminal portion. INPNC can target proteins to the cell surface (Bae, Mulchandani, and Chen 2002). The GEM of *Stenotrophomonas* sp. strain demonstrates 100% activity in suspended cultures for two weeks. This strain is capable of degrading diethyl and dimethyl OPs. Moreover, *Stenotrophomonas* sp. strain can degrade in 5 hours a mixture of 6 different OP pesticides. This strain represents a promising candidate for *in-situ* bioremediation of pesticides (Yang et al. 2010).

In recent years, genetic engineering approaches have focused on constructing GEMs with multifunctional bioremediation capacities, as we can find in the studies of Cao et al. (2012); Jiang et al. (2005), Jiang et al. (2006), Liu et al. (2006), and Yuanfan et al. (2010), among others. The constant investigation for new strategies of pesticide bioremediation is paving the way for future *in-situ* implementations. In this way, GEMs can be converted into a more realistic, efficient, and cost-effective option for bioremediation. Future investigations tend to focus on GEMs that degrade pesticides, resist the toxic environments, and can be monitored and controlled *in situ*.

A genetically engineered bacterium (GEB) that degrades pesticides and emits green fluorescence as a containment system was constructed by Li et al. (2020). This GEB contains two plasmids. The first plasmid fuses the genes for carboxylesterase B1 (CarE B1) and an enhanced green fluorescent protein (EGFP) from *Culex pipiens quinquefasciatus*. CarE B1 gene encodes an enzyme capable of an intense degradation of various pesticides such as organochloride, OPs, and the insecticides carbamates and pyrethroid (Barata, Solayan, and Porte 2004; Nishi et al. 2006; Vontas, Small, and Hemingway 2000). On the other hand, the EGFP is used in the first plasmid as a control mechanism to monitor the survival of the constructed GEM. The second plasmid contains two copies of a lethal gene which encodes a lethal nuclease for *Serratia marcescens*. The GEB can degrade pesticides, be monitored by EGFP, and could contain lethal genes. Thus, this GEM provides an effective and safe mechanism for bioremediation of pesticides *in situ*.

GEMs Against Radioactive Compounds

Low radioactive wastes from sources such as hospitals, industries, universities, and so on can accumulate long-term, leading to radiological contamination (Pant et al. 2020). Synthetic chemicals like radionuclides are highly resistant to biodegradation by native microorganisms.

The bioremediation of radioactive compounds requires GEMs because they are designed to resist the toxic environment of radioactive wastes and bioremediate a mixture of compounds, compared to native microbes. Genetically engineered radiation-resistant microbes such as the *Deinococcus radiodurans* are among the most radioresistant bacteria known so far (Pant et al. 2020).

Radionuclides

A radionuclide is an atom with an unstable nucleus containing an excess of energy. These atoms can emit radiation as they are subjected to radioactive decay through the emission of alpha particles (α), beta particles (β), or gamma rays (γ) (Hanrahan 2012). Some radionuclides like U, Rn, Th, Ra, Am, and Tc are used for nuclear medicine procedures. These are toxic for cells and usually undergo radioactive decay emitting gamma rays or subatomic particles (Tahri et al. 2013; Aquino, Barbieri, and Nascimento 2011). The released energy can be absorbed by biological materials, resulting in damage (Ansoborlo and Adam-Guillermin 2012).

The presence of radionuclides causes toxicity to bacteria. The characteristic of high resistance to ionizing radiation makes *Deinococcus radiodurans* a useful microorganism for genetic engineering approaches related to radioactive compounds. In 1998, *D. radiodurans* was utilized as a host for genes encoding toluene dioxygenase from *P. putida*. This GEM was successfully utilized to oxidize 3,4-dichloro-1-butene, chlorobenzene, indole, and toluene. The *D. radiodurans* GEM has the potential for bioremediation of wastes containing radionuclides and organic solvents (Lange et al. 1998).

Radionuclides are regularly released from nuclear power plants, reprocessing plants, hospitals, industrial facilities, research institutions, and various other places. Additionally, accidental release has occurred throughout human history, causing severe damage to the atmosphere and aquatic environment (Thorne 2012). A bacterial commonly employed as a GEM is *Staphylococcus (S) Aureus*. An arsenate detoxification pathway was assembled using the *S. aureus* arsenate resistance operon and DNA shuffling mechanisms. The resulting recombinant *E. coli* had increased arsenate resistance and the ability to detoxify arsenate by reduction (Crameri et al. 1997).

ENVIRONMENTAL BIOREMEDIATION OF GEMS

AIR

Air pollution is a mixture of solid and gaseous particles caused by residues from industrial and urban activities in fixed sources such as chimneys as well as mobile sources such as automobiles and transportation in general. Additionally, the usage of chemical products contributes to the production of pollutants in the air. Therefore, to face this type of pollution, it is necessary to apply methods and treatments directly to the sources that generate or emit pollutants.

The increasing number of pollutants in the environment is of alarming concern to ecosystems. For this reason, GEMs are employed for an urban air pollution monitoring and assessment program that evolved from a World Health Organization (WHO) urban air quality monitoring pilot project that started in 1973 (UNEP/WHO 1993). Many organic pollutants, such as polychlorinated biphenyls (PCBs), polyaromatic hydrocarbons (PAHs), and pesticides, are highly resistant to degradation (Kumar et al. 2013). In addition, the excessive use of pesticides/chemical fertilizers in modern agriculture practices has led to contamination of air, land, and water, as well as adverse consequences on the health of humans and animals in several ways (Liu et al. 2019).

The treatment of pollutants in the air has been studied and carried out by techniques such as photocatalysis that have shown a reasonable reduction of bioaerosols; ~10 to ~100 times more rapidly than the rates of natural decay (Grinshpun et al. 2007). Regarding the use of GEMs, investigations have pointed to their usefulness in biosensors using luminescent genes for environmental studies. The ability to introduce the lux phenotype, coupled with specific promoters that allow its expression only in the appropriate analyte presence in different bacterial species, provides a convenient method for rapid, simple, and sensitive environmental conditions detection. Several recombinant plasmids, which carry the lux operon constitutively expressed in many gram-negative and gram-positive bacteria, have been constructed and used to transform a wide variety of bacteria (Girotti et al. 2008). However, there are difficulties in controlling factors (such as adequate cell growth medium, oxygen, and cell concentration) to effectively contact air pollutants with bacterial cells and keep them in suitable culture conditions.

Gil, Kim, and Gu (2002) used genetically modified bioluminescent bacteria to develop whole-cell biosensors to detect toxic gaseous chemicals. A solid agar medium was used as immobilization support to measure the toxicity through the direct contact of the cells with BTEX (benzene, toluene, ethylbenzene, and xylene). Tulupov, Pavlovna, and Tulupov (2006) carried out a bioassay aimed at quantitatively determining the gas prepared in the form of a solution. The gaseous component's biological activity was determined by the response of the bioassay and depended on the concentration of the dissolved component. The biological activity of the entire gaseous medium was calculated based on the biological activity of its parts. Ecolum-5.4 was used as a luminescent bacterium to determine the toxicity of benzene in the air.

WATER

Water pollution is an issue of great concern worldwide. It can be broadly divided into three main categories of contamination: organic compounds, inorganic compounds (e.g., heavy metals), and microorganisms. Treating wastewater effluents laden with heavy metals is challenging because it greatly depends on techno-economic, environmental, and social considerations (Diep, Mahadevan, and Yakunin 2018). In recent years, the number of studies on effective processes to clean up water bodies and minimize water body pollution has increased. To this end, bioremediation methods to remove toxic metals from aqueous solutions have attracted widespread attention. Due to its environmental compatibility and possible cost-effectiveness, the use of microorganisms for bioremediation shows excellent development potential.

A wide range of microorganisms, including bacteria, fungi, yeast, and algae, can act as biologically active methylating agents capable of at least modifying toxic substances. Although microorganisms cannot destroy metals, they can change their chemical properties (Coelho et al. 2015). The latest advances in genetics provide the impetus for the use of engineered microorganisms and enzymes for bioremediation. The ability of genetically modified microorganisms has been studied for the tolerance and biotransformation of heavy metals (S. Gupta and Singh 2017).

In recent years, genetically modified microorganisms or microbial preparations have been effectively used in difficult-to-treat organic wastewater in China. To degrade pesticide pollutants in sewage, research on genetically engineered microorganisms' bioremediation has been extensively carried out (Qu and Fan 2010). For example, Yang, Liu, Guo, and Qiao, in their investigation, cloned the MPD gene, the gene encoding the organophosphorus hydrolase from the chlorpyrifos-degrading *Stenotrophomonas* sp. strain *YC-1*, and they functionally expressed it in *E. coli*, where the analysis showed the recombinants completely degraded 100 mg/L of chlorpyrifos in a liquid medium within 24 hours. On the other hand, a new gene involved in reducing synthetic was cloned from *Klebsiella* sp. The F51–1–2 strain and the deduced protein have homology with members of the aldehyde-ketoreductase (AKR) superfamily, potentially reducing the double bond connecting phosphate and sulfur in organic phosphate molecules (Qu and Fan 2010; Jiang et al. 2006).

In other cases, researchers have studied the degradation of 3-chlorobenzoic acid (3CB), 4-chlorobenzoic acid (4CB), and 4-methyl benzoic acid (4MB) as a substrate in batch and continuous culture using the genetically modified microorganism *Pseudomonas* sp. 813 FRI SN45P. This strain can fully mineralize individual compounds and substrate mixtures. Maximum specific substrate conversion rates were 0.9 g/gh for 3CB and 4CB and 1.1 g/gh for 4MB. The stability of the degradation pathways of strain *Pseudomonas* sp. B13 FRI SN45P could be demonstrated in continuous cultivation over 3.5 months (734 generation times) on 3CB, 4MB, and 4CB (Müller, Deckwer, and Hecht 2000).

SOIL

Concerns about soil contamination are on the rise in all regions because it can seriously degrade ecosystems and threaten food security by affecting agricultural yields due to toxic levels of pollutants, causing crops produced in contaminated soils to be dangerous for the consumption of animals and humans. Many pollutants are transported from the soil to surface waters and groundwater, causing environmental damage through eutrophication and direct human health problems from contaminated drinking water. Pollutants also directly harm soil microorganisms and larger organisms that live in the soil, as well as impact soil biodiversity and the service that affected organisms provide, and directly affect human health.

Soil pollution by different contaminants such as heavy metals has become an essential topic in all environmental crises today. Therefore, the remediation of contaminated soils is essential, and research is continuing to develop novel and scientific rehabilitation methods. Increasingly costly physical rehabilitation methods such as inactivation or sequestration of chemicals in landfills are being replaced by science-based biological methods such as microbial degradation carried out by natural and genetically modified microorganisms. The process of importing microorganisms to the contaminated site is called bioaugmentation, enhancing the indigenous microbiota's metabolic capacities to boost bioremediation (El Fantroussi and Agathos 2005). Genetic bioaugmentation is an *in-situ* bioremediation method that stimulates the horizontal transfer of catabolic plasmids between exogenous donor cells and indigenous bacteria to increase the biodegradation potential of contaminants. Figure 2.4 illustrates the action of bioaugmentation and genetic bioaugmentation, where GEMs are applied. Genetic bioaugmentation aims to transfer relevant genes into indigenous microorganisms that have greater fitness for survival in contaminated environments, as opposed to conventional bioaugmentation, which relies on the survival of exogenous microorganisms (Ikuma and Gunsch 2012).

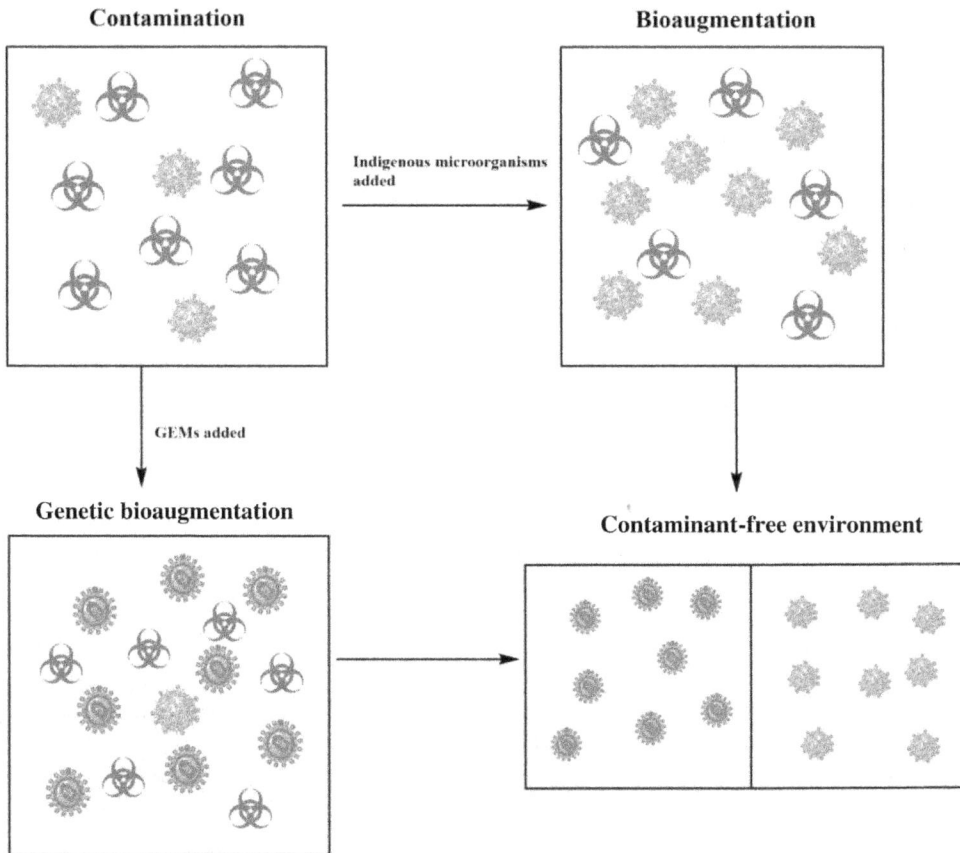

FIGURE 2.4   Bioaugmentation and genetic bioaugmentation applied to a polluted medium.

Saavedra et al. (2010), in their research, used genetically modified bacteria to degrade Polychlorobiphenyls (PCBs), a contaminant classified as "high priority" due to its toxicity, carcinogenicity, and persistence in the environment. The vast majority of microorganisms degrade PCBs incompletely, leading to the accumulation of chlorobenzoates (CBAs) as dead-end metabolites. Researchers used genetic engineering to obtain a microorganism capable of mineralizing PCB congeners, incorporating the BPH locus of *Burkholderia xenovorans* LB400, which encodes one of the most effective PCB degradation pathways, to the genome of the bacteria *Cupriavidus necator* JMP134-X3 that degrades the CBA. In one week, the genetically modified strain degraded 99% of 3-CB and 4-CB and approximately 80% of 2,4'-CB in soil. The bacterial count increased by almost two orders of magnitude in PCB-contaminated soils (Saavedra et al. 2010).

Studies on the biodegradation of aromatic compounds have also been carried out due to their use in fertilizers, herbicides, pesticides, plastics, and dyes (Duque et al. 1993). These end up in the environment due to agricultural runoff, industrial waste sites, and military operations (Duque et al. 1993). In their research, Dutta et al. (2003) genetically transferred the biodegradable plasmid pJS1 DNT (dinitrotoluene) to the *S. Meliloti* USDA strain. The results showed a reduction of more than a third of 2,4-DNT in contaminated soil of both 0.14 and 0.28 mM, and in contaminated soil of 0.55 mM it degraded 94% of the 2,4-DNT present. They concluded that enhanced bioremediation strains might not have better symbiotic efficacy than their parental strains. Still, when the toxic compound they degrade is present, they are superior to wild-type strains in terms of symbiosis and bioremediation (Dutta et al. 2003).

ACTIVATED SLUDGE ENVIRONMENTS

Sewage sludge is a slimy-looking solid, semi-solid, or liquid waste produced after old sewage (human household and industrial waste) is treated in a sewage treatment plant. The sludge can contain various organic and inorganic compounds derived from wastewater such as detergents, pesticides, oils and fats, colorants, solvents, phenols, and so on. With proper treatment, these can be used in fish farming, irrigation, forestry, horticulture, and even to produce energy in renewable biofuels (Bharathiraja et al. 2014).

Research has focused on treating hydrocarbon compounds such as phenol. The entry of a high concentration of phenol into an activated sludge process seriously damages the microbial activity if it is not previously acclimatized to the phenol. Consequently, the performance of the treatment deteriorates dramatically. It cannot be adapted to activated sludge in some cases, so it is necessary to enrich it with selected pure microbial cultures that can effectively degrade phenol to increase the rate and degree of phenol degradation (bioaugmentation). For this purpose, naturally selected microorganisms and GEMs can be used. Soda, Ike, and Fujita (1998) genetically modified *Pseudomonas putida* BH (pS10–45) in their study and analyzed it along with its unmodified counterpart to determine phenol degradation-activated sludge. The experiment was carried out in a two-phase reactor: the first in shock charges and the second in a semi-continuous way. The results showed that after the shock loads, compared to the control without GEMs, the GEM-inoculated activated sludge's phenol removal efficiency was greatly improved.

In contrast, during the semi-continuous feeds, the GEM-inoculated activated sludge settled much better than in the control process, although phenol was removed entirely in both processes. Thus, it was concluded that GEM inoculation could be a valuable means to improve activated sludge processes that treat with phenol (Soda, Ike, and Fujita 1998).

Other studies aimed to analyze the ecological stability of GEMs, that is, the period during which GEMs can survive and express helpful activity in the mixed microbial flora of the wastewater treatment process. If a GEM has low ecological stability, it can disappear from the process quickly, so the treatment's efficiency will hardly increase. For example, the survival rate of natural and genetically modified bacteria that can degrade 3-chlorobenzoate has been studied in a laboratory-scale activated sludge facility, highlighting the importance of selecting strains suitable for environmental conditions to successfully use microorganisms (McClure, Fry, and Weightman 1991b, 1991a; Fujita, Ike, and Uesugi 1994). On the other hand, the survival of *Escherichia coli* and *Pseudomonas putida* modified with the recombinant plasmid pBH500 containing the gene that encodes lacatechol 2,3-oxygenase has been studied during the activated sludge model process in the filling system for culture and extraction (FD) and continuous flow (CF) under different conditions. The results show that the mode of operation of the activated sludge process and the diversity of the native bacterial flora affect the survival of GEMs in activated sludge (Fujita, Ike, and Uesugi 1994).

OTHERS

Biodegradable Plastics

Several techniques have been studied for the degradation of plastic waste. Concern increases every moment due to the significant world production of this type of waste and its remnants in the environment. Population growth and the constant use of plastic are difficult to deal with. Genetic engineering tools have been used to manipulate microorganisms' genes and optimize their potential to biodegrade plastics in conjunction with recombinant DNA technology (Wilkes and Aristilde 2017; Jaiswal, Sharma, and Shukla 2019). Studies carried out have discovered plastic-degrading enzymes such as peroxidase and laccase, using recombinant rDNA technology. They have been incorporated into the host microorganism *E. coli* to improve its bioremediation (Sharma, Dangi, and Shukla 2018). On the other hand, research has indicated contributions to the degradation of PET plastics by the enzyme *cutinase*. This enzyme has led to genetic engineering to eradicate the formation of lumps, which is one of the main limitations in the degradation by *cutinase*, and to improve the percentage of bioremediation. The results indicated an enhanced bioremediation capacity at high temperatures by the recombined enzyme (Jaiswal, Sharma, and Shukla 2019; Shirke et al. 2018).

Textile and Food Industry

Genetic engineering can improve the generation and overexpression of enzymes in different microorganisms using recombinant DNA technology. These enzymes are more accessible to manipulate than the original ones. This technique is considered more profitable because its production can be controlled by modifying the substrate, pH, and temperature, allowing to establish specific environmental conditions for the treatment of toxic compounds (S. K. Gupta and Shukla 2016; Sharma, Dangi, and Shukla 2018). For example, recombinant peroxidase enzymes have been used in *Aspergillus* for the treatment of dyes (Sharma, Dangi, and Shukla 2018; Dua et al. 2002). Table 2.11 shows current recombinant enzyme application studies.

The application of genetic engineering in bioremediation has generated valuable advances in dealing with the environmental pollutants generated by food industry processes, which become part of the biogeochemical cycle. Among these contaminants, we have the phytosanitary compounds, which are some of the most worrisome and polluting (Thassitou and Arvanitoyannis 2001). Jaiswal, Singh, and Shukla (2019) studied the treatment of pesticides through TALEN, ZFNs, and CRISPR *Cas9* gene editing tools, as shown in Figure 2.4, obtaining the desired functions of elimination

TABLE 2.11
Applications of Recombinant Enzymes

Recombinant enzyme	Native microorganism	Engineered microorganism	Pollutants	Reference
Flavodoxin-like protein (Pst2)	*S. cerevisiae*	*E. coli*	Quinine, phenolic component, etc.	(Koch et al. 2017; Sharma, Dangi, and Shukla 2018)
Laccase (lacIIIb), Versatile (vpl2) peroxidase, Mn peroxidase, Lignin peroxidases	*T. versicolor, P. eryngii, P. chryososporium*	*P. chryososporium*	Phenolic compounds, synthetic dye, etc.	(Coconi-Linares et al. 2015; Sharma, Dangi, and Shukla 2018)
Laccase CueO	*E. coli K12*	*P. pastoris GS5115*	Synthetic dye decolorization such as malachite green, Congo red, etc.	(Ma et al. 2017; Sharma, Dangi, and Shukla 2018)
Laccase	*B. vallismortis fmb103*	*E. coli BL21*	Synthetic dye decolorization such as malachite green, Congo red, etc.	(Sun et al. 2017; Sharma, Dangi, and Shukla 2018)
Dye decolorizing peroxidase (DyP)	*G. candidum Dec1*	*A. oryzae RD005*	Dye decolorization such as malachite green, Congo red etc	(Sun et al. 2017; Sharma, Dangi, and Shukla 2018)
Mn-dependent peroxidase	*P. incarnate KUC8836*	*S. cerevisiae BY 4741*	Anthracene, phenols, dyes, etc.	(Lee et al. 2016; Sharma, Dangi, and Shukla 2018)
Mn peroxidase (Mn P)	*I. lacteus F17*	*E. coli*	Phenols, amines containing aromatic compounds, dyes, etc.	(Lin, Pan, and Cheng 2010; Sharma, Dangi, and Shukla 2018)

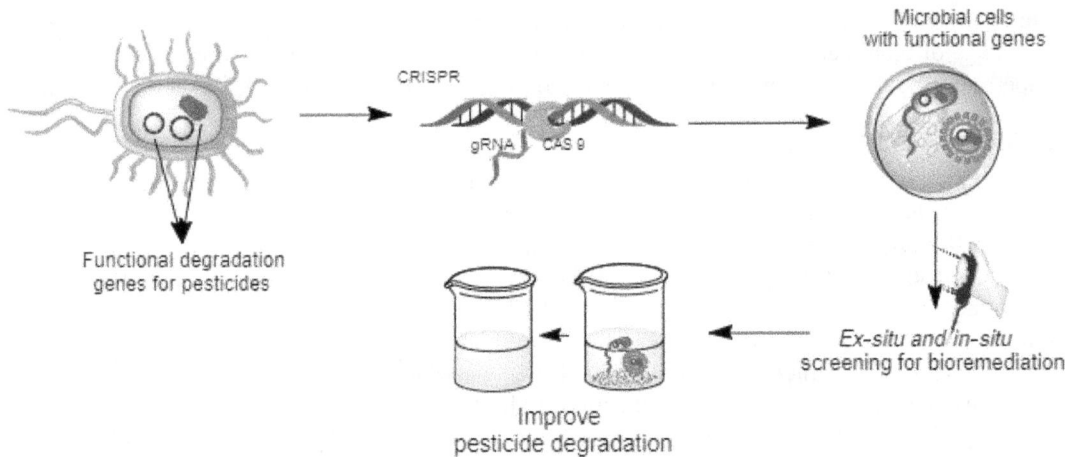

FIGURE 2.5   Using CRISPR Cas9 gene editing for pesticide bioremediation.

of the phytosanitary compound by specific microorganisms, observing admissible progress in the bioremediation of the contaminant. These tools made it possible to efficiently identify the best host microorganism for biodegradation (Jaiswal, Singh, and Shukla 2019).

Oil Industry

Bioremediation is considered to be one of the most sustainable clean technologies. Still, because it is too slow to meet the environment's urgent needs, its potential has not been fully utilized. The use of genetically modified or manipulated microorganisms to enhance the degradation of oils, especially the degradation of polyaromatic compounds and high molecular weight alkanes, has aroused great interest. Microorganisms have been developed that have an improved ability to degrade aromatic hydrocarbons and their derivatives in particular (Thomas and Ward 1992; Balba, Al-Awadhi, and Al-Daher 1998). Although technologies based on these concepts are expected to improve biore-actors' performance, experience gained from bioaugmentation tests shows that the use of GEMs will be ineffective if technologies are not developed that improve their ability to resist competition from native microbial populations (Atlas and Bartha 1992; Balba, Al-Awadhi, and Al-Daher 1998). Convenient, inexpensive, and effective methods have been developed to track genetically modified microorganisms so that their survival, transport, and ecological impact can be controlled when released into a new environment (Veal, Stokes, and Daggard 1992; Balba, Al-Awadhi, and Al-Daher 1998). Additionally, genetic engineering has provided beneficial microorganisms to report on the availability and biodegradation of polycyclic aromatic hydrocarbons through light signals that could be used as an online tool for *in-situ* monitoring of bioremediation processes (Samanta, Om, and Rakesh 2002).

POTENTIAL RISKS OF GEMS FOR BIOREMEDIATION

The main goal of constructed GEMs is to enhance the biodegradation of a wide range of pollutants. GEMs have the potential for bioremediation of air, soil, water, sludge, and other environments. Although GEMs are continuously developed for the bioremediation of polluted environments, there is a concern about their release into the environment. The survival and dispersal of GEMs without control are some of the potential risks. Further, it is crucial to consider and monitor the fate of GEMs in the environment after their use for bioremediation purposes, guaranteeing the tracking of GEMs after their release into the environment (Pandey, Paul, and Jain 2005). The release of GEMs into the environment could cause numerous potential problems such as interrupting the natural balance of ecosystems, horizontal transfer of genetically modified material to autochthonous microbes, increasing the competition for energy sources, and the gap of knowledge about the fate of byproducts created after bioremediation (Kolata 1985; Velkov 2001). The research and application of GEMs are stifled and limited due to the lack of a risk-based regulatory approach. The acquisition of governmental permissions represents a significant impediment to the application of this technology. Usually, these permissions are characterized as being complicated and time-demanding (Sayler and Ripp 2000).

Regardless of the potential environmental risks, scientists have been developing genetic engineering strategies that employ GEMs for bioremediation purposes without representing a risk to the environment. These strategies are designed to avoid environmental problems and, at the same time, enhance the biodegradation activity of GEMs. At present, GEMs are designed to be susceptible to biological containment; in other words, GEMs only perform their biodegradability function under specific selective conditions such as the presence of a specific pollutant. There is a need to design safer GEMs for environmental release that minimize potential risks (Sayler and Ripp 2000).

CONCLUSION

Conventional methods seem an unfeasible solution in current times for the remediation of pollutants. Therefore, the use of microorganisms attracts great scientific efforts. GEMs shows superior results compared to the previously established conventional remediation methods. However, despite the success, bioremediation based on GEMs is still limited to academic research. Commercial remediation currently relies on naturally occurring microbes identified at contaminated sites. Many issues surround the application of GEMs for use in bioremediation, including (1) their effectiveness compared with their counterparts present in nature; (2) their influence on indigenous microorganisms; (3) their fitness in nature; and (4) their containment. For further studies, integrating the current knowledge and improving genetic procedures would enhance hazardous contaminant biodegradation. The exploration of bacterial degradation kinetics is highly desirable. Advanced technologies applied via *in-situ* bioremediation are necessary to improve and scale the biodegradation of industrial waste and commercial-level applications (S. Varjani et al. 2020a). Currently, innovative molecular practices are needed for the characterization of new microbial strains with superior biodegradable capacities. New molecular biology techniques are starting to be probed for their application in bioremediation. Genomics, proteomics, transcriptomics, metabolomics, *in silico* models, and bioinformatic studies can lead to the obtention of strains capable of responding favorably, not only at *ex-situ* bioremediation but also *in-situ* bioremediation. These new techniques allow the study of optimal microbial strains capable of overcoming the challenges of *in-situ* environments and represent the future insights of this field (Mishra et al. 2021; Rucká, Nešvera, and Pátek 2017). Although significant progress has been made in the development and implementation of GEMs for bioremediation, much work remains to be done.

REFERENCES

Abe, Flavia R. et al. 2019. "Life history and behavior effects of synthetic and natural dyes on Daphnia magna". *Chemosphere* 236: 124390. Elsevier BV. doi:10.1016/j.chemosphere.2019.124390.

Ajaz, Mehvish, Shakeel, Sana and Rehman, Abdul. 2019. "Microbial use for azo dye degradation—a strategy for dye bioremediation". *International Microbiology* 23 (2): 149–159. Springer Science and Business Media LLC. doi:10.1007/s10123-019-00103-2.

Aksu, Zümriye and Tezer, Sevilay. 2000. "Equilibrium and kinetic modelling of biosorption of Remazol Black B by Rhizopus arrhizus in a batch system: Effect of temperature". *Process Biochemistry* 36 (5): 431–439. Elsevier BV. doi:10.1016/s0032-9592(00)00233-8.

Al-Amrani, Waheeba Ahmed et al. 2014. "Factors affecting bio-decolorization of azo dyes and COD removal in anoxic–aerobic REACT operated sequencing batch reactor". *Journal of the Taiwan Institute of Chemical Engineers* 45 (2): 609–616. Elsevier BV. doi:10.1016/j.jtice.2013.06.032.

Ali, Asjad et al. 2014. "Physiological and transcriptional responses of Baccharis halimifolia to the explosive 'composition B' (RDX/TNT) in amended soil". *Environmental Science and Pollution Research* 21 (13): 8261–8270. Springer Science and Business Media LLC. doi:10.1007/s11356-014-2764-4.

Ansoborlo, E. and Adam-Guillermin, C. 2012. "Radionuclide transfer processes in the biosphere". *Radionuclide Behaviour in the Natural Environment*: 484–513. Elsevier. doi:10.1533/9780857097194.2.484.

Aquino, Elen, Barbieri, Cleide and Oller Nascimento, Claudio Augusto. 2011. "Engineering bacteria for bioremediation". *Progress in Molecular and Environmental Bioengineering—From Analysis and Modeling to Technology Applications*. InTech. doi:10.5772/19546.

Arslan, Muhammad et al. 2015. "Plant–bacteria partnerships for the remediation of persistent organic pollutants". *Environmental Science and Pollution Research* 24 (5): 4322–4336. Springer Science and Business Media LLC. doi:10.1007/s11356-015-4935-3.

Atlas, Ronald M. and Bartha, Richard. 1992. "Hydrocarbon biodegradation and oil spill bioremediation". *Advances in Microbial Ecology*: 287–338. Springer US. doi:10.1007/978-1-4684-7609-5_6.

Auti, Asim M. et al. 2019. "Microbiome and imputed metagenome study of crude and refined petroleum-oil-contaminated soils: Potential for hydrocarbon degradation and plant-growth promotion". *Journal of Biosciences* 44 (5). Springer Science and Business Media LLC. doi:10.1007/s12038-019-9936-9.

Awad, Abdelrahman M. et al. 2019. "Adsorption of organic pollutants by natural and modified clays: A comprehensive review". *Separation and Purification Technology* 228: 115719. Elsevier BV. doi:10.1016/j.seppur.2019.115719.

Ayed, Lamia et al. 2020. "Hybrid coagulation-flocculation and anaerobic-aerobic biological treatment for industrial textile wastewater: Pilot case study". *The Journal of The Textile Institute* 112 (2): 200–206. Informa UK Limited. doi:10.1080/00405000.2020.1731273.

Azimi, Arezoo et al. 2017. "Removal of heavy metals from industrial wastewaters: A review". *ChemBioEng Reviews* 4 (1): 37–59. Wiley. doi:10.1002/cben.201600010.

Bae, Weon, Mulchandani, Ashok and Chen, Wilfred. 2002. "Cell surface display of synthetic phytochelatins using ice nucleation protein for enhanced heavy metal bioaccumulation". *Journal of Inorganic Biochemistry* 88 (2): 223–227. Elsevier BV. doi:10.1016/s0162-0134(01)00392-0.

Balba, M.T., Al-Awadhi, N. and Al-Daher, R. 1998. "Bioremediation of oil-contaminated soil: Microbiological methods for feasibility assessment and field evaluation". *Journal of Microbiological Methods* 32 (2): 155–164. Elsevier BV. doi:10.1016/s0167-7012(98)00020-7.

Barata, Carlos, Solayan, Arun and Porte, Cinta. 2004. "Role of B-esterases in assessing toxicity of organophosphorus (chlorpyrifos, malathion) and carbamate (carbofuran) pesticides to Daphnia magna". *Aquatic Toxicology* 66 (2): 125–139. Elsevier BV. doi:10.1016/j.aquatox.2003.07.004.

Basutkar, Madhuri R. and Shivannavar, Channappa T. 2019. "Decolorization study of reactive Red-11 by using dye degrading bacterial strain Lysinibacillus boronitolerans CMGS-2". *International Journal of Current Microbiology and Applied Sciences* 8 (06): 1135–1143. Excellent Publishers. doi:10.20546/ijcmas.2019.806.140.

Bencheqroun, Zineb et al. 2019. "Removal of basic dyes from aqueous solutions by adsorption onto Moroccan clay (Fez City)". *Mediterranean Journal of Chemistry* 8 (3): 158. Mediterranean Journal of Chemistry. doi:10.13171/10.13171/mjc8319050803hz.

Benkhaya, Said, M'rabet, Souad and El Harfi, Ahmed. 2020. "Classifications, properties, recent synthesis and applications of azo dyes". *Heliyon* 6 (1): e03271. Elsevier BV. doi:10.1016/j.heliyon.2020.e03271.

Bernardes, Mariana Furio Franco et al. 2015. "Impact of pesticides on environmental and human health". *Toxicology Studies—Cells, Drugs and Environment*. InTech. doi:10.5772/59710.

Bharathiraja, B. et al. 2014. Biofuels from sewage sludge-a review. *Recent Trends in Biotechnology and Chemical Engineering* 6 (9): 4417–4427. www.researchgate.net/publication/266853355_Biofuels_from_sewage_sludge-_A_review.

Bhatia, Deepika et al. 2017. "Biological methods for textile dye removal from wastewater: A review". *Critical Reviews in Environmental Science and Technology* 47 (19): 1836–1876. Informa UK Limited. doi:10.1080/10643389.2017.1393263.

Bhattacharjee, Gargi et al. 2020. "Microbial bioremediation of industrial effluents and pesticides". *Bioremediation of Pollutants*: 287–302. Elsevier. doi:10.1016/b978-0-12-819025-8.00013-2.

Cao, Xiangyu et al. 2012. "Simultaneous degradation of organophosphate and organochlorine pesticides by Sphingobium japonicum UT26 with surface-displayed organophosphorus hydrolase". *Biodegradation* 24 (2): 295–303. Springer Science and Business Media LLC. doi:10.1007/s10532-012-9587-0.

Chaube, P., Indurkar, H. and Moghe, S. 2010. "Biodegradation and decolorisation of dye by mix consortia of bacteria and study of toxicity on Phaseolus mungo and Triticum aestivum". *Asiatic Journal of Biotechnology Resources*: 45–56. Pacific Journals. www.researchgate.net/profile/Sandhya-Moghe/

publication/228486921_Biodegradation_and_decolorisation_of_Direct_Violet_51_and_Tatrazine_dye_from_isolated_fungus_TYPE_I/links/53e1eb630cf24f90ff659f71/Biodegradation-and-decolorisation-of-Direct-Violet-51-and-Tatrazine-dye-from-isolated-fungus-TYPE-I.pdf.

Chen, Dan, Wang, Hongyu and Yang, Kai. 2016. "Effective biodegradation of nitrate, Cr(VI) and p-fluoronitrobenzene by a novel three dimensional bioelectrochemical system". *Bioresource Technology* 203: 370–373. Elsevier BV. doi:10.1016/j.biortech.2015.12.059.

Coconi-Linares, Nancy et al. 2015. "Recombinant expression of four oxidoreductases in Phanerochaete chrysosporium improves degradation of phenolic and non-phenolic substrates". *Journal of Biotechnology* 209: 76–84. Elsevier BV. doi:10.1016/j.jbiotec.2015.06.401.

Coelho, L. et al. 2015. "Bioremediation of polluted waters using microorganisms". *Advances in Bioremediation of Wastewater and Polluted Soil*. InTech. doi:10.5772/60770.

Conko, Gregory et al. 2016. "A risk-based approach to the regulation of genetically engineered organisms". *Nature Biotechnology* 34 (5): 493–503. Springer Science and Business Media LLC. doi:10.1038/nbt.3568.

Crameri, Andreas et al. 1997. "Molecular evolution of an arsenate detoxification pathway by DNA shuffling". *Nature Biotechnology* 15 (5): 436–438. Springer Science and Business Media LLC. doi:10.1038/nbt0597-436.

Crini, Grégorio and Lichtfouse, Eric. 2018. "Advantages and disadvantages of techniques used for wastewater treatment". *Environmental Chemistry Letters* 17 (1): 145–155. Springer Science and Business Media LLC. doi:10.1007/s10311-018-0785-9.

Crowley, D.E. et al. 1991. "Mechanisms of iron acquisition from siderophores by microorganisms and plants". *Plant and Soil* 130 (1–2): 179–198. Springer Science and Business Media LLC. doi:10.1007/bf00011873.

Das, Adya and Mishra, Susmita. 2017. "Removal of textile dye reactive green-19 using bacterial consortium: Process optimization using response surface methodology and kinetics study". *Journal of Environmental Chemical Engineering* 5 (1): 612–627. Elsevier BV. doi:10.1016/j.jece.2016.10.005.

de Lorenzo, V. 2009. "Recombinant bacteria for environmental release: What went wrong and what we have learnt from it". *Clinical Microbiology and Infection* 15: 63–65. Elsevier BV. doi:10.1111/j.1469-0691.2008.02683.x.

Diep, Patrick, Mahadevan, Radhakrishnan and Yakunin, Alexander F. 2018. "Heavy metal removal by bioaccumulation using genetically engineered microorganisms". *Frontiers in Bioengineering and Biotechnology* 6. Frontiers Media SA. doi:10.3389/fbioe.2018.00157.

Dong, Hao et al. 2019. "Biochemical characterization of a novel azoreductase from Streptomyces sp.: Application in eco-friendly decolorization of azo dye wastewater". *International Journal of Biological Macromolecules* 140: 1037–1046. Elsevier BV. doi:10.1016/j.ijbiomac.2019.08.196.

Douglas, Thomas A. et al. 2011. "Desorption and transformation of nitroaromatic (TNT) and nitramine (RDX and HMX) explosive residues on detonated pure mineral phases". *Water, Air, and Soil Pollution* 223 (5): 2189–2200. Springer Science and Business Media LLC. doi:10.1007/s11270-011-1015-2.

Dua, M. et al. 2002. "Biotechnology and bioremediation: Successes and limitations". *Applied Microbiology and Biotechnology* 59 (2–3): 143–152. Springer Science and Business Media LLC. doi:10.1007/s00253-002-1024-6.

Duque, E. et al. 1993. "Construction of a pseudomonas hybrid strain that mineralizes 2,4,6-trinitrotoluene". *Journal of Bacteriology* 175 (8): 2278–2283. American Society for Microbiology. doi:10.1128/jb.175.8.2278-2283.1993.

Dutta, Sisir K. et al. 2003. "Enhanced bioremediation of soil containing 2,4-dinitrotoluene by a genetically modified Sinorhizobium meliloti". *Soil Biology and Biochemistry* 35 (5): 667–675. Elsevier BV. doi:10.1016/s0038-0717(03)00016-6.

Dzantor, E. Kudjo. 2007. "Phytoremediation: The state of rhizosphere 'engineering' for accelerated rhizodegradation of xenobiotic contaminants". *Journal of Chemical Technology & Biotechnology* 82 (3): 228–232. Wiley. doi:10.1002/jctb.1662.

El Fantroussi, Saïd and Agathos, Spiros N. 2005. "Is bioaugmentation a feasible strategy for pollutant removal and site remediation?" *Current Opinion in Microbiology* 8 (3): 268–275. Elsevier BV. doi:10.1016/j.mib.2005.04.011.

Exbrayat, J.-M. 2016. "Microscopy: Light microscopy and histochemical methods". *Encyclopedia of Food and Health*: 715–723. Elsevier. doi:10.1016/b978-0-12-384947-2.00460-8.

Feng, Nai-Xian et al. 2017. "Efficient phytoremediation of organic contaminants in soils using plant–endophyte partnerships". *Science of The Total Environment* 583: 352–368. Elsevier BV. doi:10.1016/j.scitotenv.2017.01.075.

Freeman, H.S. and Mock, G.N. 2012. "Dye application, manufacture of dye intermediates and dyes". *Handbook of Industrial Chemistry and Biotechnology*: 475–548. Springer US. doi:10.1007/978-1-4614-4259-2_13.

Fu, Y. and Viraraghavan, T. 2002. "Dye biosorption sites in aspergillus niger". *Bioresource Technology* 82 (2): 139–145. Elsevier BV. doi:10.1016/s0960-8524(01)00172-9.

Fujita, Masanori, Ike, Michihiko and Uesugi, Kazuya. 1994. "Operation parameters affecting the survival of genetically engineered microorganisms in activated sludge processes". *Water Research* 28 (7): 1667–1672. Elsevier BV. doi:10.1016/0043-1354(94)90235-6.

Gammon, Derek W. et al. 2019. "Pyrethroid neurotoxicity studies with bifenthrin indicate a mixed Type I/II mode of action". *Pest Management Science*75 (4): 1190–1197. Wiley. doi:10.1002/ps.5300.

Garbisu, Carlos and Alkorta, Itziar. 1999. "Utilization of genetically engineered microorganisms (GEMs) for bioremediation". *Journal of Chemical Technology & Biotechnology* 74 (7): 599–606. Wiley. doi:10.1002/(sici)1097-4660(199907)74:7<599::aid-jctb82>3.0.co;2-g.

Gerhardt, Karen et al. 2017. *Phytoremediation of Salt-Impacted Soils and Use of Plant Growth-Promoting Rhizobacteria (PGPR) to Enhance Phytoremediation*. Springer International Publishing. doi:10.1007/978-3-319-52381-1.

Ghattas, Ann-Kathrin et al. 2017. "Anaerobic biodegradation of (emerging) organic contaminants in the aquatic environment". *Water Research* 116: 268–295. Elsevier BV. doi:10.1016/j.watres.2017.02.001.

Gil, Geun Cheol, Kim, Young Joon and Gu, Man Bock. 2002. "Enhancement in the sensitivity of a gas biosensor by using an advanced immobilization of a recombinant bioluminescent bacterium". *Biosensors and Bioelectronics* 17 (5): 427–432. Elsevier BV. doi:10.1016/s0956-5663(01)00305-0.

Gilden, Robyn C., Huffling, Katie and Sattler, Barbara. 2010. "Pesticides and health risks". *Journal of Obstetric, Gynecologic & Neonatal Nursing* 39 (1): 103–110. Elsevier BV. doi:10.1111/j.1552-6909.2009.01092.x.

Girotti, Stefano et al. 2008. "Monitoring of environmental pollutants by bioluminescent bacteria". *Analytica Chimica Acta* 608 (1): 2–29. Elsevier BV. doi:10.1016/j.aca.2007.12.008.

Glenn, Jeffrey K., Akileswaran, Lakshmi and Gold, Michael H. 1986. "Mn(II) oxidation is the principal function of the extracellular Mn-peroxidase from phanerochaete chrysosporium". *Archives of Biochemistry and Biophysics* 251 (2): 688–696. Elsevier BV. doi:10.1016/0003-9861(86)90378-4.

Gregory, Peter. 2009. "Dyes and dye intermediates". *Kirk-Othmer Encyclopedia of Chemical Technology*. John Wiley & Sons, Inc. doi:10.1002/0471238961.0425051907180507.a01.pub2.

Grinshpun, Sergey A. et al. 2007. "Control of aerosol contaminants in indoor air: Combining the particle concentration reduction with microbial inactivation". *Environmental Science and Technology* 41 (2): 606–612. American Chemical Society (ACS). doi:10.1021/es061373o.

Guo, Ying et al. 2018. "Study on the degradation mechanism and pathway of benzene dye intermediate 4-methoxy-2-nitroaniline via multiple methods in Fenton oxidation process". *RSC Advances* 8 (20): 10764–10775. Royal Society of Chemistry (RSC). doi:10.1039/c8ra00627j.

Gupta, Dharmendra Kumar and Walther, Clemens. 2014. *Radionuclide Contamination and Remediation Through Plants. Radionuclide Contamination and Remediation Through Plants*. Springer, Cham. doi:10.1007/978-3-319-07665-2.

Gupta, Sanjeev K. and Shukla, Pratyoosh. 2015. "Advanced technologies for improved expression of recombinant proteins in bacteria: Perspectives and applications". *Critical Reviews in Biotechnology* 36 (6): 1089–1098. Informa UK Limited. doi:10.3109/07388551.2015.1084264.

Gupta, Saurabh and Singh, Daljeet. 2017. "Role of genetically modified microorganisms in heavy metal bioremediation". *Advances in Environmental Biotechnology*: 197–214. Springer, Singapore. doi:10.1007/978-981-10-4041-2_12.

Hanrahan, Grady. 2012. "Surface/groundwater quality and monitoring". *Key Concepts in Environmental Chemistry*: 109–152. Elsevier. doi:10.1016/b978-0-12-374993-2.10004-4.

Herrero, M., de Lorenzo, V. and Timmis, K.N. 1990. "Transposon vectors containing non-antibiotic resistance selection markers for cloning and stable chromosomal insertion of foreign genes in gram-negative bacteria". *Journal of Bacteriology* 172 (11): 6557–6567. American Society for Microbiology. doi:10.1128/jb.172.11.6557-6567.1990.

Hong, Yiguo et al. 2007. "Respiration and growth of Shewanella decolorationis S12 with an azo compound as the sole electron acceptor". *Applied and Environmental Microbiology* 73 (1): 64–72. American Society for Microbiology. doi:10.1128/aem.01415-06.

Hussain, Imran et al. 2018. "Microbe and plant assisted-remediation of organic xenobiotics and its enhancement by genetically modified organisms and recombinant technology: A review". *Science of The Total Environment* 628–629: 1582–1599. Elsevier BV. doi:10.1016/j.scitotenv.2018.02.037.

Ikuma, Kaoru and Gunsch, Claudia K. 2012. "Genetic bioaugmentation as an effective method for in situ bioremediation". *Bioengineered* 3 (4): 236–241. Informa UK Limited. doi:10.4161/bioe.20551.

Jaiswal, Shweta, Sharma, Babita and Shukla, Pratyoosh. 2019. "Integrated approaches in microbial degradation of plastics". *Environmental Technology and Innovation* 17: 100567. Elsevier BV. doi:10.1016/j.eti.2019.100567.

Jaiswal, Shweta, Singh, Dileep Kumar and Shukla, Pratyoosh. 2019. "Gene editing and systems biology tools for pesticide bioremediation: A review". *Frontiers in Microbiology* 10. Frontiers Media SA. doi:10.3389/fmicb.2019.00087.

Janssen, Dick B. and Stucki, Gerhard. 2020. "Perspectives of genetically engineered microbes for groundwater bioremediation". *Environmental Science: Processes & Impacts* 22 (3): 487–499. Royal Society of Chemistry (RSC). doi:10.1039/c9em00601j.

Jiang, Jian-Dong et al. 2005. "Construction of multifunctional genetically engineered pesticides-degrading bacteria by homologous recombination". *Sheng wu gong cheng xue bao= Chinese Journal of Biotechnology* 21 (6): 884–891.

Jiang, Jiandong et al. 2006. "Simultaneous biodegradation of methyl parathion and carbofuran by a genetically engineered microorganism constructed by mini-Tn5 transposon". *Biodegradation* 18 (4): 403–412. Springer Science and Business Media LLC. doi:10.1007/s10532-006-9075-5.

Jing, Ran, Fusi, Soliver and Kjellerup, Birthe V. 2018. "Remediation of polychlorinated biphenyls (PCBs) in contaminated soils and sediment: State of knowledge and perspectives". *Frontiers in Environmental Science* 6: 1–17. Frontiers Media SA. doi:10.3389/fenvs.2018.00079.

Kamthan, Ayushi et al. 2016. "Genetically modified (GM) crops: Milestones and new advances in crop improvement". *Theoretical and Applied Genetics* 129 (9): 1639–1655. Springer Science and Business Media LLC. doi:10.1007/s00122-016-2747-6.

Kang, Jun Won. 2014. "Removing environmental organic pollutants with bioremediation and phytoremediation". *Biotechnology Letters* 36 (6): 1129–1139. Springer Science and Business Media LLC. doi:10.1007/s10529-014-1466-9.

Keasling, Jay D. and Bang, Sang-Weon. 1998. "Recombinant DNA techniques for bioremediation and environmentally-friendly synthesis". *Current Opinion in Biotechnology* 9 (2): 135–140. Elsevier BV. doi:10.1016/s0958-1669(98)80105-5.

Kinnunen, Anu et al. 2017. "Improved efficiency in screening for lignin-modifying peroxidases and laccases of basidiomycetes". *Current Biotechnology* 6 (2): 105–115. Bentham Science Publishers Ltd. doi:10.2174/2211550105666160330205138.

Koch, Karin et al. 2017. "Structure, biochemical and kinetic properties of recombinant Pst2p from saccharomyces cerevisiae, a FMN-dependent NAD(P)H:quinone oxidoreductase". *Biochimica et Biophysica Acta (BBA)—Proteins and Proteomics* 1865 (8): 1046–1056. Elsevier BV. doi:10.1016/j.bbapap.2017.05.005.

Köhler, S., Belkin, S. and Schmid, R.D. 2000. "Reporter gene bioassays in environmental analysis". *Fresenius' Journal of Analytical Chemistry* 366 (6–7): 769–779. Springer Science and Business Media LLC. doi:10.1007/s002160051571.

Kolata, G. 1985. "How safe are engineered organisms?" *Science* 229: 34–36.

Kumar, Arvind et al. 2020. "Genetically engineered bacteria for the degradation of dye and other organic compounds". *Abatement of Environmental Pollutants*: 331–350. Elsevier. doi:10.1016/b978-0-12-818095-2.00016-3.

Kumar, Narasimhan Manoj et al. 2017. "Genetically modified organisms and its impact on the enhancement of bioremediation". *Energy, Environment, and Sustainability*: 53–76. Springer, Singapore. doi:10.1007/978-981-10-7485-1_4.

Kumar, Sandeep et al. 2013. "Genetically modified microorganisms (GMOs) for bioremediation". *Biotechnology for Environmental Management and Resource Recovery*: 191–218. Springer India. doi:10.1007/978-81-322-0876-1_11.

Kumar Singh, Anand et al. 2012. "A current review of cypermethrin-induced neurotoxicity and nigrostriatal dopaminergic neurodegeneration". *Current Neuropharmacology* 10 (1): 64–71. Bentham Science Publishers Ltd. doi:10.2174/157015912799362779.

Kunz, A., Mansilla, H. and Durán, N. 2002. "A degradation and toxicity study of three textile reactive dyes by ozone". *Environmental Technology* 23 (8): 911–918. Informa UK Limited. doi:10.1080/09593332308618358.

Lange, Cleston C. et al. 1998. "Engineering a recombinant Deinococcus radiodurans for organopollutant degradation in radioactive mixed waste environments". *Nature Biotechnology* 16 (10): 929–933. Springer Science and Business Media LLC. doi:10.1038/nbt1098-929.

Lee, Aslan Hwanhwi et al. 2016. "Heterologous expression of a new manganese-dependent peroxidase gene from Peniophora incarnata KUC8836 and its ability to remove anthracene in Saccharomyces cerevisiae". *Journal of Bioscience and Bioengineering* 122 (6): 716–721. Elsevier BV. doi:10.1016/j. jbiosc.2016.06.006.

Lellis, Bruno et al. 2019. "Effects of textile dyes on health and the environment and bioremediation potential of living organisms". *Biotechnology Research and Innovation* 3 (2): 275–290. Elsevier BV. doi:10.1016/j. biori.2019.09.001.

Li, Qin et al. 2020. "A safety type of genetically engineered bacterium that degrades chemical pesticides". *AMB Express* 10 (1). Springer Science and Business Media LLC. doi:10.1186/s13568-020-00967-y.

Lin, Ta-Chen, Pan, Po-Tsen and Cheng, Sheng-Shung. 2010. "Ex situ bioremediation of oil-contaminated soil". *Journal of Hazardous Materials* 176 (1–3): 27–34. Elsevier BV. doi:10.1016/j.jhazmat.2009.10.080.

Lin, Ziqiu et al. 2020. "Degradation of acephate and its intermediate methamidophos: Mechanisms and biochemical pathways". *Frontiers in Microbiology* 11. Frontiers Media SA. doi:10.3389/fmicb. 2020.02045.

Liu, Lina et al. 2019. "Mitigation of environmental pollution by genetically engineered bacteria—Current challenges and future perspectives". *Science of The Total Environment* 667: 444–454. Elsevier BV. doi:10.1016/j.scitotenv.2019.02.390.

Liu, Zhi et al. 2006. "Construction of a genetically engineered microorganism for degrading organophosphate and carbamate pesticides". *International Biodeterioration & Biodegradation* 58 (2): 65–69. Elsevier BV. doi:10.1016/j.ibiod.2006.07.009.

Lorenzo, Victor et al. 1998. "Mini-transposons in microbial ecology and environmental biotechnology". *FEMS Microbiology Ecology* 27 (3): 211–224. Oxford University Press (OUP). doi:10.1111/j.1574-6941.1998. tb00538.x.

Lushchak, Volodymyr et al. 2018. "Pesticide toxicity: A mechanistic approach". *EXCLI Journal* 17: 1101. doi:10.17179/excli2018-1710.

Ma, Xiaojian et al. 2017. "High-level expression of a bacterial laccase, CueO from Escherichia coli K12 in Pichia pastoris GS115 and its application on the decolorization of synthetic dyes". *Enzyme and Microbial Technology* 103: 34–41. Elsevier BV. doi:10.1016/j.enzmictec.2017.04.004.

Mandal, Tamal, Dasgupta, Dalia and Datta, Siddhartha. 2010. "A biotechnological thrive on COD and chromium removal from leather industrial wastewater by the isolated microorganisms". *Desalination and Water Treatment* 13 (1–3): 382–392. Informa UK Limited. doi:10.5004/dwt.2010.996.

Martin, Belinda C. et al. 2016. "Citrate and malonate increase microbial activity and alter microbial community composition in uncontaminated and diesel-contaminated soil microcosms". *SOIL* 2 (3): 487–498. Copernicus GmbH. doi:10.5194/soil-2-487-2016.

McClure, N.C., Fry, J.C. and Weightman, A.J. 1991a. "Genetic engineering for wastewater treatment". *Water and Environment Journal* 5 (6): 608–616. Wiley. doi:10.1111/j.1747-6593.1991.tb00678.x.

McClure, N.C., Fry, J.C. and Weightman, A.J. 1991b. "Survival and catabolic activity of natural and genetically engineered bacteria in a laboratory-scale activated-sludge unit". *Applied and Environmental Microbiology* 57 (2): 366–373. American Society for Microbiology. doi:10.1128/ aem.57.2.366-373.1991.

Mishra, Sandhya et al. 2021. "Recent advanced technologies for the characterization of xenobiotic-degrading microorganisms and microbial communities". *Frontiers in Bioengineering and Biotechnology* 9. Frontiers Media SA. doi:10.3389/fbioe.2021.632059.

Müller, R., Deckwer, W.-D. and Hecht, V. 2000. "Degradation of chloro- and methyl-substituted benzoic acids by a genetically modified microorganism". *Biotechnology and Bioengineering* 51 (5): 528–537. Wiley. doi:10.1002/(sici)1097-0290(19960905)51:5<528::aid-bit4>3.0.co;2-e.

Nahurira, Ruth et al. 2017. "Degradation of Di(2-Ethylhexyl) phthalate by a novel gordonia alkanivorans strain YC-RL2". *Current Microbiology* 74 (3): 309–319. Springer Science and Business Media LLC. doi:10.1007/s00284-016-1159-9.

Nicolia, Alessandro et al. 2013. "An overview of the last 10 years of genetically engineered crop safety research". *Critical Reviews in Biotechnology* 34 (1): 77–88. Informa UK Limited. doi:10.3109/073885 51.2013.823595.

Nikfar, S. and Jaberidoost, M. 2014. "Dyes and colorants". *Encyclopedia of Toxicology*: 252–261. Elsevier. doi:10.1016/b978-0-12-386454-3.00602-3.

Nishi, Kosuke et al. 2006. "Characterization of pyrethroid hydrolysis by the human liver carboxylesterases hCE-1 and hCE-2". *Archives of Biochemistry and Biophysics* 445 (1): 115–123. Elsevier BV. doi:10.1016/j.abb.2005.11.005.

Ojuederie, Omena and Babalola, Olubukola. 2017. "Microbial and plant-assisted bioremediation of heavy metal polluted environments: A review". *International Journal of Environmental Research and Public Health* 14 (12): 1504. MDPI AG. doi:10.3390/ijerph14121504.

O'Mahony, T., Guibal, E. and Tobin, J.M. 2002. "Reactive dye biosorption by Rhizopus arrhizus biomass". *Enzyme and Microbial Technology* 31 (4): 456–463. Elsevier BV. doi:10.1016/s0141-0229(02)00110-2.

Paitan, Yossi et al. 2004. "Monitoring aromatic hydrocarbons by whole cell electrochemical biosensors". *Analytical Biochemistry* 335 (2): 175–183. Elsevier BV. doi:10.1016/j.ab.2004.08.032.

Pandey, Gunjan, Paul, Debarati and Jain, Rakesh K. 2005. "Conceptualizing 'suicidal genetically engineered microorganisms' for bioremediation applications". *Biochemical and Biophysical Research Communications* 327 (3): 637–639. Elsevier BV. doi:10.1016/j.bbrc.2004.12.080.

Pant, Gaurav et al. 2020. "Biological approaches practised using genetically engineered microbes for a sustainable environment: A review". *Journal of Hazardous Materials* 405: 124631. Elsevier BV. doi:10.1016/j.jhazmat.2020.124631.

Pathiraja, Gathanayana et al. 2019. "Solubilization and degradation of polychlorinated biphenyls (PCBs) by naturally occurring facultative anaerobic bacteria". *Science of The Total Environment* 651: 2197–2207. Elsevier BV. doi:10.1016/j.scitotenv.2018.10.127.

Paul, Debarati, Pandey, Gunjan and Jain, Rakesh K. 2005. "Suicidal genetically engineered microorganisms for bioremediation: Need and perspectives". *BioEssays* 27 (5): 563–573. Wiley. doi:10.1002/bies.20220.

Pratush, Amit, Kumar, Ajay and Hu, Zhong. 2018. "Adverse effect of heavy metals (As, Pb, Hg, and Cr) on health and their bioremediation strategies: A review". *International Microbiology* 21 (3): 97–106. Springer Science and Business Media LLC. doi:10.1007/s10123-018-0012-3.

Purg, Miha et al. 2016. "Probing the mechanisms for the selectivity and promiscuity of methyl parathion hydrolase". *Philosophical Transactions of the Royal Society A: Mathematical, Physical and Engineering Sciences* 374 (2080): 20160150. The Royal Society. doi:10.1098/rsta.2016.0150.

Qu, Jiuhui and Fan, Maohong. 2010. "The current state of water quality and technology development for water pollution control in China". *Critical Reviews in Environmental Science and Technology* 40 (6): 519–560. Informa UK Limited. doi:10.1080/10643380802451953.

Ripp, Steven et al. 2000. "Controlled field release of a bioluminescent genetically engineered microorganism for bioremediation process monitoring and control". *Environmental Science & Technology* 34 (5): 846–853. American Chemical Society (ACS). doi:10.1021/es9908319.

Roy, Dipankar Chandra et al. 2018. "Biodegradation of Crystal Violet dye by bacteria isolated from textile industry effluents". *PeerJ* 6: 1–15. doi:10.7717/peerj.5015.

Rucká, Lenka, Nešvera, Jan and Pátek, Miroslav. 2017. "Biodegradation of phenol and its derivatives by engineered bacteria: Current knowledge and perspectives". *World Journal of Microbiology and Biotechnology* 33 (9): 1–8. Springer Science and Business Media LLC. doi:10.1007/s11274-017-2339-x.

Saavedra, Juan Matías, Acevedo, Francisca, González, Myriam and Seeger, Michael. 2010. "Mineralization of PCBs by the genetically modified strain Cupriavidus necator JMS34 and its application for bioremediation of PCBs in soil". *Applied Microbiology and Biotechnology* 87 (4): 1543–1554. Springer Science and Business Media LLC. doi:10.1007/s00253-010-2575-6.

Samaksaman, Ukrit et al. 2016. "Thermal treatment of soil co-contaminated with lube oil and heavy metals in a low-temperature two-stage fluidized bed incinerator". *Applied Thermal Engineering* 93: 131–138. Elsevier BV. doi:10.1016/j.applthermaleng.2015.09.024.

Samanta, Sudip K., Singh, Om V. and Jain, Rakesh K. 2002. "Polycyclic aromatic hydrocarbons: Environmental pollution and bioremediation". *Trends in Biotechnology* 20 (6): 243–248. Elsevier BV. doi:10.1016/s0167-7799(02)01943-1.

Saxena, Gaurav et al. 2020. "Genetically modified organisms (GMOs) and their potential in environmental management: Constraints, prospects, and challenges". *Bioremediation of Industrial Waste for Environmental Safety*: 1–19. Springer, Singapore. doi:10.1007/978-981-13-3426-9_1.

Sayler, G.S. and Ripp, S. 2000. "Field applications of genetically engineered microorganisms for bioremediation processes." 11 (3): 286–289. Elsevier BV. doi:10.1016/s0958-1669(00)00097-5. www.sciencedirect.com/science/article/pii/S0958166900000975.

Serdar, Cüneyt M. and Gibson, David T. 1985. "Enzymatic hydrolysis of organophosphates: Cloning and expression of a parathion hydrolase gene from pseudomonas diminuta". *Nature Biotechnology* 3 (6): 567–571. Springer Science and Business Media LLC. doi:10.1038/nbt0685-567.

Sharma, Babita, Dangi, Arun Kumar and Shukla, Pratyoosh. 2018. "Contemporary enzyme based technologies for bioremediation: A review". *Journal of Environmental Management* 210: 10–22. Elsevier BV. doi:10.1016/j.jenvman.2017.12.075.

Shimazu, Mark, Mulchandani, Ashok and Chen, Wilfred. 2001. "Simultaneous degradation of organophosphorus pesticides and p-nitrophenol by a genetically engineered Moraxella sp. with surface-expressed organophosphorus hydrolase". *Biotechnology and Bioengineering* 76 (4): 318–324. Wiley. doi:10.1002/bit.10095.

Shimizu, Hiroshi. 2002. "Metabolic engineering—integrating methodologies of molecular breeding and bio-process systems engineering". *Journal of Bioscience and Bioengineering* 94 (6): 563–573. Elsevier BV. doi:10.1016/s1389-1723(02)80196-7.

Shirke, Abhijit N. et al. 2018. "Stabilizing leaf and branch compost cutinase (LCC) with glycosylation: Mechanism and effect on PET hydrolysis". *Biochemistry* 57 (7): 1190–1200. American Chemical Society (ACS). doi:10.1021/acs.biochem.7b01189.

Shivananju, B.N. et al. 2019. "Optical biochemical sensors based on 2D materials". *Fundamentals and Sensing Applications of 2D Materials*: 379–406. Elsevier. doi:10.1016/b978-0-08-102577-2.00010-5.

Singh, Brajesh K. and Walker, Allan. 2006. "Microbial degradation of organophosphorus compounds". *FEMS Microbiology Reviews* 30 (3): 428–471. Oxford University Press (OUP). doi:10.1111/j.1574-6976.2006.00018.x.

Singh, Jay Shankar et al. 2011. "Genetically engineered bacteria: An emerging tool for environmental remediation and future research perspectives". *Gene* 480 (1–2): 1–9. Elsevier BV. doi:10.1016/j.gene.2011.03.001.

Singh, Pardeep et al. 2020. "Bioremediation". *Abatement of Environmental Pollutants*: 1–23. Elsevier. doi:10.1016/b978-0-12-818095-2.00001-1.

Soda, Satoshi, Ike, Michihiko and Fujita, Masanori. 1998. "Effects of inoculation of a genetically engineered bacterium on performance and indigenous bacteria of a sequencing batch activated sludge process treating phenol". *Journal of Fermentation and Bioengineering* 86 (1): 90–96. Elsevier BV. doi:10.1016/s0922-338x(98)80040-8.

Sponza, Delia Teresa. 2006. "Toxicity studies in a chemical dye production industry in Turkey". *Journal of Hazardous Materials* 138 (3): 438–447. Elsevier BV. doi:10.1016/j.jhazmat.2006.05.120.

Strauss, Steven H. and Sax, Joanna K. 2016. "Ending event-based regulation of GMO crops". *Nature Biotechnology* 34 (5): 474–477. Springer Science and Business Media LLC. doi:10.1038/nbt.3541.

Sun, Jianna et al. 2017. "Heterologous production of a temperature and pH-stable laccase from Bacillus vallismortis fmb-103 in Escherichia coli and its application". *Process Biochemistry* 55: 77–84. Elsevier BV. doi:10.1016/j.procbio.2017.01.030.

Tahri, Nezha et al. 2013. "Biodegradation: Involved microorganisms and genetically engineered microorganisms". *Biodegradation—Life of Science*. InTech. doi:10.5772/56194.

Thassitou, P.K. and Arvanitoyannis, I.S. 2001. "Bioremediation: A novel approach to food waste management". *Trends in Food Science & Technology* 12 (5–6): 185–196. Elsevier BV. doi:10.1016/s0924-2244(01)00081-4.

Then, Christoph, Kawall, Katharina and Valenzuela, Nina. 2020. "Spatiotemporal controllability and environmental risk assessment of genetically engineered gene drive organisms from the perspective of European Union genetically modified organism regulation". *Integrated Environmental Assessment and Management* 16 (5): 555–568. Wiley. doi:10.1002/ieam.4278.

Thomas, J.M. and Ward, C.H. 1992. "Subsurface microbial ecology and bioremediation". *Journal of Hazardous Materials* 32 (2–3): 179–194. Elsevier BV. doi:10.1016/0304-3894(92)85091-e.

Thorne, M. 2012. "Modelling radionuclide transport in the environment and calculating radiation doses". *Radionuclide Behaviour in the Natural Environment*: 517–569. Elsevier. doi:10.1533/9780857097194.3.517.

Timmis, Kenneth N. and Pieper, Diemar H. 1999. "Bacteria designed for bioremediation". *Trends in Biotechnology* 17 (5): 201–204. Elsevier BV. doi:10.1016/s0167-7799(98)01295-5.

Tulupov, P.E., Pavlovna, N.S. and Tulupov, A.P. 2006. Method of determining biological activity (toxicity) of solid metal components and action thereof on living organism specimens. RU 2281496 C1 20060810, issued 2006. https://patents.google.com/patent/RU2281495C9/en.

UNEP/WHO. 1993. *GEMS/AIR—A Global Programme for Urban Air Quality Monitoring and Assessment*. Nairobi: Urban Air Quality Monitoring and Assessment.

Van Eenennaam, A.L. and Young, A.E. 2014. "Prevalence and impacts of genetically engineered feedstuffs on livestock populations1". *Journal of Animal Science* 92 (10): 4255–4278. Oxford University Press (OUP). doi:10.2527/jas.2014-8124.

Vargas, J.M. 1975. "Pesticide degradation". *Journal of Arboriculture* 1 (12): 232–233.

Varjani, Sunita J. et al. 2020a. "Microbial degradation of dyes: An overview". *Bioresource Technology* 314: 123728. Elsevier BV. doi:10.1016/j.biortech.2020.123728.

Varjani, Sunita J. et al. 2020b. "Microbial degradation of dyes: An overview". *Bioresource Technology* 314: 123728. Elsevier BV. doi:10.1016/j.biortech.2020.123728.

Varjani, Sunita J. and Upasani, Vivek N. 2017a. "A new look on factors affecting microbial degradation of petroleum hydrocarbon pollutants". *International Biodeterioration & Biodegradation* 120: 71–83. Elsevier BV. doi:10.1016/j.ibiod.2017.02.006.

Varjani, Sunita J. and Upasani, Vivek N. 2017b. "Critical review on biosurfactant analysis, purification and characterization using rhamnolipid as a model biosurfactant". *Bioresource Technology* 232: 389–397. Elsevier BV. doi:10.1016/j.biortech.2017.02.047.

Veal, Duncan A., Stokes, H.W. and Daggard, Grant. 1992. "Genetic exchange in natural microbial communities". *Advances in Microbial Ecology*: 383–430. Springer US. doi:10.1007/978-1-4684-7609-5_8.

Velkov, Vassili V. 2001. "Stress-induced evolution and the biosafety of genetically modified microorganisms released into the environment". *Journal of Biosciences* 26 (5): 667–683. Springer Science and Business Media LLC. doi:10.1007/bf02704764.

Verma, Samakshi and Kuila, Arindam. 2019. "Bioremediation of heavy metals by microbial process". *Environmental Technology & Innovation* 14: 100369. Elsevier BV. doi:10.1016/j.eti.2019.100369.

Vontas, J.G., Small, G.J. and Hemingway, J. 2000. "Comparison of esterase gene amplification, gene expression and esterase activity in insecticide susceptible and resistant strains of the brown planthopper, Nilaparvata lugens (Stal)". *Insect Molecular Biology* 9 (6): 655–660. Wiley. doi:10.1046/j.1365-2583.2000.00228.x.

Wang, Yunxiao et al. 2020. "Treatment of azo dye wastewater by the self-flocculating marine bacterium Aliiglaciecola lipolytica". *Environmental Technology & Innovation* 19: 100810. Elsevier BV. doi:10.1016/j.eti.2020.100810.

Wenzel, Walter W. et al. 2003. "Chelate-assisted phytoextraction using canola (Brassica napus L.) in outdoors pot and lysimeter experiments". *Plant and Soil* 249 (1): 83–96. Springer Science and Business Media LLC. doi:10.1023/a:1022516929239.

Whitfield Åslund, Melissa L. et al. 2010. "Effects of amendments on the uptake and distribution of DDT in Cucurbita pepo ssp pepo plants". *Environmental Pollution* 158 (2): 508–513. Elsevier BV. doi:10.1016/j.envpol.2009.08.030.

Wijnhoven, S. et al. 2007. "Heavy-metal concentrations in small mammals from a diffusely polluted floodplain: Importance of species- and location-specific characteristics". *Archives of Environmental Contamination and Toxicology* 52 (4): 603–613. Springer Science and Business Media LLC. doi:10.1007/s00244-006-0124-1.

Wilkes, R.A. and Aristilde, L. 2017. "Degradation and metabolism of synthetic plastics and associated products by pseudomonas sp.: Capabilities and challenges". *Journal of Applied Microbiology* 123 (3): 582–593. Wiley. doi:10.1111/jam.13472.

Wozniak, Chris A. and McHughen, Alan. 2012. *Regulation of Agricultural Biotechnology: The United States and Canada. Regulation of Agricultural Biotechnology: The United States and Canada.* Vol. 9789400721. doi:10.1007/978-94-007-2156-2.

Wu, Gang et al. 2010. "A critical review on the bio-removal of hazardous heavy metals from contaminated soils: Issues, progress, eco-environmental concerns and opportunities". *Journal of Hazardous Materials* 174 (1–3): 1–8. Elsevier BV. doi:10.1016/j.jhazmat.2009.09.113.

Xu, Renjie et al. 2016. "A new method for extraction and heavy metals removal of abalone visceral polysaccharide". *Journal of Food Processing and Preservation* 41 (4): e13023. Wiley. doi:10.1111/jfpp.13023.

Yang, Chao et al. 2010. "Genetic engineering of Stenotrophomonas strain YC-1 to possess a broader substrate range for organophosphates". *Journal of Agricultural and Food Chemistry* 58 (11): 6762–6766. American Chemical Society (ACS). doi:10.1021/jf101105s.

Yang, Ting et al. 2018. "Biodegradation of Di-(2-ethylhexyl) phthalate by Rhodococcus ruber YC-YT1 in contaminated water and soil". *International Journal of Environmental Research and Public Health* 15 (5): 964. MDPI AG. doi:10.3390/ijerph15050964.

Yu, Xiaolong et al. 2019. "Oxidation degradation of tris-(2-chloroisopropyl) phosphate by ultraviolet driven sulfate radical: Mechanisms and toxicology assessment of degradation intermediates using flow cytometry analyses". *Science of the Total Environment* 687: 732–740. Elsevier BV. doi:10.1016/j.scitotenv.2019.06.163.

Yuanfan, Hong et al. 2010. "Characterization of a Fenpropathrin-degrading strain and construction of a genetically engineered microorganism for simultaneous degradation of methyl parathion and Fenpropathrin". *Journal of Environmental Management* 91 (11): 2295–2300. Elsevier BV. doi:10.1016/j.jenvman.2010.06.010.

Zhang, Wenping et al. 2020. "Insights into the biodegradation of lindane (γ-hexachlorocyclohexane) using a microbial system". *Frontiers in Microbiology* 11: 1–12. Frontiers Media SA. doi:10.3389/fmicb.2020.00522.

3 Electroporation for the Production of Genetically Modified Microorganisms

Sumeyra Gurkok

INTRODUCTION

Environmental pollution has increased rapidly in the last few decades due to improper human activities, agricultural policies, and industrialization. A wide variety of treatment methods such as physical, chemical, and thermal are used to remove contaminants such as hydrocarbons, polycyclic aromatic hydrocarbons, polychlorinated biphenyls, chlorinated volatile organic compounds, dyes, and pesticides. At present, biological treatment, which can be applied directly to targeted contaminated areas, is considered an effective alternative to traditional methods. Bioremediation involves processes that use biological systems, such as plants, microorganisms, or their enzymes to decompose and/or detoxify pollutants by metabolic or enzymatic means.

Bioremediation is actually a natural phenomenon that occurs with microorganisms that already exist locally without any human contribution and/or intervention. However, this is a slow process under normal conditions as most of the chemicals in pollutants are not readily biodegraded by microorganisms. Therefore, various bioremediation techniques are employed to accelerate this phenomenon. One of the most used techniques is bio-stimulation, where oxygen and nutrients are supplied to the contaminated sites to promote microbial growth. Another technique is bio-augmentation, where microorganisms that efficiently break down pollutants are developed in the laboratory and inoculated into contaminated sites.

Remediation of polluted sites using microbial processes has proven to be effective, reliable, cost-effective, and environmentally friendly, but the contribution to improved bioremediation can be much greater with the use of GMOs. Pollutants that can be eliminated in a long time with naturally existing microorganisms can be removed faster with genetic modifications of the microorganisms. Advanced recombinant DNA technologies allow organisms to acquire desired properties or to further improve the properties they already possess. GMOs are frequently used in bioremediation due to the superior properties they have acquired. Modifications in the genetic material of microorganisms allow the constitutive production of hydrolytic enzymes that break down hydrocarbons and the production of catabolic enzymes with increased chemical specificity to degrade contaminants. The design of microorganisms that can use organic pollutants as a carbon and energy source can be driven by genetic modifications. *In-situ* monitoring of microbial viability and enabling microbial growth under adverse or extreme conditions can also be achieved by genetic modifications.

METHODS FOR GMO CONSTRUCTION

Plasma membrane forms a selectively permeable barrier between the inside of the cell and the external environment. This hydrophobic barrier is impermeable to many hydrophilic macromolecules such as DNA, RNA, proteins, antibodies, chemicals, and drugs. Most molecular techniques require a foreign material to be inserted into a host cell. Therefore, unique ways of introducing these molecules are necessary.

DOI: 10.1201/9781003188568-3

The introduction of particular genetic materials into the target cell in an efficient, repeatable, and safe manner to manufacture GMOs is the ultimate objective of most genetic engineering investigations. However, DNA is a highly charged macromolecule that cannot diffuse through the membrane of a cell. As a result, a variety of biological, chemical, and physical techniques for delivering foreign genetic elements have been devised.

Agrobacterium tumefaciens-mediated transformation [1,2], protoplast transformation employing cell wall destroying enzymes [3], and viral infections [4] are all biological techniques. Chemical methods include lipofection [5] and the use of cationic polymers such as calcium phosphate, polyethylenimine, dimethylaminoethyl-dextran, chitosan, and polyphosphoesters for gene delivery into cells [4,6,7,8].

Although biological and chemical methods of genetic transformation are still widely employed, physical approaches like as electroporation, microinjection, biolistic (gene gun), vacuum infiltration, glass bead agitation, silicon carbide fibers, laser micro beams, ultrasound, and shock-wave-mediated transformation have become more popular in recent years [9,10,11,12]. Physical approaches relate to introducing DNA into a cell by enhancing the permeability of the cell membrane using physical force, while avoiding cytotoxic effects and safety concerns associated with bacterial and viral gene delivery strategies, as well as toxic effects from large doses of chemical vectors.

Electroporation is a widely applicable physical technique that creates transient holes on the surface of the cell by causing a change in polarity on cell membranes through electrical pulses, and thus, the entrance of DNA into cells is facilitated. With the electroporation method, in addition to DNA, various exogenous molecules with varied molecular masses and forms can be transferred to cells during the electropermeabilized state.

A pulse of high intensity electric field is applied for a brief period of time to transiently permeabilize the cell membrane. This transient permeability is caused by the applied electric field inducing a polarity shift in the membrane, resulting in a dipolar movement inside the cells and a potential differential across the plasma membrane. If the external electric field is not excessive, the cell can survive by resealing the permeabilized membrane once the external electric field is withdrawn.

At present, electroporation is widely employed in molecular and medical research, including DNA transfection or transformation, direct transfer of plasmids between cells, transdermal drug delivery, and cancer tumor electrochemotherapy.

HISTORY OF ELECTROPORATION

The electroporation method has been used for decades. Earlier research has reported the rupture of erythrocytes and bacterial protoplasts due to the irreversible permeability of cell membranes by high electric fields [13]. Immediately after, several researchers reported that brief electrical pulses in the μs and kV/cm range might trigger the temporary cell membrane permeability, allowing the transfer of small substances into the cells. Application of intense electric pulse for DNA transfer was first performed on mouse fibroblasts by Wong and Neumann [14]. Neumann and his colleagues [14,15] discovered that fibroblasts receive external DNA upon exposure to electric shock. Later, this technique was adapted to all cell types. Unlike fibroblasts, it can even be applied to lymphocytes that cannot be transfected with other alternative procedures [16]. At that time, the mechanism of DNA transport through cell membranes was unknown, but a simple physical model for improved DNA penetration into cells in high electric fields was proposed [17]. Since then, many advances have been made in the field and currently, it is known that electroporation creates pores on the cell surface by causing polarity changes on cell membrane through electrical pulses, thus facilitating the entry of DNA into cells [18].

Chang and Reese [18] studied the formation of membrane pores in electro-permeable cells using rapid freezing electron microscopy in human erythrocytes subjected to radio frequency electric field. They showed that electroporation was associated with the formation of volcano-shaped breaches on the membrane. The formation of the membrane pore is basically divided into three stages: in the first stage, the pores are created and expand rapidly after the first few milliseconds following the electric pulse. In the second stage, pore structures remain essentially constant for a few milliseconds to several seconds, although some of the smaller pores may continue to grow. In the final stage, the pores decrease in size due to the sealing-up process.

DNA transfer via electroporation has advanced fast over the last decade, and currently, it is a well-established technology that can be applied to all types of cells such as bacteria, fungi, plants, and animals [19].

MECHANISM OF ELECTROPORATION

The term "electroporation" implies a temporary increase in the membrane permeability when exposed to an electric field, but the processes behind the permeability of the cell membrane and concomitant transfer of genetic materials are still not fully elucidated. Several models have been proposed to elucidate the mechanism of transient membrane permeabilization and the associated access of non-permeant molecules into the cell [20,21]. A thermodynamic approach supports the pore formation, which enables the system's electrostatic energy to be reduced. While line tension tends to suppress pore formation, which is an activated process, surface tension promotes it. Applied electric field pulses enhance surface tension and hence decrease the activation energy barrier [21].

Application of pulses of a high-intensity electric field to cells causes structural rearrangement of the membrane. Transient aqueous pathways, in the form of pores, are created by the rearrangements. The electric pulses both induce the formation of pores and provide impetus for ionic and molecular transport across the pores locally [22].

Exposure to an electric field in an electroporation cuvette (Figure 3.1.a) creates a charge in the form of a transmembrane potential in the cell membrane (Figure 3.1.b). The electric pulse causes the dipoles of the molecules to orient themselves according to the electric field. It is then distributed in and around the cell such that the side facing the cathode is depolarized and the other side of the

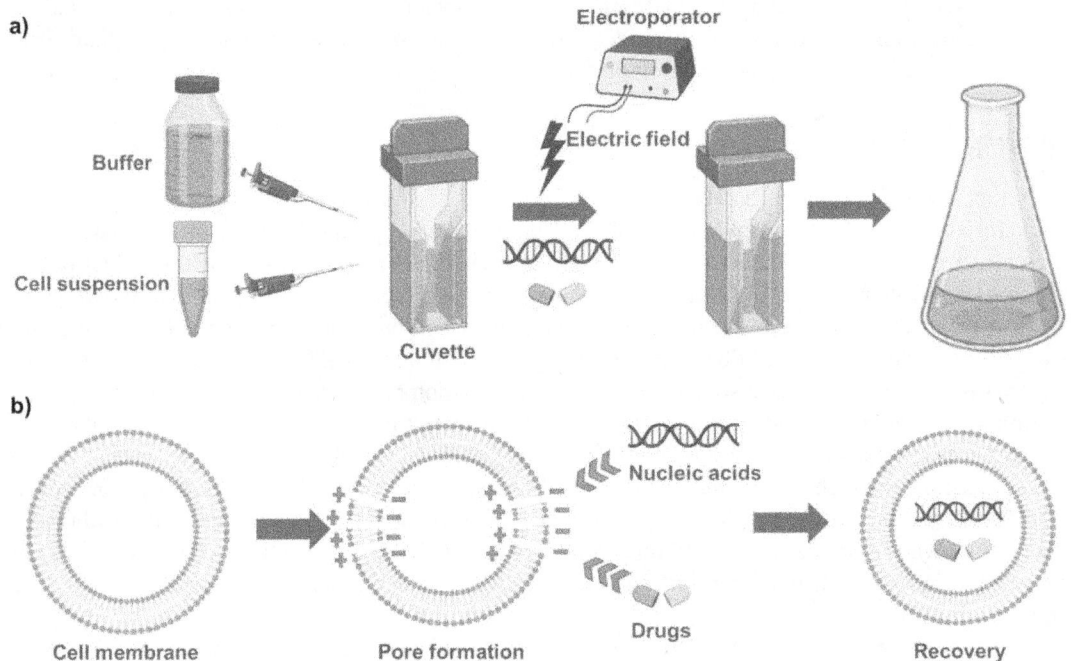

FIGURE 3.1 Demonstration of basic electroporation mechanism. Pore formation (a); uptake of exogenous molecules by electropermeabilized cell (b).

cell towards the anode is hyperpolarized due to the different charge accumulation on either side of the plasma membrane. When the transmembrane potential pass over the dielectric strength of the membrane (around 500 mV), the membrane is subjected to a permeation event causing the formation of hydrophobic pores allowing water movement and ion flow [23,24].

Small compounds enter primarily by simple diffusion, whereas macromolecules, including DNA, enter via a multistage process that involves the electrophoretically oriented engagement of the molecule with the destabilized and permeabilized membrane during the electric field followed by its movement through the cell membrane. Hence, efficient transfer of DNA into cells relies both on membrane permeabilization and the manner in which DNA engages with the plasma membrane. The DNA molecules move to the nucleus of the cell after entering the cytoplasm [21]. Aside from allowing ions, DNA, and drugs to enter the cytoplasm, the electropermeabilized plasma membrane also permits cell fusion or the introduction of proteins into cells [25].

Using appropriately selected and optimized electric field parameters, the membrane's electropermeability is a reversible process, and a cell can recover to its normal physiological condition when the electric field is removed. If the parameters related to the electric field exceed a certain value, the cells become irreversibly permeable and lose their vitality [25].

PARAMETERS AFFECTING THE ELECTROPORATION EFFICIENCY

Parameters affecting the electroporation efficiency include the surface concentration and shape of DNA. Free-ended linear DNA is more prone to recombination and, most probably, integration into the chromosomal DNA, resulting in durable transformants. Supercoiled DNA is easier to pack into chromatin and is more efficient for transient gene expression in general [26]. Cell-associated parameters include the type, size, and shape of the cell, cell density at the time of transfection, cell resistance to membrane permeation, and regeneration capacity of the cells.

In addition to biological parameters, the effectiveness of the electroporation process is strongly reliant on parameters related to the electric field. The intensity, number, and length of pulses, extent and length of membrane permeability, pattern and length of the molecular flow, and frequency of pulse repeats and pulse waveforms are among the most important parameters [27,28].

In electroporation, the external electrical field is used in the form of pulses. In order to influence intracellular structures or produce a high number of tiny holes in membranes, electrical field lengths between 10 ns to 1000 ns are frequently employed. The two most commonly used waveforms in the electroporation of nucleic acids are exponential-decay pulse and square pulse.

EXPONENTIAL-DECAY PULSE

For exponential-decay pulse, the capacitor releases the set voltage (V_0) into the cells and the voltage delivered to the cells drops exponentially (ms) over time. The voltage V at a given time t is given by $V = V_0 e^{-(t/RC)}$. In the particular situation where t = T, then $V = V_0/e$. The CR value is termed as the voltage decay's time constant. The decay rate increases as the time constant decreases (Figure 3.2.a). The exponential-decay pulses are easily standardized in most cell types for highly efficient transfection. The released pulse is defined by the time constant (t) and the field strength (kV/cm), which are adjusted by varying voltage and capacitance settings to accomplish a wide pulse gradient [21].

SQUARE WAVE PULSE

Sensitive cells can be damaged by exponential-decay waves, and square wave pulses may then be preferred for increased efficiency and viability. In addition, square waveform pulses are preferred for *in-vivo* applications. In square (or rectangular) wave pulse, the pulse delivered from the capacitor is characterized by the voltage applied, the duration of the pulse, the quantity of pulses, and the duration of the interval between pulses [21]. The electroporator delivers pulses to cells with a set

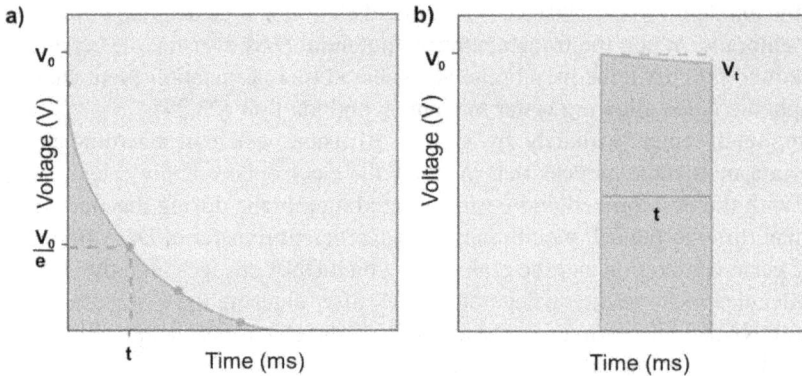

FIGURE 3.2 Pulse wave forms; exponential-decay pulse (a) and square wave pulse (b).

voltage for a defined period of time. The time of pulses varies between 100 μs and 100 ms. The voltage again decreases during the pulse exponentially, and at the end of the pulse, the voltage (V_t) is slightly less than the initial voltage (V_0) (Figure 3.2.b). Square wave generators can produce rapidly repetitive pulses.

The permeability of the cell membrane is accomplished by exposing the cell to an intense electric field over a short time. The principal amount behind this process is probably the difference in transmembrane potential induced in the first approximation, which is proportional to the cell radius and the product of the delivered electric field intensity [28,29]. It has also been revealed that the electric field affects the permeability of the plasma membrane in two different ways. First, the electric field starts the permeability of the cell membrane in spots at which the transmembrane potential difference surpasses the 200–300 mV threshold value [30]. Second, the size of the permeable region of the cell membrane is determined by the intensity of the electric field [28,31], which means that the permeability of the cell membrane will only take place once the delivered electric field is greater than the threshold value. Because the generated transmembrane potential difference is proportional to the cell radius, it is obvious that the threshold value of the electric field changes according to the size of the cell, implying that big cells are more susceptible to lower electric field strengths than tiny cells [32]. Likewise, the generated transmembrane potential difference has been demonstrated to be dependent on cell density, shape, position, and arrangement [33]. Therefore, it is inconvenient to generalize the parameters related to the electric field for different cell types (bacteria, fungi, plant, or animal) and experimental conditions. As a result, different cell types and applications require pulses with different waveforms, strengths, and duration [28].

ADVANTAGES AND DISADVANTAGES OF ELECTROPORATION

Electroporation has a number of advantages over alternative gene delivery methods. It is easy to apply, fast, low cost, non-viral, and non-toxic. It is an adaptable approach that may be safely employed to all cell types and cells in any form, whether *in vivo, in vitro,* or *ex vivo* [16]. The electric field is applied as a very short pulse on the membrane that causes only temporary pores and does not permanently damage the cells; then the cells can heal by resealing these pores. Electroporation is reproducible and efficient so that a high frequency of permanent transformants can be obtained. It is not limited by the size of the plasmid; various types of DNA constructs can be transferred without size limitation of DNA; and uptake is immediate. Electroporation can be performed at a small scale and requires smaller amounts of cells and DNA than other techniques. In addition, multiple plasmids can be introduced into the same cell by electroporation [34].

Drawbacks of electroporation include the requirement of laborious protocols for process and post-transformation regeneration. It is used to transform mainly protoplasts. Plant cells must be converted to protoplast before electroporation to enable DNA transfer into the cell. If the pulses are of the wrong intensity and length, some pores may become too large or fail to reseal after membrane discharge, causing cell damage or rupture [22]. The transport of materials into and out of the cell during electropermeabilization is relatively unspecific, which may cause ion unbalance and result in inappropriate cell function and finally, cell loss.

MATERIALS AND METHODS FOR ELECTROPORATION

ELECTROPORATION BUFFER

The ionic composition and conductivity of the buffer should be designated to improve cell transfection rates by minimizing cell death while ensuring the efficient transfer of nucleic acids. It should be nuclease-free to prevent degradation of DNA and capable of preserving cell viability after electroporation [35]. Buffers used in electroporation have different compositions such as HEPES as well as phosphate- and saline-based buffers. The conductivity is adjusted by the salt and the osmolality is adjusted with an osmotic factor, which is generally a sugar or an inert protein [26,36].

ELECTROPORATOR

An electroporator (or electropulsator) is a device used for conducting genetic transformation via electroporation. The electroporator consists of a chamber (80–800 µl) with a slot (4 mm) housing two parallel plate electrodes made of aluminum or stainless steel. The electrodes are in contact with an electrolyte solution that contains the cell suspension and the DNA that will be transferred into the cell.

ELECTRICAL PULSE

Although the voltage required for electroporation depends on the size and shape of the cell [37], each application generally consists of one or more electrical pulse applications ranging from 1.6–2.0 kV and lasting from 10^{-6} to 10^{-2} s for small cells like bacteria. This applied voltage causes cellular permeability owing to an electrical imbalance in the cell membrane once the potential difference is greater than 0.5 V (membrane voltage threshold is between 0.5–1 V) under normal pressure and temperature conditions. In this way, DNA may be transferred into the cells without altering membrane integrity or cellular functions [35]. Subsequently, the resealing of the cellular membranes and the isolation of gene-transferred cells is encouraged [11].

TARGET CELL

The electroporation technique is preferably performed with electrocompetent cells that are resistant to high-voltage electrical fields and quick-to-recover. Cells in the mid-to-late-log phase are preferred for electroporation. The cells can be in suspension or adherent to a substratum. Adherent cells are trypsinized before electroporation and then the trypsin is inactivated by serum.

POST-ELECTROPORATION MEDIUM

Since cells may be damaged during the electroporation process, a nutrient-rich microbial broth medium is used in the recovery step for maximizing the transfection efficiency. A SOC medium containing yeast extract (0.5%), tryptone (2%), 2.5 mM KCl, 10 mM NaCl, 10 mM $MgCl_2$, 10 mM $MgSO_4$, and 20 mM glucose (or dextrose) is the most preferred medium due to its contents of peptides, amino acids, water-soluble vitamins, and glucose, as well as its low salt content. SOC medium has been often used for higher transformation efficiencies of *Escherichia coli* competent cells [38].

ELECTROPORATION CUVETTE

Special cuvettes are used in electroporation, which must comply with protocols for biocompatibility, sterility, and engineering tolerances. In addition, they must provide optimum and reproducible impedance measurements and be suitable for generating uniform pulses for improved gene delivery.

In brief, the target cell suspension is placed in an electroporation cuvette containing the electroporation buffer, as shown in Figure 3.1.a. The DNA to be transferred is loaded and the cuvette is connected to the power supply of the electroporator. The cells are subsequently subjected to a brief exposure to a pulsed electrical field. Electropermeabilized and transformed cells are incubated briefly in a nutrient-rich growth medium for resealing before transfer to a selective growth medium.

A SAMPLE EXPERIMENTAL PROTOCOL FOR THE TRANSFORMATION OF BACTERIA BY ELECTROPORATION

1. Sterile culture tubes (17x100 mm) are prepared at room temperature.
2. SOC medium is kept ready in a 37 °C water bath to be used in the recovery step of competent cell transformation.
3. Selective agar plates are preheated at 37 °C for 1 hour.
4. Electroporation cuvettes (1 mm) and microcentrifuge tubes are placed on ice.
5. Electrocompetent cells are thawed on ice for approximately 10 min and mixed by gently tapping.
6. 25 μl of the cells ($1-10x10^6$) are transferred to the chilled microcentrifuge tube.
7. 1 μl of DNA solution (final concentration of 5–10 μg/ml is sufficient) is added to the microcentrifuge tube and mixed gently with electrocompetent cells.
8. The DNA-cell mixture is carefully transferred into a cooled electroporation cuvette without creating air bubbles.
9. Electroporation is applied under conditions of 2.0 kV, 100 Ω, 25 μF, and typical time constant ~ 2.6 milliseconds.
10. Immediately after electroporation, 975 μl SOC medium preheated at 37 °C is supplied to the cuvette and gently mixed with the DNA-cell mixture.
11. The mixture is transferred to the 17x100 mm sterile culture tube and incubated at 37 °C for 1 hour at 250 rpm.
12. Properly diluted 50–250 μl of mixture is spread on pre-warmed selective agar plates and incubated at 37 °C overnight.

The protocol was modified according to the method in "New England Biolabs: Electroporation Protocol (C2986)" [39].

Essentially, the universal phospholipid bilayer membrane makes electroporation practically applicable to different cell types in addition to bacterial cells. Electroporation has also been applied successfully in the *in-vivo* transfer of plasmid DNA to a number of tissues [40]. Plasmid DNA is injected into the tissues such as muscle or skin and electrodes are positioned around the site of injection. Target tissue cells are exposed to a defined pulsed electrical field for a predetermined time. The animal is allowed to reseal the pores and recover. For plant cells, the procedure is slightly modified by removing the cell wall and applying it to the protoplasts [26].

BIOREMEDIATION BY GENETICALLY MODIFIED MICROORGANISMS

The history of improved bioremediation with the use of GMOs dates back to the 1980s. GMO refers to any organism whose genetic material has been changed via the use of genetic engineering methods. The specific targeted modification of genetic material using recombinant DNA technologies has allowed scientists to improve the genetic makeup of the organism.

The gene to be inserted into the host organism must be isolated and coupled with additional genetic components, including a promoter and terminator sequence, as well as a selectable marker. The isolated gene must then be safely and properly transferred to the host genome using one of the available techniques. Various molecular techniques are available for the construction of GMOs. Among them, electroporation has shown to be an effective, fast, and safe technique in gene transfer, with a significant potential for the production of GMOs for bioremediation.

Most GMOs are constructed for use in a defined task. In bioremediation, constructing GMOs also has several goals which include:

i: increasing the microorganism's ability to break down a wide variety of chemical contaminants by introducing new genes that encode superior enzymes or by introducing new pathways for degradation of pollutants,

ii: allowing the microorganism to constitutively produce a hydrocarbon degrading enzyme by combining the gene to be inserted with a strong promotor of a housekeeping gene,

iii: producing microorganisms carrying hybrid metabolic pathways capable of degrading recalcitrant compounds by combining DNA fragments from different origins and placing the resulting ensemble of genes under the control of strong promoters,

iv: allowing microorganisms to produce hydrolytic/catabolic enzymes with increased chemical specificity and affinity for a substrate to be degraded,

v: directing the microorganism to use an organic pollutant as a carbon and energy source,

vi: developing microorganisms that can survive and grow under adverse/extreme conditions,

vii: accelerating biodegradation by combining catabolic fragments from different organisms,

vi: monitoring the viability of microorganisms *in situ* by introducing bioluminescent reporter genes.

DESIGN AND CONSTRUCTION OF GENETICALLY MODIFIED ORGANISMS

Before starting the design and construction of GMOs, it is necessary to identify the desired gene (or DNA fragments) with the desired trait in nature. For bioremediation, a microorganism that has the ability to naturally degrade pollutants is selected first. Since the organism has enzymes (or a set of enzymes) that allow it to use environmental contaminants as substrates, this microorganism is used as the source of the gene(s) encoding the desired enzymes. To obtain the desired gene, genomic DNA or plasmid DNA of the microorganism is purified by molecular techniques. The capability of certain bacteria to degrade hydrocarbons is a plasmid DNA-encoded trait [41, 42]. The desired genes (or DNA fragments) are cut by restriction enzymes and amplified using a specified set of primers by a polymerase chain reaction. The amplified gene is placed into the vector containing a strong promoter and is ready for transfer to the new host organism by electroporation. The host organism to which the desired gene will be transferred can be a microorganism or a plant [43].

Luo et al. [44] constructed a recombinant pCom8-GPo1 vector by inserting an alkane hydroxylase (alkB) gene of diesel oil-degrading *Pseudomonas putida* GPo1 into pCom8 expression vector. They transformed *E. coli* DH5 with the recombinant pCom8-GPo1 vector using electroporation and obtained increased diesel oil degradation rates.

Another approach is the transfer of the degradative plasmid directly into the host organism. Vasudevan et al. [41] studied with *P. fluorescens* (containing a 1.8 kb plasmid) which was able to utilize hexadecane as the sole carbon source. They showed that the hexadecane degradation ability was lost when the plasmid was removed from *P. fluorescens* (plasmid curing). In addition, they transformed *E. coli* cells with *P. fluorescens* plasmid and showed that *E. coli* also gained the ability to break down hexadecane. Results of plasmid curing and transformation of *E. coli* proved that plasmid DNA was responsible for degradation of hexadecane and this trait can be transferred from one cell to another.

The introduction of new genes encoding enzymes or new pathways for the degradation of recalcitrant pollutants has been used for years for increasing the capability of microorganisms to degrade a wide variety of chemical contaminants [45,47]. Tsoi et al. [46] cloned oxygenolytic ortho-dehalogenation (ohb) genes from 2-chlorobenzoate (2-CBA)- and 2,4-dichlorobenzoate (2,4-dCBA)-degrading *P. aeruginosa* 142 into *E. coli* cells by using electrotransformation. They observed DH5αF′ (pOD22) and DH5αF′ (pOD33) strains of recombinant *E. coli* converted 2-CBA to catechol, as well as 2,4-dCBA and 2,5-dCBA to 4-chlorocatechol. A subclone of *E. coli* pOD33, plasmid pE43, having the minimized ohb DNA region, enabled *P. putida* PB2440 to grow on 2-CBA as the sole carbon source.

The local population of hydrocarbon-degrading microorganisms is of paramount importance for *in-situ* bioremediation of oil-contaminated sites. Therefore, augmentation of the population is an effective technique for improved soil reclamation [47]. However, microbial population changes during bioremediation should be regularly checked [48]. For tracking and monitoring the living microorganisms in the environment, the use of bioluminescent reporter genes has been extensively studied [49]. Mishra et al. [50] transferred the luciferase gene as a bioluminescent reporter gene into a hydrocarbon-degrading *Acinetobacter* sp. by using electroporation. They used 25 μF capacitance, 1000 Ω (ohms) resistance, 2.5 kV, and 0.1 cm cuvette gap conditions in electroporation to transform *A. baumannii* with the recombinant plasmid construct containing the luxCDABE fragment, and survival of the lux-tagged strain could be monitored in crude oil-contaminated sites.

CONCLUSION

Environmental pollution remains a global problem that directly affects public health and the eco-system, and traditional treatment strategies are insufficient to combat it. Bioremediation stands out as an effective alternative method to deal with environmental pollution. To further support biore-mediation, genetic engineering is also used to improve the biodegradation capabilities of microor-ganisms. The targeted specific modification of genetic material using advanced recombinant DNA technologies has allowed scientists to improve the organism's genetic makeup so that it can be used more effectively in bioremediation. For the construction of GMOs with high bioremediation potential, several DNA transfer methods are available. Among them, electroporation provides an effective, fast, and safe gene delivery and is used frequently in the development of GMOs to be used in bioremediation.

Electroporation has become a well-established physical technique that temporarily increases the permeability of the cell membrane by using an intense electric field. Exposure to high-voltage electric fields creates temporary pores that facilitate the entry of DNA into cells. It is a versatile approach that can be used effectively on all cell types, yielding a high frequency of stable trans-formants. Recently, the widespread use of electroporation has greatly increased with the advent of commercial devices that are safe and straightforward to use, and provide highly reproducible outputs. It is clear that in the future, electroporation will be further improved with advancing tech-nologies and will continue to be a promising technique for obtaining GMOs with superior bioreme-diation abilities.

REFERENCES

[1] Frandsen, R.J., A guide to binary vectors and strategies for targeted genome modification in fungi using *Agrobacterium tumefaciens*-mediated transformation. *J. Microbiol. Methods.*, 87(3), 247–62, 2011.

[2] Li, X., Li, H., Yuzhu Z., Zong, P., Zhan, Z., Piao, Z., Establishment of a simple and efficient *Agrobacterium*-mediated genetic transformation system to Chinese Cabbage (Brassica rapa L. ssp. pekinensis). *Hortic. Plant J.*, 7(2), 117–128, 2021.

[3] Wang, Q., Yu, G., Chen, Z., Han, J., Hu, Y., Wang, K., Optimization of protoplast isolation, transforma-tion and its application in sugarcane (*Saccharum spontaneum* L). *Crop J.*, 9(1), 133–142, 2021.

[4] Graham, F.L., van der Eb, A.J., A new technique for the assay of infectivity of human adenovirus 5 DNA. *Virology*, 52(2), 456–467, 1973.

[5] Kawata, Y., Yano, S., Kojima, H., *Escherichia coli* can be transformed by a liposome-mediated lipofection method. *Biosci. Biotechnol. Biochem.*, 67(5), 1179–1181, 2003.

[6] Chowdhury, E.H., Kunou, M., Nagaoka, M., Kundu, A.K., Hoshiba, T., Akaike, T., High-efficiency gene delivery for expression in mammalian cells by nanoprecipitates of Ca-Mg phosphate. *Gene*, 341, 77–82, 2004.

[7] Miyazaki, M., Obata, Y., Abe, K., Furusu, A., Koji, T., Tabata, Y., Kohno, S., Gene transfer using non-viral delivery systems. *Perit. Dial. Int.*, 26(6), 633–40, 2006.

[8] Khan, N.T., Gene delivery system: Non-viral mediated chemical approaches. *J. Tissue Sci. Eng.*, 08(03), 2017.

[9] Graessmann, M., Graessmann A., Microinjection of tissue culture cells using glass capillaries. In *Microinjection and Organelle Transplantation Techniques* (eds. J.E. Celis, A. Graessmann, A. Loyter), pp. 3–13, Academic Press, New York, 1986.

[10] Mellott, A.J., Forrest, M.L., Detamore, M.S., Physical non-viral gene delivery methods for tissue engineering. *Ann. Biomed. Eng.*, 41(3), 446–468, 2013.

[11] Rivera, A.L., Loske, A., Gómez Lim, M., Fernández, F., Genetic transformation of cells using physical methods. *J. Genet. Syndr. Gene Ther.*, 5, 4, 2014.

[12] Ismagul, A., Yang, N., Maltseva, E., Iskakova, G., Mazonka, I., Skiba, Y., Bi, H., Eliby, S., Jatayev, S., Shavrukov, Y., Borisjuk, N., Langridge, P., A biolistic method for high-throughput production of transgenic wheat plants with single gene insertions. *BMC Plant Biol.*, 18, 135, 2018.

[13] Sale, A.J., Hamilton, W.A., Effects of high electric fields on micro-organisms: III. Lysis of erythrocytes and protoplasts. *Biochim. Biophys. Acta—Biomembr.*, 163, 37–43, 1968.

[14] Wong, T.K., Neumann, E., Electric field mediated gene transfer. *Biochem. Biophys. Res. Commun.*, 107(2), 584–587, 1982.

[15] Neumann, E., Schaefer-Ridder, M., Wang, Y., Hofschneider, P.H., Gene transfer into mouse lyoma cells by electroporation in high electric fields. *EMBO J.*, 1(7), 841–845, 1982.

[16] Potter, H., Weir, L., Leder, P., Enhancer-dependent expression of human κ immunoglobulin genes introduced into mouse pre-B lymphocytes by electroporation. *Proc Natl Acad Sci USA*, 81, 7161–7165, 1984.

[17] Neumann, E., Rosenheck, K., Permeability induced by electric impulses in vesicular membranes. *J. Memb. Biol.*, 10, 279–290, 1972.

[18] Chang, D.C., Reese, T.S., Changes in membrane structure induced by electroporation as revealed by rapid-freezing electron microscopy. *Biophys. J.*, 58, 1–12, 1990.

[19] Joersbo, M., Brunstedt, J., Electroporation and transgenic plant production. In *Electrical Manipulation of Cells* (eds. P.T. Lynch, M.R. Davey), Springer, Boston, MA, 1996.

[20] Escoffre, J.M., Dean, D.S., Hubert, M., Rols, M.P., Favard, C., Membrane perturbation by an external electric field: A mechanism to permit molecular uptake. *Eur. Biophys. J.*, 36, 973–983, 2007.

[21] Escoffre, J.M., Portet, T., Wasungu, L., Teissié, J., Dean, D., Rols, M.P., What is (still not) known of the mechanism by which electroporation mediates gene transfer and expression in cells and tissues. *Mol. Biotechnol.*, 41(3), 286–295, 2009.

[22] Weaver, J.C., Electroporation theory. In *Plant Cell Electroporation and Electrofusion Protocols. Methods in Molecular Biology™* (ed. J.A. Nickoloff), Vol. 55, Springer, Totowa, NJ, 1995.

[23] Hibino, M., Itoh, H., Jr Kinosita, K., Time courses of cell electroporation as revealed by submicrosecond imaging of transmembrane potential. *Biophys. J.*, 64, 1789–1800, 1993.

[24] Young, J.L., Dean, D.A., Electroporation-mediated gene delivery. *Adv. Genet.*, 49–88, 2015.

[25] Neumann, E., Sowers, A.E., Jordan, C.A., *Electroporation and Electrofusion in Cell Biology*, Plenum, New York, 1989.

[26] Potter, H., Transfection by electroporation. *Current Protocols in Molecular Biology*, Chapter 9, Unit–9.3, 2003.

[27] Pucihar, G., Mir, L.M., Miklavčič, D., The effect of pulse repetition frequency on the uptake into electropermeabilized cells in vitro with possible analysis and it application. *Bioelectrochem. Bioenerg.*, 57, 167–172, 2002.

[28] Puc, M., Čorović, S., Flisar, K., Petkovšek, M., Nastran, J., Miklavčič, D., Techniques of signal generation required for electropermeabilization. *Bioelectrochemistry*, 64(2), 113–124, 2004.

[29] Canatella, P.J., Karr, J.F., Petros, J.A., Prausnitz, M.R., Quantitative study of electroporation-mediated molecular uptake and cell viability. *Biophys. J.*, 80(2), 755–764, 2001.

[30] Kotnik, T., Pucihar, G., Induced transmembrane voltage—theory, modeling, and experiments. In *Advanced Electroporation Techniques in Biology and Medicine* (eds. A.G. Pakhomov, D. Miklavčič, M.S. Markov), pp. 51–70, CRC Press, Boca Raton, 2010.

[31] Teissié, J., Rols, M.P., An experimental evaluation of the critical potential difference inducing cell membrane electropermeabilization. *Biophys. J.*, 65, 409–413, 1993.

[32] Kotnik, T., Bobanovic, F., Miklavčič, D., Sensitivity of trans-membrane voltage induced by applied electric fields—a theoretical analysis. *Bioelectrochem. Bioenerg.*, 43, 285–291, 1997.

[33] Valic, B., Golzio, M., Pavlin, M. Schatz, A., Faurie, C., Gabriel, B., Teissié, J., Rols, M.P., Miklavcic, D., Effect of electric field induced transmembrane potential on spheroidal cells: Theory and experiments. *Eur. Biophys. J.*, 32, 519–528, 2003.

[34] Matsuda, T., Cepko C.L., Controlled expression of transgenes introduced by in vivo electroporation. *P.N.A.S.*, 104(3), 1027–1032, 2007.

[35] Djuzenova, C.S., Zimmermann, U., Frank, H., Sukhorukov, V.L., Richter, E., Fuhr, G., Effect of medium conductivity and composition on the uptake of propidium iodide into electropermeabilized myeloma cells. *Biochim. Biophys. Acta. Biomembr.*, 1284, 143–152, 1996.

[36] Sherba, J.J., Hogquist, S., Lin, H., Shan, J.W., Shreiber, D.I., Zahn, J.D., The effects of electroporation buffer composition on cell viability and electro-transfection efficiency. *Sci. Rep.*, 10, 3053, 2020.

[37] Wells, D.J., Electroporation and ultrasound enhanced non-viral gene delivery in vitro and in vivo. *Cell Biol. Toxicol.*, 26(1), 21–28, 2010.

[38] Hanahan, D., Studies on transformation of *Escherichia coli* with plasmids. *J. Mol. Biol.*, 166(4), 557–580, 1983.

[39] New England biolabs: Electroporation protocol (C2986). protocols.io https://dx.doi.org/10.17504/protocols.io.crgv3v.

[40] Heller, L.C., Heller, R., In vivo electroporation for gene therapy. *Hum. Gene Ther.*, 17(9), 890–897, 2006.

[41] Vasudevan, N., Bharathi, S., & Arulazhagan, P., Role of plasmid in the degradation of petroleum hydrocarbon by *Pseudomonas fluorescens* NS1. *J. Environ. Sci. Health A.*, 42(8), 1141–1146, 2007.

[42] Lanka, U., Kandisa, R.V., Kannangara, S., Sirisena, D.M., Plasmid encoded toluene and xylene degradation by phyllosphere bacteria. *J. Environ. Anal. Toxicol*, 08(02), 2018.

[43] Darbani, B., Farajnia, S., Toorchi, M., Zakerbostanabad, S., Noeparvar, S., Neal Stewart Jr., C., DNA delivery methods to produce transgenic plants. *Sci. Alert.*, 10(4), 323–340, 2011.

[44] Luo, Q., He, Y., Hou, D.-Y., Zhang, J.-G., Shen, X.-R., GPo1 alkB gene expression for improvement of the degradation of diesel oil by a bacterial consortium. *Braz. J. Microbiol.*, 46(3), 649–657, 2015.

[45] Hrywna, Y., Tsoi, T.V., Maltseva, O.V. Quensen, J.F., Tiedje, J.M., Construction and characterization of two recombinant bacteria that grow on ortho- and para-substituted chlorobiphenyls. *Appl. Environ. Microbiol.*, 65(5), 2163–2169, 1999.

[46] Tsoi, T.V., Plotnikova, E.G., Cole, J.R., Guerin, W.F., Bagdasarian, M., Tiedje, J.M., Cloning, expression, and nucleotide sequence of the pseudomonas aeruginosa 142 ohb genes coding for oxygenolytic ortho dehalogenation of halobenzoates. *Appl. Environ. Microbiol.*, 65(5), 2151–2162, 1999.

[47] Andreolli, M., Lampis, S., Brignoli, P., Vallini, G., Bioaugmentation and biostimulation as strategies for the bioremediation of a burned woodland soil contaminated by toxic hydrocarbons: A comparative study. *J. Environ. Manage.*, 153, 121–131, 2015.

[48] MacNaughton, S.J., Stephen, J.R., Venosa, D.A., Davis, G.A., Chang, Y., White, D.C., Microbial population changes during bioremediation of an experimental oil Spill. *Appl. Environ. Microbiol.*, 65, 3566–3574, 1999.

[49] Nunnes-Halldorson, V., Norma-Leticia, D., Bioluminescent bacteria: Lux genes as environmental biosensors. *Braz. J. Microbiol.*, 34, 91–96, 2003.

[50] Mishra, S., Sarma, P.M., Lal. B., Crude oil degradation efficiency of a recombinant *Acinetobacter baumannii* strain and its survival in crude oil-contaminated soil microcosm. *FEMS Microbiol. Lett.*, 235(2), 323–331, 2004.

4 Application of Fermentation Techniques in the Production of Genetically Engineered Microorganisms (GMOs)

Ankita Sharma, Anchal Rana,
Nandita Sharma and Samriti Sharma

INTRODUCTION

Biotechnology is a term that first appeared in the early 20th century, but its applications are as old as social human practices. Initially, the term "biotechnology" was limited to the field of biochemical engineering, especially industrial microbiology and enzyme technology. In any field of application, biotechnological tools developed from basic sciences such as recombinant DNA technology, metabolic engineering, and guided evolution; process growth, bioprocess modeling, scale-up rationale, and bioreactor design are critical as well (Joshi et al., 2019). The processing of raw materials from plant or animal origin to produce foods dates back more than two million years and achieved a significant milestone with the regulation of fire. The processing of fermented foods such as wine, bread, and cheese is the first type of biotechnology that can be traced back to ancient times. The use of fermentation in conjunction with the heating of foods resulted in major changes in food safety and quality. Bioprocesses are expected to replace chemical manufacturing processes in the 21st century because they are environmentally sustainable and dependent on natural activities carried out by living things, ensuring the industrial sector's long-term viability. New production systems are integrated production systems (IPS) with the goal of processing foods, oil, and industrial compounds with the least amount of energy and the least amount of environmental impact, allowing for long-term sustainability.

Fermentation and genetic engineering are two main components of biotechnology used in the food processing, chemical, and pharmaceutical industries (Yokoi et al., 1993). Microorganisms operate as mini-factories, converting raw materials into finished goods, and they can be genetically modified to meet the needs of a specific industry. Genetically modified microbes are created by transplanting a gene of interest into microbes using molecular biology and recombinant DNA technologies. These engineered microbes must become a part of a useful industrial process to have an effect on society at large, and fermentation is the one that is credited with biotechnology's economic success.

HISTORY OF FERMENTATION

Fermentation has been used to make products for everyday use from the beginning of human history. Fermentation is considered one of the oldest forms of food processing known today. For thousands of years, microbial processes have been primarily used to retain or reformeatable products for human consumption. Notable examples include cereal fermentation, beer produced from hops, and wine, all of which conserve a large part of nutrition while preserving the products. Bread and beverages which are produced from natural yeast contain large numbers of vitamins and other nutrients.

DOI: 10.1201/9781003188568-4

For instance, the fermentation of milk to yogurt and cheese through lactic acid increases the shelf life of dairy products. In addition, through the process of fermentation, many food products can be preserved and their flavors enhanced, such as tea leaves, coffee beans, and pickled vegetables.

Up to the second half of the 19th century, the process of fermentation was considered more of an art than a science. During that time, a meager portion of the leftover batch was used to make bacterial culture. The theory that bacteria are responsible for fermentation did not gain popularity until 1857, when Louis Pasteur released research describing the cause of failed commercial alcohol fermentations. He also popularised the technique of pasteurisation (heat treatment) to increase wine storage quality. It was the initial stage in the process of sterilising a fermentation medium in order to prevent contamination. Emil Christian Hansen proposed the idea of using pure yeast culture in beer manufacturing in Denmark in 1883.

At the beginning of the 18th century, the production of beer and wine was relatively large to satisfy the demand of urban populations. Furthermore, it was the era of the industrial revolution, which resulted in growing populations in large cities. As a result, there was a considerable shift from small-scale to large-scale food production, which was necessary to fulfil the demands of expanding and more distant markets.

In Great Britain and Germany, acetone-butanol fermentation was carried out during World War I, and is considered the first aseptic large-scale fermentation. The main objective is to make butanol as a precursor to butanediol, which would be used to make synthetic rubber. Following the onset of the war, Great Britain's antecedent of acetone for weapons manufacturing, which had been imported from Germany previously, was cut off, causing the focus to move to acetone. As the method was developed and expanded, competing bacteria were introduced which overtook *Clostridium acetobutylicum*. The culture media had to be sterilised, and the procedure had to be carried out under aseptic circumstances.

The United States Constitution's 18th Amendment prohibited the sale of alcoholic drinks in 1919, enabling premises to be utilised for other purposes. Organic acids, notably citric acid and lactic acid, which are generated from non-sterile beer and wine corn fermenters, became the focus of fermentation. Apart from this, various products from agriculture such as corn steep liquor and the use of subside culture from fungus were also being used in fermentation. In 1930, penicillin was created in the United Kingdom, and in 1942, marketed in the United States. Penicillin was lauded as one of the first medicines to successfully treat infection from bacteria that had resulted in serious sickness. During the time of World War II and the years that followed, its demand was very high. Penicillin was first grown in quart milk bottles as a surface culture, but its production is hindered by cost, availability, and bottle handling. Submerged culture fermentation was also invented by scientists. Using stirred tank bioreactors, this procedure not only boosts productivity per unit volume, but may also boost production scale. The huge success of penicillin in the 1940s and 1950s encouraged pharmaceutical firms to find and develop a slew of new antibiotics. The majority of the fermentation processes were aerobic, requiring a lot of aeration and agitation. Biochemical engineering evolved at this time as a unique area that allows researchers to investigate difficulties in the fermentation process and to develop fermenters with high transfer rates.

In the 1960s, amino acid fermentation was discovered in Japan. Monosodium glutamate (MSG) was created to replace MSG derived from natural sources as a taste enhancer. Another fermentation from amino acids was produced employing glutamic acid bacteria cultures. In the 1970s, enzyme commercialization and application in industrial processes began, with 80% of all enzymes in use today.

In 1970, genetic engineering opened up new possibilities for fermented products. In 1977, insulin emerged as the first genetically modified compound. Following that, a vast number of genetically engineered products have been produced.

BIOREACTORS

A bioreactor is a vessel in which a biological reaction occurs in order to transform any material (raw material or substrate) into a product. It is utilised in a variety of bioprocesses, including the production of soy sauce, the treatment of domestic and industrial waste, the manufacturing of vaccines and antibiotics, and the production of important compounds.

Several biological processes rely on bioreactors. In order to contemplate a bioreactor, firstly we have to identify the biological process, which can be determined by market demand or the biotransformation process. However, advancement in recombinant DNA techniques and genome sequencing through different biological processes can be achieved: microorganisms, plant cells, animal cells, or enzymes. Various processes, such as genetic expressions, metabolic modification, and bioreaction pathways, must be comprehended. Secondly, medium requirement for particular biological process which will enhance the performance. Moreover, the physical environment (pH, temperature, mixing and shear stress, mass transfer) also affects performance of bioprocesses. A suitable bioreactor type can be chosen after examining the physical and biological requirements for a certain biological process. It is also crucial to consider various factors (i.e., transfer of oxygen, mixing, shear, operational stability and dependability, scale-up, and cost) in a balanced way. Bioreactor optimization and characterisation are critical because they influence biological performance. Control and support systems are essential for a successful bioreactor system.

BIOREACTOR TYPES

Suspension and immobilisation systems are the two primary types of bioreactors. Immobilisation systems include packed bed, membrane, and fluidized bed bioreactors, whereas suspension cultures include air-lift, stirred tank, and bubble column bioreactors. Some bioreactors go within each of these categories. For instance, cells or enzymes immobilized on carriers, might be suspended in air-lift/bubble column or stirred tank bioreactors with the appropriate carriers.

Although each type of bioreactor has its unique design and selection criteria, certain fundamental principles are to be followed. For example, supplies of supplements and waste products must be eliminated. It is necessary to design an operational model, optimise environmental parameters, and measure cell growth and product generation kinetics. And various environmental parameters (i.e., pH, dissolved oxygen concentration, temperature) and substrate concentrations need to be monitored. Furthermore, bioreactor design should be economical, simple, feasible, and free from microbial contagion. Commonly used bioreactors used in industrial fermentation are discussed subsequently.

Stirred Tank Bioreactor (STBR)

STBR is one of the most widely used conventional bioreactors (i.e., good mixing, transferability of oxygen, scale-up, and substitute impellers, and easy compliance with cGMP). The essential component of the STBR is the impeller/agitator, which performs functions like heat and mass transmission, aeration, and homogenization mixing. However, there are some drawbacks to this form of bioreactor, including significant shear stress, issues regarding shaft sealing and stability, and immense power consumption. Animal and plant cells were more perceptive than microbes, which led to adapting and improving the impeller system by balancing mixing and mass transfer. By designing novel impeller designs, several alterations to traditional STRs have been produced (Yokoi et al., 1993; Ogbonna et al., 1996). Various researchers have studied the multiple impeller systems hydrodynamics (Cronin et al., 1989). A number of biopharmaceutical companies have successfully used STRs on a 10000–20000-l scale for large-scale animal cell cultures (Butler, 2005; Kretzmer, 2002).

Pneumatically Agitated Bioreactors

The pneumatic bioreactor is a gas-liquid dispersion reactor that contains a cylindrical tank with nozzles, perforated plates, and a ring sparger that introduces gas at the bottom of the vessel without the use of mechanical elements. Air-lift and bubble column reactors are two different forms of pneumatically agitated bioreactors. They normally include a main body, an air-bubbling device, a steam generator for sterilisation, an air vent system and air inputs, as well as control systems for monitoring temperature, oxygen, and pH, and systems for transporting steam, medium, air, and secondary products.

Air-lift bioreactors provide a number of advantages, including a gentler distribution of shear stress, as well as being simple to build and scale up at minimal cost. However, they have a number of drawbacks, including poor mixing of fluid and foaming. For a range of fermentation processes such as cell culture and biological wastewater treatment, an air-lift bioreactor with various configurations has been created (Margaritis and Sheppard, 1981).

Membrane Bioreactors

A membrane bioreactor is a form of flow reactor in which cells or enzymes are separated from product streams using specialised membranes. These membranes are made of cellulose, acetate, nitrate, polyvinylidene difluoride, polysulfone, polypropylene, polytetrafluoroethylene (PTFE), and polyacrylonitrile. For microfiltration and ultrafiltration, these membranes are commonly used. The pore size of microfiltration membranes ranges from 0.1 to 0.5 μm, whereas the pore size of ultrafiltration membranes ranges from 20 to 1000Å.

Microfiltration membranes are used to confine cells inside a reactor while imposing little restriction on the flow of soluble nutrients and products. Ultrafiltration membranes are used to retain or reject macromolecules, whereas microfiltration membranes are used to confine cells inside a reactor while imposing little restriction on the flow of soluble nutrients and products. Cell density, low shear stress, and large volumetric output are the most significant characteristics of membrane bioreactors. Large-scale applications are limited by the method's constraints, which include low cell viability, product inhomogeneity, diffusion gradient, and poor process stability. Despite the constraints, numerous membrane configurations (such as flat sheet and rotating bioreactors) have been examined, with the hollow-fiber configuration being one of the more intriguing. For biocatalysis, fermentation, cell cultures, and wastewater treatment, membrane reactors have been commonly used.

Fixed-Bed Bioreactors

Immobilization systems are frequently used in fixed (packed) bed reactors. The ease of operation and rapid reaction speeds of this type of bioreactor are well known. Large solid-liquid-specific interfacial contact areas result from the immobilisation of enzymes or cells in suitable carriers, and the velocity of liquid decreases the film barrier to mass transfer dramatically. Fixed bed bioreactors have poor mass and heat transfer coefficients because of low liquid velocities. However, effective gas-liquid interaction and removal of carbon dioxide are crucial in aerobic biological systems. By collecting stagnant gas pockets, a fixed-bed reactor causes flooding and affects the distribution of liquid. As a result, aerobic microbial fermentation methods do not consider this type of bioreactor (Nielsen and Villadsen, 1994).

Fluidized Bed Bioreactors

Another typical bioreactor for immobilisation systems is the fluidized bed reactor. A fluidized-bed reactor can provide an intermediate level of mixing between the packed bed reactor and the STBR (Varley and Birch, 1999). On one side, the fluid's upward flow moves the immobilised cells higher; the particles climb until gravity compels them to descend. The only way to achieve fluidization for particles is by movement (upward and downward). The fluidized-bed bioreactor has several advantages, including a homogenous system that can be easily monitored and balanced operational parameters (i.e., high mass transfer, efficient mixing and heat transfer, and the ability to scale up). Back-mixing and fluidization patterns are a few of the drawbacks. The use of a fluidized-bed reactor with immobilised cells is mostly for the treatment of wastewater. Microcarrier cultures have also been grown in fluidized reactors (Nienow, 1998).

Wave Bioreactors

In recent years, wave bioreactors, which generate a wave motion by mechanically swinging a culture-containing bag back and forth, have become popular. These waves offer aeration and transfer of mass, which results in an optimal environment for plant and animal cell suspension growth. The technique has been tested with a variety of cell lines, for example, Chinese hamster ovary (CHO) cells, hybridoma cells, insect cells, NS0 cells, and anchorage-dependent cells) (Chisti, 1999). Furthermore, it also has a sterilized, disposable chamber for convenience of use and contamination prevention. The chamber is mounted on a platform that adheres to the regulations of cGMP. Wave bioreactor systems of this type have recently been scaled up to 1000l capacity, with cell densities as high as 107 cells per ml.

EFFECTS OF PROCESS PARAMETERS ON BIOLOGICAL PERFORMANCES

For successful biotechnological processes, bioreactor analysis is critical. The main goal of a bioreactor is to create the best possible environment for biological activity. Various factors like pH, temperature, mixing, transfer of oxygen, and substrate concentration need to be optimised in the bioreactor. For instance, the ability of the bioreactor to balance substrate concentration is a vital function. However, in batch or fed-batch processes, substrate concentration may be susceptible to geographic change as well as time fluctuation, either purposefully or unintentionally.

Temperature

In a bioreactor, temperature is one of the most important factors to regulate. Microorganisms are categorized according to their temperature as thermophiles (> 50 °C); mesophiles (> from 20 °C to 50 °C) and psychrophiles (< 20 °C). Microorganisms require an optimal temperature for development; if the temperature falls below this, the microorganisms' development is hampered, accruing lower rates of cellular development and synthesis. However, if the temperature rises beyond optimum, it will not only kill microorganisms, but it will also alter protein expression and metabolite production, decreasing product yield and quality.

Effect of pH

The ideal pH ranges for each biological system are varied and pH 5 and 7 are considered ideal for most bacteria. pH can fluctuate throughout the fermentation process. Substrate consumption and numerous metabolites cause change in pH as the cell grows. For example, ammonia is utilised in fermentation which it includes nitrogen and lowers the pH. As a result, pH must be monitored and optimised using a base or acid. Elmahdi et al. (2003) reported the pH influence on the growth and synthesis of erythromycin by *Saccharopolyspora erythraea CA340*. Optimal pH resulted in an increase in the production of erythromycin.

Mixing

Proper mixing in a bioreactor is essential for providing an adequate supply of nutrients and preventing the accumulation of harmful metabolites. Various factors such as fluid hydrodynamics, impeller type, power input, fluid rheology, and vessel size are responsible for influencing mixing conditions. It was also reported that biomass affects the efficiency of mixing at low broth viscosity levels.

Oxygen Transfer

In aerobic biological systems, the transfer of oxygen is of major concern. A well-mixed bioreactor can provide an appropriate and timely supply of several nutrients required for cellular development and metabolism. However, the transfer of oxygen becomes a limiting step for biological systems' optimal functioning as well as scaling up. When there is a lack of oxygen, both cell development and product creation are severely hampered. For example, it was discovered that even a few minutes of excessive aeration during the fermentation of penicillin had a significant influence on the cells' capacity to generate the antibiotic.

Shear Force

Several factors such as shear stress, shear duration, power dissipation, and the cell's development phase influence shear-induced cell death. Shear damage can be due to sparging, which can occur in the bioreactor, including the bubble formation zone, suspension's surface, and rising zone.

Furthermore, bubble-free aeration can eliminate the problem of rising air bubbles for indirect aeration by utilising membranes, in which oxygen supply is regulated by diffusion. In addition, sensitive cell cultures minimising shear stress intensity is a general approach. Furthermore, with proper bioreactor design and management, shear damage might be reduced.

TYPES OF FERMENTATION

On the basis of the inoculum, fermentor used, product isolated, and various growth parameters, there are eight types of fermentation systems, (Figure 4.1) namely:

1. Batch system
2. Fed-batch system
3. Continuous system
4. Aerobic system
5. Anaerobic system
6. Surface system
7. Submerged system
8. Solid state system

Each one of them is described as follows:

Batch System

The batch fermentation system deals with the closed fermentation process where a limited amount of medium is added along with fixed volume of inoculum. The fermentation process takes place for

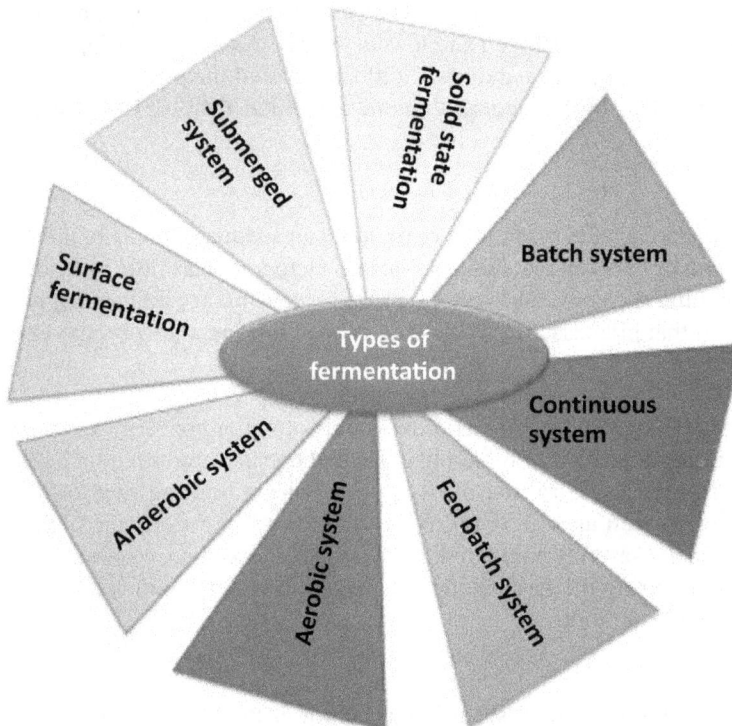

FIGURE 4.1 Types of fermentation systems.

a specific interval of time under optimum conditions. During this process, the cell undergoes various cycles of multiplication and growth which in turn leads to the alteration in the composition of medium and metabolites (Butler, 2005). After a specific interval of time the product is harvested and further subjected to purification.

Advantages:

- Used for production metabolites (primary and secondary metabolites)
- This is a simple process
- Low risk of contamination
- Low risk of mutation

Disadvantages

- The fermentor after every use needs to be cleaned and sterilized
- It is only applicable for reactions with high substrate concentration
- For every fermentation process, new culture and fresh medium are to be added, which can be quite a cumbersome process
- Labour-intensive process

FED-BATCH SYSTEM

This is the modified version of batch fermentation where the substrate is added in a specific period of time repeatedly, which maintains the substratum at an optimum level. Fed-batch fermentation plays a crucial role where this process avoids the catabolic repression of the secondary metabolites. This process is employed in such a way that the critical nutrients are added in low amounts to the medium to initiate the fermentation and then the substrate is added step by step in varied intervals of time (Liu et al., 2018). Fed-batch culture is also divided into sub-parts depending upon the volume of the culture added. These are:

Fixed volume cultures: little amount of the substrate including the limiting factor is added which leads to an insignificant increase in the medium

Variable volume cultures: the volume is increased by the addition of the same amount of medium.

Advantages:

- Evaporation enables the replacement of water loss in this system
- Controlled and well-managed conditions for the substrate during fermentation
- No extra or special equipment is required from outside
- Antibiotics like penicillin are favourably harvested in fed-batch systems where they are produced at an optimum level

Disadvantages:

- A detailed analysis of the microorganisms being used is necessary to correlate its philosophy with the desired productivity
- It requires a skillful operator to handle the bioreactor
- Direct methods like chromatography are inefficient in measuring the concentration of the feeding substrate.

CONTINUOUS SYSTEM

As the name suggests, a continuous system deals with fermentation for an indefinite period of time. The nutrients are added freshly after specific intervals of time continuously and the used medium with the product is harvested simultaneously. With the continuous flow of nutrients, the volume of the medium is maintained. Initially, the medium and inoculums are added to the point where culture growth is maintained and thereafter the supply of nutrients and simultaneous removal of broth takes place (Liu et al., 2018). Continuous fermentation can be of three types: single-stage fermentation, recycle-stage fermentation, and multiple-stage fermentation, depending on the addition of the medium at different stages of fermentation.

Advantages:

- It is a continuous process with little or no break
- It generates a high yield compared to other systems
- It is used mainly for the production of organic solvents, single-cell proteins, and antibiotics.

Disadvantages:

- Chances of mutation and contamination are high
- Requires efficient knowledge and skillful labour
- Wastage of nutrients can occur.

ANAEROBIC SYSTEM

Anaerobic fermentation refers to the system where fermentation takes place in the absence of oxygen. It requires two types of microorganisms: a) obligate anaerobes like *Clostridium* sp. that survive only in the absence of oxygen, b) facultative anaerobes like lactic acid bacteria which are active where minimal amounts of oxygen are present. The already present oxygen is removed by using exhaust pumps or by flushing it out while pumping in gases like carbon dioxide or hydrogen. Additionally, anaerobic conditions can be created by first sterilising the medium and then immediately adding inoculum at the base. If the medium is viscous or at a stationary position, then oxygen flow is also retarded.

Advantages:

- Many beneficial byproducts like hydrogen and carbon dioxide are produced which can be of great economic importance to manufacturers.

Disadvantages:

- Requires extra equipment such as pumps for maintaining anaerobic conditions
- Special media (viscous media) is required which is prepared by comparatively costly chemicals.

AEROBIC SYSTEM

The employment of oxygen for the fermentation process is called aerobic fermentation. Most of the fermentation takes place in aerobic conditions for the production of commercially important products. Aerobic fermentation can be of three types: static, submerged, or surface.

STATIC SYSTEM

In this type of fermentation, the microorganisms grow on slightly wet/moist solid material in the absence of water or minimal water. This technique is widely used in recent years for many bioprocesses. Here, the substrate is the only carbon source in the absence of water.

SUBMERGED SYSTEM

In this type of fermentation, the nutrients inserted into the medium are in liquid form, also termed broth. Most industrial bioprocessors deal with liquid broth medium (Liu et al., 2018). The microorganisms grow inside the substratum and extract nutrients from it. Impellors and agitators are employed for the aeration.

SURFACE SYSTEM

As the name suggests, surface fermentation is the type of aerobic fermentation where the substratum is either in the form of solid or liquid. The microorganisms grow on the surface and extract nutrients from the substratum. This is of most importance for products that rely on sporulation. The major drawback of this process is that uniform conditions are not maintained.

GENETICALLY ENGINEERED MICROBES (GEMS)

The food processing, chemical, and pharmaceutical industries rely on two major pillars of biotechnology: fermentation and genetic engineering. Microorganisms form the miniature factories that convert raw materials into end-products, and using genetic engineering they can be exploited to suit an industry's demands. Transplantation of genes of interest into microbes using molecular biology and recombinant DNA technologies leads to the formation of genetically engineered microbes. To impact society at large, these engineered microbes must become part of useful industrial processes accredited with the economic success of biotechnology.

ADVANTAGES OF USING GEMs

The laboratory work for an industrial product to be produced by fermentation does not end with finding an organism that naturally produces it. Many challenges are faced along the course of making this organism suitable for industrial-level production of the desired product. These challenges are an answer to why we need GEMs.

1. **Increased production**
   Industrial production is synonymous with mass or large-scale production. The organism naturally producing a substance may not be capable of meeting the needs of mass production. Therefore, through genetic manipulation microbes can be equipped to produce substances beyond their natural capacities leading to higher productivity that eventually makes the process economically competent.
2. **Eliminating harmful byproducts**
   During the process of fermentation, microbes produce many byproducts other than the desired product. These byproducts can act as inhibitors and slow down the whole process. In certain cases, the removal of these byproducts from the desired product is cumbersome, thereby resulting in low yields or inefficient products. Genetic engineering of microbes has proved to be successful in limiting the production or conversion of inhibitors such that they do not interfere with the production of the final substance (Joshi et al., 2019).
3. **Eliminating undesirable properties**
   Many microbes possess properties that are part of their environmental survival strategies, such as objectionable color, odor, and production of slime and spores. However, these properties are not required in a controlled environment and are difficult to handle. For example, the formation of spores could lead to airborne spread of the microbe. These properties can be modified or suppressed by genetic modification of the microbe without compromising the production of the substance required.

4. **Procuring useful substances from pathogenic microbes**

Certain useful enzymes, antibiotics, and other substances are produced by pathogenic microbes that cannot be easily inoculated or allowed to be mass-produced. In such cases, the gene responsible for the desired substance can be transferred to an innocuous and genetically tractable microbe to be safely produced.

APPLICATION OF FERMENTATION IN THE PRODUCTION OF GEMs

Genetic engineering is a technology used at the laboratory level. To expand the scope and efficiency of GEMs, various fermentation techniques are utilized. GEMs may be used as whole living organisms in certain cases, or the substances produced by them may prove beneficial to humankind. They have great potential to solve major problems of human society; however, without good production techniques like fermentation, it's just a single wheel of a cart. Fermentation along with GEMs can be applied for:

1. **Utilization of abundant, potentially useful raw materials**

Certain byproducts produced on a large scale can be used as raw materials for a fermentation process and converted into useful materials. Fermentation processes were developed to convert n-alkanes produced as petroleum refinery byproducts into single-cell proteins for use in animal feed.

2. **Diminishing the scarcity of an established product**

Certain products once obtained by other processes can be made by fermentation in large quantities. The production of human hormones, organic compounds, and inorganic chemicals which were difficult to obtain through natural processes was made simpler using GEMs and fermentation. Examples include insulin, human growth hormone, lactic acid, and citric acid.

3. **Mass production of newly identified products**

The field of antibiotics is a never-ending search. The finding that microorganisms can produce antibiotics has expanded the reach of microbiology and biotechnology to healthcare. Thousands of antibiotics have been discovered and a lot of them have been proven to be clinically useful. Strain improvement strategies and genetic modifications have been employed for the enhancement and discovery of novel variants of antibiotics. The list is very long, including doxorubicin, rifamycin, chloramphenicol, tetracycline, and erythromycin, to name a few (Demain and Adrio, 2008).

4. **Solution to environmental concerns**

Increasing anthropogenic activities result in causing great damage to the environment. Constant efforts are being made to find solutions to such problems. The environment is greatly influenced by the diversity of microorganisms present in it. Biodegradation capabilities of microorganisms have been discovered and utilized for bioremediation, sewage treatment, plastic decomposition, and more. The discovery of bacterial hemoglobin (VHb) has proven useful in bioremediation under hypoxic conditions (Urgun-Demirtas et al., 2006).

5. **Addressing scarcity of raw materials**

Mounting global energy, climate, and environmental challenges have increased the demand for renewable energy sources over traditional fossil fuels. The production of bioethanol derived from grain and sugar crops is restricted due to increasing demand for food supplies for the ever-increasing population. Lignocellulosic feedstocks and agricultural residues can also be used, as they contain cellulose that can be converted into bioethanol, but only after biomass pretreatment, enzymatic saccharification, and sugar fermentation. Many microbes like *Clostridium cellulovorans* have been metabolically engineered to produce biofuels, as they possess cellulolytic activity (Liu et al., 2018).

REFERENCES

Butler, M. 2005. Industrial processes with animal cells. *Applied Microbiology and Biotechnology* 68: 283–291.

Chisti, Y. 1999. Modern systems of plant cleaning. In *Encyclopedia of Food Microbiology* 1806–1815. London: Academic Press.

Demain, A.L. and Adrio, J.L. 2008. Strain improvement for production of pharmaceuticals and other microbial metabolites by fermentation. *Progress in Drug Research* 65: 252–289.

Elmahdi, I., Baganz, F., Dixon, K., Harrop, T., Sugden, D. and Lye, G.J. 2003. pH control in microwell fermentations of S. Erythraea CA340: Influence on biomass growth kinetics and erythromycin biosynthesis. *Biochemical Engineering* 16: 299–310.

Joshi, B., Joshi, J., Bhattarai, T. and Sreerama, L. 2019. Currently used microbes and advantages of using genetically modified microbes for ethanol production. In *Bioethanol Production from Food Crops*. Elsevier Inc.

Kretzmer, G. 2002. Industrial processes with animal cells. *Applied Microbiology and Biotechnology* 59: 135–142.

Liu, H., Sun, J., Chang, J.S. and Shukla, P. 2018. Engineering microbes for direct fermentation of cellulose to bioethanol. *Critical Reviews in Biotechnology* 38(7): 1089–1105.

Margaritis, A. and Sheppard, J.D. 1981. Mixing time and oxygen transfer characteristics of a double draft tube airlift fermentor. *Biotechnology and Bioengineering* 23: 2117–2135.

Nielsen, J. and Villadsen, J. 1994. *Bioreaction Engineering Principles*. New York, NY: Plenum Press.

Nienow, A.W. 1998. Hydrodynamics of stirred bioreactors. *Applied Mechanics Reviews* 51: 3–32.

Ogbonna, J.C., Yada, H., Masui, H. and Tanaka, H. 1996. A novel internally illuminated stirred tank photobioreactor for large scale cultivation of photosynthetic cells. *Fermentation Bioengineering* 82(61): 67.

Urgun-Demirtas, M., Stark, B. and Pagilla, K. 2006. Use of genetically engineered microorganisms (GEMs) for the bioremediation of contaminants. *Critical Reviews in Biotechnology* 26(3): 145–164.

Varley, J. and Birch, J. 1999. Reactor design for large scale suspension animal cell culture. *Cytotechnology* 29: 177–205.

Yokoi, H., Koga, J., Yamamura, K., Seike, Y. and Tanaka, H. 1993. High density cultivation of plant cells in a new aeration-agitation type fermentor, Maxblend Fermentor. *Fermentation Bioengineering* 75: 48–52.

5 Application of Genetically Modified Microorganisms as Biosorbents for Polluted Environments

Nagma Parveen, Amrita Kumari Panda,
Rojita Mishra, Aseem Kerketta and Satpal Singh Bisht

INTRODUCTION

Biosorption is a passive process in which various substances including heavy metals adsorb to biological material such as dead and live microbes, plants and seaweeds through electrostatic interaction, ion exchange, precipitation and chemical interactions (Mukkata et al., 2018). Microorganisms utilize pollutants as nutrition sources because of their adsorptive abilities. Biosorption technology is considered one of the most economical cleanup technologies. The negatively charged plasma membrane of microbes binds easily to positively charged heavy metals (Fomina and Gadd, 2014). For example, the outer murein layer of gram-positive bacteria contains anionic muramic acid, teichoic acid and meso-di-aminopimelic acid; similarly, the membrane of the gram-negative bacteria contains phospholipids, lipopolysaccharides, glycoproteins and lipoproteins that act as the active sites in biosorption processes (Gupta et al., 2015; Ayangbenro and Babalola, 2017). Dead biomass is preferred over live biomass as they do not require growth media and maintenance of growth conditions (Volesky, 2007). Many researchers have reviewed and published articles on the biosorption properties of microorganisms (Ilyas et al., 2017; Diep et al., 2018). Some studies have reported that genetically engineered microbes express various cell surface proteins that bind heavy metals with greater specificity (Ueda, 2016).

WHY BIOSORPTION?

The following are the key benefits of this technology over traditional techniques: biosorption is inexpensive, requires no expensive growth media, is quick to process, has a high metal binding performance, specificity for metal ions of interest, regenerates biomass, uses fewer chemicals, recovers metal ions in condensed form and produces no toxic waste (Figure 5.1) (Aryal, 2020).

MECHANISM OF BIOSORPTION

The mechanism of biosorption includes surface chemical modifications, intracellular adsorption and surface adsorption methods. Intracellular adsorption occurs when target metal ions are transferred to organelles and accumulates. Cell-surface adsorption methods include heterologous expression of metal-binding peptides. Out of all these methods, cell-surface adsorption methods have several advantages: (i) no need for cell disruption; (ii) no need for protein purification; (iii) recycling of biosorbents; and (iv) specific adsorption of target metals. The heavy metals biosorption process is biphasic. The first phase is the Rapid Initial Phase; this reversible phase is metabolism-independent and unaffected by

DOI: 10.1201/9781003188568-5

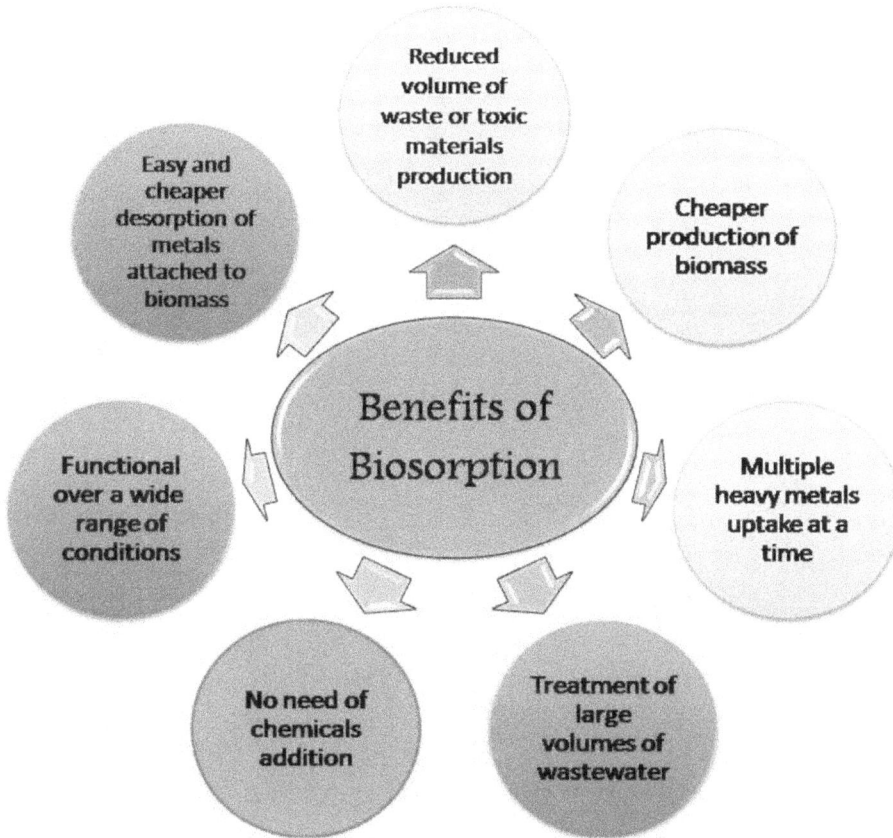

FIGURE 5.1 Benefits of biosorption over traditional processes.

temperature. The second phase is the Slower Phase; this phase is metabolism-dependent and affected by environmental factors such as temperature (temperature can damage the target site and lead to lower biosorption potential) and metabolic inhibitors (Perpetuo et al., 2011). Heavy metal ion connection to bacterial cell walls (peptidoglycan) may be reliant or autonomous to metabolism (Shamim, 2018).

METABOLISM-DEPENDENT AND -INDEPENDENT BIOSORPTION

Living biological material exhibits metabolism-dependent biosorption (Shamim, 2018), with transport through cell membrane and precipitation playing a key role (Perpetuo et al., 2011). The metabolism-independent mechanism occurs most often in biomass formed from dead cells. Biosorption involves transport through the cell membrane, complexation, ion exchanges, precipitation and physical adsorption (Figure 5.2). It is fundamental to recognize the functional groups liable for the binding of target metals to better understand how metals adsorb to biological material (Javanbakht et al., 2014).

1. **Transport across cell membranes:** Diffusion across the cell walls allows different cations to enter the cell. The permeability of the cell membrane would have a substantial effect on cation adsorption (Javanbakht et al., 2014). Transport of target metals across cell membranes may use the same mechanism that transports metabolically vital cations like Na^+, K^+ and Mg^{2+}. The incidence of target metal ions with similar valency and ionic radius as vital ions may cause confusion in metal transport systems (Perpetuo et al., 2011). The molecular mechanism of biosorption of copper by *Aspergillus niger* is primarily due to an energetic dynamic process that results in intracellular copper absorption (Javanbakht et al., 2014).

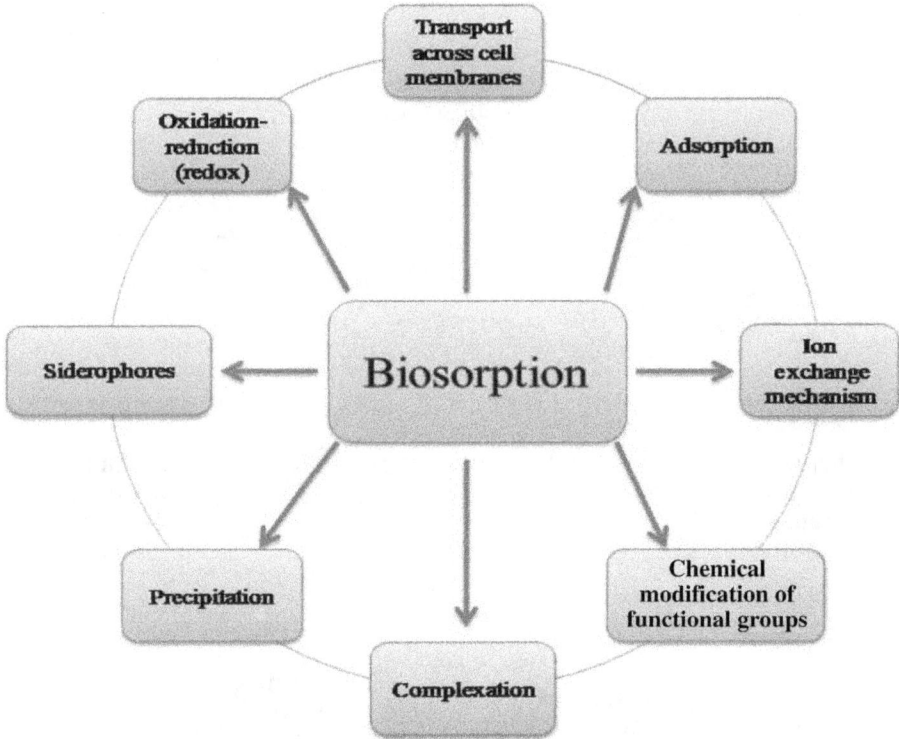

FIGURE 5.2   Different mechanisms involved in biosorption.

2. **Adsorption:** Adsorption is a mechanism in which a metal ion in a solution attaches to polyelectrolytes in the microbial cell wall to attain electroneutrality through van der Waals forces and electrostatic interactions, redox interaction, covalent bonding and biomineralization. Ionic interactions or physiochemical adsorption are two types of adsorption processes. The presence of anionic ligands on bacterial cell walls is also important in heavy metal adsorption (Shamim, 2018). Carboxylic groups found in *Staphylococcus xylosus* biomass are primarily liable for Fe (III) biosorption, with 2 moles of carboxylic groups needed for 1 mole of Fe (III) ion biosorption (Aryal et al., 2010; Aryal, 2020). Chojnacka et al. (2005) investigated the potential binding sites on the surface of *Spirulina* sp. and determined the adsorptive surface by measuring the amount of adsorption of methylene blue onto *Spirulina* sp. biomass. Surprisingly, it was discovered that physical adsorption played a negligible role in the entire biosorption process, with Cr (III) physical adsorption accounting for just 3.7% of the overall biosorption process in all morphological forms of cells. This process is metabolically independent and mutable as it offers numerous benefits, particularly for the treatment of wastewater with small pollutant concentrations (Perpetuo et al., 2011).

In yeasts and fungi, biosorption is mediated by ionic interactions between the ligands/functional groups present on the cell walls of microbial biomass and target metal ions in solutions. Copper biosorption by *Zoogloea ramigera* and *Chiarella vulgaris* has been shown to be due to electrostatic interactions (Aksu et al., 1992), except for the biosorption of chromium by the fungi *Aspergillus niger*. The occurrence of Pb^{2+}, Cu^{2+} and Cr^{2+} alters zeta potential on cell surfaces, suggesting that metal absorption is linked to the ionic interaction of metals with anionic functional groups on the *Rhodococcus opacus* surface (Bueno et al., 2008). Dead biomass has a lesser isoelectric point, which magnifies biosorption ability, whereas living cells have a much greater isoelectric point, which decreases biosorption capability (Huang et al., 2013). Whereas, Malik (2004), Gadd (2009) and Hajdu et al. (2010) concluded that living cells are capable of continuous adsorption of metal and self-sustainability, and living biomass is preferred over dead biological material (Shamim, 2018).

3. **Ion Exchange Mechanism**

The divalent metal ions interchange with the functional groups of the hetero-polysaccharides found on the cell walls of microbes. For example, alginates are salts of K^+, Na^+, Ca^{2+} and Mg^{2+} found in the cell walls of marine algae (Perpetuo et al., 2011). They exchange with counter ions including Cu^{2+}, CO^{2+}, Zn^{2+} and Cd^{2+}, resulting in heavy metal biosorption (Kuyucak and Volesky, 1989). Ion exchange is an alterable stoichiometric process in which an ion from an electrolyte is replaced by an equally charged ion immobilized on a surface, thereby upholding total electroneutrality. Zeolites were the earliest ion-exchange matrix used for the elimination of waste effluents. Metal sulfides bonded covalently with heavy metal ions act as a greatly effective and inexpensive ion exchange matrix. Various sulfide-based matrices have been considered for the removal of heavy-metal pollution from wastewater. Layered metal thiosulphate has been reported as an efficient ion exchange matrix for Pb^{2+} sequestration and is able to remove 99% (393 mg/g) of Pb2+ from solutions (Rathore et al., 2017). Pb^{2+} and Cd^{2+} biosorption on *Amanita rubescens* and *Lactarius scrobiculatus* biomass were found to be accelerated by chemical ion exchange (Javanbakht et al., 2014).

4. **Chemical Modification of Functional Groups**

There is some evidence that the O– (hydroxyl, carboxyl groups linked to Pb^{2+} sequestration in *Aeromonas hydrophila*), N– (amide group involved in Fe^{3+} sequestration in *Cicer arientinum*), S– (Sulfate, thiol), or P– (Phosphate group involved in copper ion sequestration in *Mucor rouxii*) containing functional groups directly contribute in metal binding (Javanbakht et al., 2014). Researchers discovered that the biosorption of Cr^{3+} ions was hampered by the esterification of functional groups, demonstrating their importance in the biosorption mechanism (Chojnacka et al., 2005). Thus, it can be established that modification of certain functional groups can accelerate the process of biosorption.

5. **Complexation:** Metal can be removed from the solution by complex development on the surface of the cell. Complexation is essential in metal-ligand and sorbate-sorbent interfaces (Javanbakht et al., 2014). *C. vulgaris* biosorbs copper by forming coordination bonds between metals and carboxyl/amino groups of polysaccharides (Aksu et al., 1992). *Pseudomonas syringae* sequesters calcium, cadmium, magnesium, copper, zinc and mercury solely due to complexation (Perpetuo et al., 2011; Javanbakht et al., 2014).

6. **Precipitation:** Precipitation can be both metabolism-independent or -dependent. Precipitation occurs due to the reactivity among the destination metal(s) and the substance(s) generated by microbes as part of their defense systems. Microorganisms' active defense systems react in the impendence of toxic metal-delivering compounds, favouring the precipitation method (Perpetuo et al., 2011; Javanbakht et al., 2014). Metal uptake occurs either on the surface of the cell or in electrolytes in precipitation (Javanbakht et al., 2014).

7. **Siderophores:** Some microorganisms when growing in an iron-lacking medium generate effective iron chelators, known as siderophores. They contribute significantly to the complex formation of toxic substances and radioactive elements by enhancing their solubility (Perpetuo et al., 2011).

8. **Oxidation-reduction (redox):** Microorganism life forms can proficiently immobilise toxic metals by decreasing heavy metal ions, lowering their oxidation state and producing metal ions (load zero) that are less bioactive (Gadd, 2004; Perpetuo et al., 2011). Many advanced oxidation methods have been used for the removal of drug toxicants from municipal wastewater, for example. Angosto et al. (2020) reported amalgamations of ultraviolet radiation with hydrogen peroxide for the removal of Diclofenac from effluent.

Desorption: Retrieval of metal ions from metal-loaded biological material and recycling of biomass are critical for an effective biosorption practice. Acids, alkalis, EDTA, H_2O_2, sodium carbonate, sodium bicarbonate, Potassium chloride, Calcium carbonate, Calcium nitrate, Sodium citrate, nitrilotriacetic acid and other desorbing agents have been used to study the desorption of heavy metals from metal-loaded biomass (Aryal, 2020). The most effective desorbent was found to be nitric acid (at a concentration of 0.1M), which extracted much of the metallic ions bound to the biomass (98%) while causing no reduction.

CHEMICAL MODIFICATION

Surface chemical modifications and internal modifications alter the structure and organization of live cells, and increase the adsorption ability and specificity of biological materials for exact heavy metals. The negative charge of the cells repels the anionic metal species; thus, by chemical modification, the biosorption ability of biological materials can be improved (Yang et al., 2015).

SURFACE MODIFICATION

Simple soaking methods (Yin et al., 2019) or complex grafting methods (Bayramoglu et al., 2016) can be used to greatly expand the biosorption capacity by the addition of various functional groups. Cationic surfactants are often used for the removal of non-biodegradable oxyanions of As, W, B and Mo in wastewater as they change the surface charge. The surface-modified *Trichoderma asperellum* BPL MBT1 proved an efficient biosorbent for cationic dyes (Shanmugam et al., 2021); polyethylenimine-grafted *Shewanella haliotis* was reported to be very effective for gold biosorption and have 4.2 times more adsorption ability (Zhou et al., 2017).

CELL SURFACE DISPLAY OF HETEROLOGOUS PEPTIDES

Display of heterologous fusion short peptides and an anchor motif or microbial outer membrane protein for increasing the biosorption ability of microbial cells to inorganic metals is an advanced tool for the evolution of cell surface-centered microbial biosorbents (Eskandari et al., 2015). Many studies have been carried out recently on innovative genetically modified microbial biosorbents.

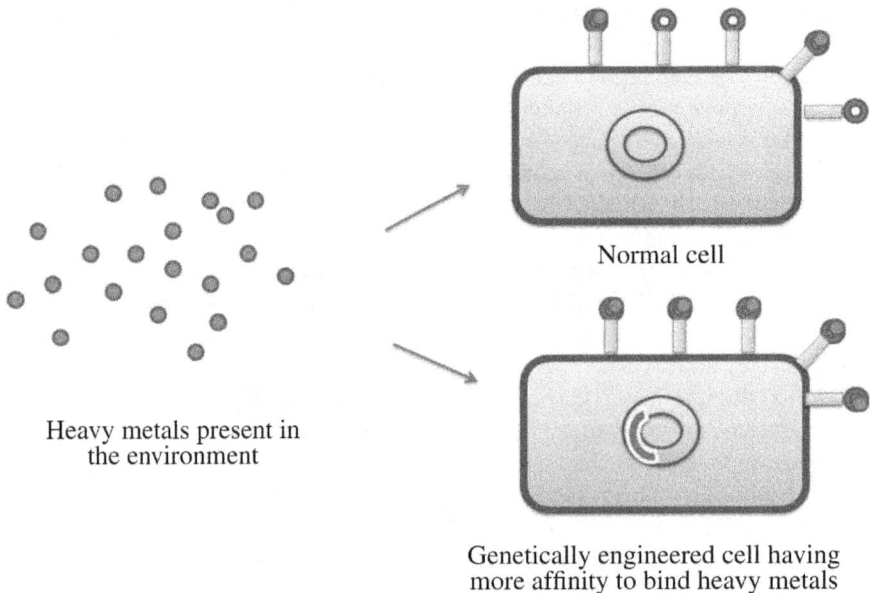

FIGURE 5.3 Expression of metal-binding peptides on bacterial surfaces improves metal biosorption.

Phytochelatin is one of the naturally occurring metal-binding peptide containing 2–11 repeats of γGlu-Cys residue ended by a Gly moiety and are non-ribosomally synthesized (Singh et al., 2008). There are reports that show synthetic phytochelatin EC20 on *E. coli* cells absorbs Cd approximately 60 mole/mg dry wt. of cells (Bae et al., 2000).

GENETICALLY MODIFIED MICROORGANISMS AS INNOVATIVE BIOSORBENTS

Several bacterial species such as *Bacillus, Enterobacter, Micrococcus* and *Pseudomonas* have been utilized as biosorbents due to their size, ubiquity and pliability to changing environmental conditions (Srivastava et al., 2015). The reason for their exceptional sorption capacity is the presence of several chemosorption sites and high surface-to-volume ratios (Mosa et al., 2016). Several microorganisms alter the redox state of heavy metals and detoxify them from the environment by secreting a variety of enzymes. For example *Pseudomonas* sp. B50A removed 86% mercury from a medium, reducing Hg(II) to Hg^0 by secreting mercuric reductase (Giovanella et al., 2016). Genetically engineered microbes with the mercuric reductase enzyme (MerA) reduce toxic Hg^{2+} to volatile non-toxic Hg^0, providing an eco-friendly solution to mercury toxicity. Reports are also there for the heterologous expression of a membrane-bound protein (MerT) to enhance Hg^{2+} uptake (Naguib et al., 2018). There are studies reported that expression of genes through genetic engineering enhances the removal efficacy of toxic heavy metals from polluted environment and wastewater (Table 5.1).

Genetic engineering of microorganisms with various heavy metal-specific binding peptides is an efficient strategy for the removal of toxic heavy metals from polluted environments. Genetic engineering practices to extract heavy metals from polluted locations have garnered substantial attention. For example, a photosynthetic bacterium *Rhodopseudomonas palustris* was genetically engineered to display the membrane transport machinery for Hg^{2+} and metallothionine to eliminate Hg^{2+} from polluted wastewater (Deng and Jia, 2011). The surface expression of metal binding peptides CP2 and HP3 on *Saccharomyces cerevisiae* enhances their Cd^{2+} and Zn^{2+} biosorption ability (Vinopal et al., 2007). Few studies also reported that the surface display of peptides through genetic engineering act as an efficient approach for remediation strategies; for example, PbrR, a lead-binding protein along with a specific promoter from lead precise operon of *Cupriavidus metallidurans* CH34 was expressed in *E. coli* (Wei et al., 2014). This surface-engineered *E. coli* adsorbs lead selectively and protects *Arabidopsis thaliana* seed germination from lead toxicity. The expression of a *Lentinula edodes* cadmium-binding protein promisingly improved the cadmium biosorption ability of engineered *E. coli* (Table 5.1). To increase the biosorption ability of living biomass genetic engineering is also a suitable tool. For instance, the deletion of a Cu^{2+}/Cd^{2+} pump P-type ATPase CrpA gene in *Aspegillus nidulans* exhibited 2.7 times greater Cd biosorption capability and enabled the expansion of novel fungal biomass-based biosorption technology to excerpt toxic Cd^{2+} (Boczonádi et al., 2020). Transformation of *E. coli* BL21 by metallothionein from *Corynebacterium glutamicum* results in considerably superior Pb^{+2} biosorption (Jafarian and Ghaffari, 2017).

EFFICIENCY OF BIOSORBENTS FOR REMOVAL OF TOXIC HEAVY METALS FROM POLLUTED ENVIRONMENTS

An ideal biosorbent should have characteristics such as accessibility, non-toxicity, abundant metal sequestration capability and significant usability and re-usability (Bilal et al., 2018). It is essential to investigate the effectiveness of metal absorption by microbial populations for large-scale applications. The parameter (q: mg/g) mg of metal accumulated per g of biosorbent content is commonly used to calculate metal uptake, while "qH" is described as a variable of the metal biosorbed, the sorptive material utilized, and the operational parameters (Perpetuo et al., 2011). According to several reports, many species of both prokaryotic organisms and eukaryotes, have the capabilities to biosorb hazardous metal ions. Metal toxicity affects eukaryotes rather than bacteria (Perpetuo et al., 2011). Temperature

TABLE 5.1

Examples of Various Surface Engineering Techniques for Metal Biosorption (PbrR, PbrR691 and PbrD: Pb-binding Proteins; AtHMA4: *Arabidopsis thaliana* Heavy Metal ATPases; MerP: Metal-binding Proteins; PbrD: Lead (II) Binding Protein)

S.N	Gene expressed	Microorganism	Target metal	Biosorption capacity	Reference
1	Phytochelatin EC20	*Escherichia coli*	Cd Pt	absorbed 37.5% of Cd^{2+} 112.67 mg/g	de Souza and Vicente, 2020 Tan et al., 2019
		Saccharomyces cerevisiae	Cd	168 mg/g	Yang et al., 2017
2	PbrR, PbrR691 and PbrD	*Escherichia coli* BL21	Pb	942.1 µmol/g, 754.3 µmol/g and 864.8-µmol/g cell dry weight	Jia et al., 2020
3	Cysteine-rich methylmercury-binding peptide	*E. coli*	methylHg.	129.5 µmol/g dry cells	Liu et al., 2019
4	non-metallothioneins cadmium-binding protein	*E. coli*	Cd	Increased an average of 10.31-fold	Dong et al., 2019
5	Cd and Zn transporter (AtHMA4)	*Chlamydomonas reinhardtii*	Cd and Zn	Microalgae calcium alginate beads biosorption capacity 5–10 times higher than free-swimming transgenic cells	Ibuot et al., 2020
6	Hg^{2+}-binding peptide CysLysCysLysCysLysCys (CL)	*E. coli*	Hg	344.93 µmol/g Dry cells	Liu et al., 2019
7	Pb^{2+} binding domain (PbBD)	*E. coli*	Pb	34.4 µmol/g dry cells (1.92-fold higher)	Hui et al., 2018
8	Metallothionein	*Saccharomyces cerevisiae*	Cr	8.27 mg/g	Zhang and Yi, 2017
9	Chromium adsorption protein (MerP)	*Escherichia coli* BL21	Cr	2.38 mmol/g	Wang et al., 2021
10	*pbrD*	*Escherichia coli* BL21 (DE3)	Pb	1628 µg/g from Dubinin– Radushkevich isotherm model	Keshav et al., 2019

and pH are the two important parameters that affect the rate of the biosorption process. As most of the metal ions present in electrolytes or solutions are cations, the negative charge of the biosorbent matters for higher amounts of metal biosorption. The optimum pH for metal adsorption is 7–8. At higher pHs, metal precipitation occurs and at lesser pHs, H^+ contends with metal ions to interact with binding sites, thus limiting the biosorption capacity. The optimum pH for anionic metals is 2–4 as at this pH the cells have more positive charges (Torres, 2020). Similarly, many studies reported that in cases of living biomass with a rise in temperature, the extent of metal biosorbed increases. A study reported that the cell wall structure of a biomass affects biosorption of many metals; for example, Pb (II) biosorption varies with different microbial biosorent used. Wang et al., 2019 compared lead

TABLE 5.2

Biosorption Using Different Microbial Species and their Efficiency

Toxic metal	Microbial group	Microbial biosorbent	pH	Temperature (°C)	Biosorption capacity	Reference
Cadmium	Fungus	*Penicillium simplicissimum*	5.0	28	52.5 mg/g	Fan et al., 2008
		Rhizopus arrhizus	4.0–7.0	—	26.8 mg/g	Fourest and Roux, 1992
		Penicillium digitatum	5.5	25	18.0 mg/g	Galun et al., 1987
		Mucor rouxii	6.0	—	7.67 mg/g	Yan and Viraraghavan, 2008
		Beauveria bassiana	5.0	30	46.2 mg/g	Hussein et al., 2011
		Metarhizium anisopliae	5.0	30	44.22 mg/g	Hussein et al., 2011
		Phanerochaete chrysosporium	4.5	27	15.2 mg/g	Li et al., 2004
		N. clavispora ASU1	7.0	30	185mg/g	Hassan et al., 2018
		Mucor rouxii (treated with NaOH)	5.0	—	8.46%	Yan and Viraraghavan, 2003; Javanbakht et al., 14
	Bacteria	*Pseudomonas veronii 2E*	7.5	32	50%	Vullo et al., 2008
		Citrobacter strain MCMB-181	6.0	28	43.48 mg/g	Puranik and Paknikar, 1999
Zinc	Fungus	*Penicillium simplicissimum*	5.0	28	65.5 mg/g	Fan et al., 2008
		Phanerochaete chrysosporium	7.0	25	28.9 mg/g	Arica et al., 2003
		N. clavispora ASU1	6.0	30	153 mg/g	Hassan et al., 2018
		Fusarium spp.	6.0	40	50%	Velmurugan et al., 2010
		Mucor rouxii (treated with NaOH)	5.0	—	7.75%	Yan and Viraraghavan, 2003
		Penicillium digitatum	5.5	25	14.1mg/g	Galun et al., 1987
		Phomopsis sp.	5.0	—	10mg/g	Saiano et al., 2005
	Bacteria	*Pseudomonas putida*	5.0	30	27.4 mg/g	Chen et al., 2005
		Thiobacillus ferrooxidans	6.0	25	82.6 mg/g	Celaya et al., 2000
		Escherichia coli	—	28	65.9 mg/g	Kao et al., 2008
		Pseudomonas aeruginosa AT18	7.0	26–30	77.5 mg/g	Silva et al., 2009
		Shewanella putrefaciens	—	22	22 mg/g	Chubar et al., 2008
		P. aeruginosa ASU 6a	6.0	30	83.3 mg/g	Joo et al., 2010
		B. cereus AUMC B52	6.0	30	66.6 mg/g	Joo et al., 2010
Copper	Bacteria	*Bacillus sphaericus*	7.0	37	51.3%	Al-Daghistani, 2012
		B. pumilus	7.0	37	66%	
		Paenibacillus alvei	7.0	37	54%	
Nickel	Fungus	*Mucor rouxii* (treated with NaOH)	5.0	—	11.09%	Yan and Viraraghavan, 2003
	Bacteria	*Bacillus sphaericus*	7.0	37	49.4%	Al-Daghistani, 2012
		B. pumilus	7.0	37	33.1%	
		Paenibacillus alvei	7.0	37	27.7%	
Chromium	Bacteria	*Bacillus sphaericus*	7.0	37	53.2%	Al-Daghistani, 2012
		B. pumilus	7.0	37	74%	Al-Daghistani, 2012
		Paenibacillus alvei	7.0	37	42.2%	Al-Daghistani, 2012
		B. cereus	2.0	37	86.79%	Sultan et al., 2012
		B. pumilis	2.0	37	87.79%	Sultan et al., 2012
		Pantoea agglomerans	3.0	37	83.64%	Sultan et al., 2012

(Continued)

TABLE 5.2 *(Continued)*

Biosorption Using Different Microbial Species and their Efficiency

Toxic metal	Microbial group	Microbial biosorbent	pH	Temperature (°C)	Biosorption capacity	Reference
Lead	Fungus	*Mucor rouxii* (treated with NaOH)	5.0	—	35.69 mg/g	Yan and Viraraghavan, 2003
		Aspergillus niger modified with Oxalic Acid	7.0	25°C	92.84%	Awofolu et al., 2006
		Penicillium austurianum	7.0	25°C	94.21%	Awofolu et al., 2006

biosorption among *Microbacterium* sp. OLJ1, *Pseudomonas putida* I3 and *Talaromyces amestolkiae* and observed adsorption capacities of 237.02, 345.02 and 199.02 mg/g, respectively. They found high potential of these biosorbents for the sequestration of trace metals from wastewater. Zhang et al., 2020 evaluated the efficiency of dead biomass of *Saccharomyces cerevisiae* immobilized on alginate-boric acid beads and found 113.4 μmol/g of uranium biosorption ability. Alginate-immobilized live cells of *Azotobacter nigricans* NEWG-1 remove 82% of copper after an incubation of 6 h (Ghoniem et al., 2020). Table 5.2 summarizes a few examples of microbial biosorbents, optimum conditions and their biosorption efficiency.

REFERENCES

Aksu, Z., Sag, Y., & Kutsal, T. 1992. The biosorption of copper (II) by *C. vulgaris* and *Z. ramigera*. *Environmental Technology*, 13: 579–586.

Al-Daghistani, H. I. 2012. Bio-remediation of Cu, Ni and Cr from rotogravure wastewater using immobilized, dead, and live biomass of indigenous thermophilic *Bacillus* species. *The Internet Journal of Microbiology*. doi: 10.5580/2a7f.

Angosto, J. M., Roca, M. J., & Fernández-López, J. A. 2020. Removal of diclofenac in wastewater using biosorption and advanced oxidation techniques: Comparative results. *Water*, 12(12): 3567.

Arica, M. Y., Arpa, C., Ergene, A., Bayramoglu, G. U., & Genc, O. 2003. Ca- alginate as a support for Pb (II) and Zn (II) biosorption with immobilized *Phanerochaete chrysosporium*. *Carbohydrate Polymers*, 52: 167–174.

Aryal, M. 2020. A comprehensive study on the bacterial biosorption of heavy metals: Materials, performances, mechanisms, and mathematical modellings. *Reviews in Chemical Engineering*. https://doi.org/10.1515/revce-2019-0016.

Aryal, M., Ziagova, M., & Liakopoulou-Kyriakides, M. 2010. Study on arsenic biosorption using Fe (III)-treated biomass of *Staphylococcus xylosus*. *Chemical Engineering Journal*, 162: 178–185.

Awofolu, O., Okonkwo, J., Roux-Van Der Merwe, R., Badenhorst, J., & Jordaan, E. 2006. A new approach to chemical modification protocols of *Aspergillus niger* and sorption of lead ion by fungal species. *Electronic Journal of Biotechnology*, 9(4): 340–348.

Ayangbenro, A. S., & Babalola, O. O. 2017. A new strategy for heavy metal polluted environments: A review of microbial biosorbents. *International Journal of Environmental Research and Public Health*, 14: 94. doi: 10.3390/ijerph14010094.

Bae, W., Chen, W., Mulchandani, A., & Mehra, R. K. 2000. Enhanced bioaccumulation of heavy metals by bacterial cells displaying synthetic phytochelatins. *Biotechnology & Bioengineering*, 70: 518–524.

Bayramoglu, G., Akbulut, A., & Arica, M. Y. 2016. Aminopyridine modified *Spirulina platensis* biomass for chromium (VI) adsorption in aqueous solution. *Water Science and Technology*, 74: 914–926.

Bilal, M., Rasheed, T., Sosa-Hernández, J. E., Raza, A., Nabeel, F., & Iqbal, H. M. N. 2018. Biosorption: An interplay between marine algae and potentially toxic elements—a review. *Marine Drugs*, 16(2): 65. https://doi.org/10.3390/md16020065.

Boczonádi, I., Török, Z., Jakab, Á., Kónya, G., Gyurcsó, K., Baranyai, E., . . . Pócsi, I. 2020. Increased Cd^{2+} biosorption capability of *Aspergillus nidulans* elicited by crpA deletion. *Journal of Basic Microbiology*, 60(7): 574–584.

Bueno, B. Y. M., Torem, M. L., Molina, F., & de Mesquita, L. M. S. 2008. Biosorption of lead(II), chromium(III) and copper(II) by R. opacus: Equilibrium and kinetic studies. *Minerals Engineering*, 21: 65–75.

Celaya, R. J., Noriega, J. A., Yeomans, J. H., Ortega, L. J., & Ruiz-Manriquez, A. 2000. Biosorption of Zn (II) by *Thiobacillus ferrooxidans*. *Bioprocess Engineering*, 22: 539–542.

Chen, X. C., Wang, Y. P., Lin, Q., Shi, J. Y., Wu, W. X., & Chen, Y. X. 2005. Biosorption of copper (II) and zinc (II) from aqueous solution by *Pseudomonas putida* CZ1. *Colloids and Surfaces B: Biointerfaces*, 46: 101–107.

Chojnacka, K., Chojnacki, A., & Górecka, H. 2005. Biosorption of Cr^{3+}, Cd^{2+} and Cu^{2+} ions by blue-green algae *Spirulina* sp.: Kinetics, equilibrium and the mechanism of the process. *Chemosphere*, 59(1): 75–84. doi: 10.1016/j.chemosphere.2004.10.005. PMID: 15698647.

Chubar, N., Behrends, T., & Cappellen, P. V. 2008. Biosorption of metals (Cu^{2+}, Zn^{2+}) and anions (F^-, $H_2PO_4^-$) by viable and autoclaved cells of the Gram-negative bacterium *Shewanella putrefaciens*. *Colloids and Surfaces B: Biointerfaces*, 65: 126–133.

de Souza, C. B., & Vicente, E. J. 2020. Expression of synthetic Phytochelatin EC20 in *E. Coli* increases its biosorption capacity and cadmium resistance. *Bioscience Journal*, 36(2).

Deng, X., & Jia, P. 2011. Construction and characterization of a photosynthetic bacterium genetically engineered for Hg^{2+} uptake. *Bioresource Technology*, 102(3): 3083–3088.

Diep, P., Mahadevan, R., & Yakunin, A. F. 2018. Heavy metal removal by bioaccumulation using genetically engineered microorganisms. *Frontiers in Bioengineering and Biotechnology*, 6: 157.

Dong, X. B., Huang, W., Bian, Y. B., Feng, X., Ibrahim, S. A., Shi, D. F., . . . Liu, Y. 2019. Remediation and mechanisms of cadmium biosorption by a cadmium-binding protein from *Lentinula edodes*. *Journal of Agricultural and Food Chemistry*, 67(41): 11373–11379.

Eskandari, V., Yakhchali, B., Sadeghi, M., Karkhane, A. A., & Ahmadi-Danesh, H. 2015. Efficient cadmium bioaccumulation by displayed hybrid CS3 Pili: Effect of heavy metal binding motif insertion site on adsorption capacity and selectivity. *Applied Biochemistry and Biotechnology*, 177: 1729–1741.

Fan, T., Liu, Y., Feng, B., Zeng, G., Yang, C., Zhou, M., Zhou, H., Tan, Z., & Wang, X. 2008. Biosorption of cadmium(II), zinc(II) and lead(II) by *Penicillium simplicissimum*: Isotherms, kinetics and thermodynamics. *Journal of Hazardous Materials*, 160: 655–661.

Fomina, M., & Gadd, G. M. 2014. Biosorption: Current perspectives on concept, definition and application. *Bioresource Technology*, 160: 3–14. doi: 10.1016/j.biortech.2013.12.102.

Fourest, E., & Roux, J. C. 1992. Heavy metal biosorption by fungal mycelia by-products: Mechanisms and influence of pH. *Applied Microbiology and Biotechnology*, 37: 399–403.

Gadd, G. M. 2004. Microbial influence on metal mobility and application for bioremediation. *Geoderma*, 122(2–4): 109–119.

Gadd, G. M. 2009. Biosorption: Critical review of scientific rationale, environmental importance and significance for pollution treatment. *Journal of Chemical Technology and Biotechnology*, 84: 13–28.

Galun, M., Galun, E., & Siegel, B. Z. 1987. Removal of metal ions from aqueous solutions by *Penicillium biomass*: Kinetic and uptake parameters. *Water, Air, and Soil Pollution*, 33: 359–371.

Ghoniem, A. A., El-Naggar, N. E., Saber, W. I. A., El-Hersh, M. S., & El-Khateeb, A. Y. 2020. Statistical modeling-approach for optimization of Cu(2+) biosorption by *Azotobacter nigricans* NEWG-1; characterization and application of immobilized cells for metal removal. *Scientific Reports*, 10: 9491.

Giovanella, P., Cabral, L., Bento, F. M., Gianello, C., & Camargo, F. A. O. 2016. Mercury (II) removal by resistant bacterial isolates and mercuric (II) reductase activity in a new strain of *Pseudomonas* sp. B50A. *New Biotechnology*, 33(1): 216–223.

Gupta, V. K., Nayak, A., & Agarwal, S. 2015. Bioadsorbents for remediation of heavy metals: Current status and their future prospects. *Environmental Engineering Research*, 20: 1–18.

Hajdu, R., Pinheiro, J. P. R., Galceran, J., & Slaveykova, V. I. 2010. Modeling of Cd uptake and efflux kinetics in metal-resistant bacterium Cupriavidus metallidurans. *Environmental Science and Technology*, 44: 4597–4602. doi: 10.1021/es100687h.

Hassan, S., Koutb, M., Nafady, N. A., & Hassan, E. A. 2018. Potentiality of Neopestalotiopsis clavispora ASU1 in biosorption of cadmium and zinc. *Chemosphere*, 202: 750–756. https://doi.org/10.1016/j.chemosphere.2018.03.114.

Huang, W., & Liu, Z. M. 2013. Biosorption of Cd(II)/Pb(II) from aqueous solution by biosurfactant-producing bacteria: Isotherm kinetic characteristic and mechanism studies. *Colloids and Surfaces B*, 105: 113–119.

Hui, C. Y., Guo, Y., Yang, X. Q., Zhang, W., & Huang, X. Q. 2018. Surface display of metal binding domain derived from PbrR on *Escherichia coli* specifically increases lead (II) adsorption. *Biotechnology Letters*, 40(5): 837–845.

Hussein, K. A., Hassan, S. H., & Joo, J. H. 2011. Potential capacity of *Beauveria bassiana* and *Metarhizium anisopliae* in the biosorption of Cd^{2+} and Pb^{2+}. *The Journal of General and Applied Microbiology*, 57: 347–355.

Ibuot, A., Webster, R. E., Williams, L. E., & Pittman, J. K. 2020. Increased metal tolerance and bioaccumulation of zinc and cadmium in *Chlamydomonas reinhardtii* expressing a AtHMA4 C-terminal domain protein. *Biotechnology and Bioengineering*, 117(10): 2996–3005.

Ilyas, S., Kim, M.-S., Lee, J.-C., Jabeen, A., & Bhatti, H. N. 2017. Bio-reclamation of strategic and energy critical metals from secondary resources. *Metals*, 7: 1–17. doi: 10.3390/met7060207.

Jafarian, V., & Ghaffari, F. (2017) A unique metallothionein-engineered in *Escherichia coli* for biosorption of lead, zinc, and cadmium; absorption or adsorption? *Microbiology*, 86: 73–81.

Javanbakht, V., Alavi, S. A., & Zilouei, H. 2014. Mechanisms of heavy metal removal using microorganisms as biosorbent. *Water Science and Technology*, 69(9): 1775–1787. doi: 10.2166/wst.2013.718. PMID: 24804650.

Jia, X., Li, Y., Xu, T., & Wu, K. 2020. Display of lead-binding proteins on *Escherichia coli* surface for lead bioremediation. *Biotechnology & Bioengineering*, 117: 3820–3834.

Joo, J. H., Hassan, S. H., & Oh, S. E. 2010. Comparative study of biosorption of Zn2+ by *Pseudomonas aeruginosa* and *Bacillus cereus*. *International Biodeterioration & Biodegradation*, 64: 734–741.

Kao, W. C., Huang, C. C., & Chang, J. S. 2008. Biosorption of nickel, chromium and zinc by MerP-expressing recombinant *Escherichia coli*. *Journal of Hazardous Materials*, 158: 100–106.

Keshav, V., Achilonu, I., Dirr, H. W., & Kondiah, K. 2019. Recombinant expression and purification of a functional bacterial metallo-chaperone PbrD-fusion construct as a potential biosorbent for Pb (II). *Protein Expression and Purification*, 158: 27–35.

Kuyucak, N., & Volesky, B. 1989. Desorption of cobalt-laden algal biosorbent. *Biotechnology & Bioengineering*, 33(7): 815–822.

Li, Q., Wu, S., Liu, G., Liao, X., Deng, X., Sun, D., Hu, Y., & Huang, Y. 2004. Simultaneous biosorption of cadmium (II) and lead (II) ions by pretreated biomass of *Phanerochaete chrysosporium*. *Separation and Purification Technology*, 34: 135–142.

Liu, M., Lu, X., Khan, A., Ling, Z., Wang, P., Tang, Y., . . . Li, X. 2019. Reducing methylmercury accumulation in fish using *Escherichia coli* with surface-displayed methylmercury-binding peptides. *Journal of Hazardous Materials*, 367: 35–42.

Malik, A. 2004. Metal bioremediation through growing cells. *Environment International*, 30: 261–278. doi: 10.1016/j.envint.2003.08.001.

Mosa, K. A., Saadoun, I., Kumar, K., Helmy, M., & Dhankher, O. P. 2016. Potential biotechnological strategies for the cleanup of heavy metals and metalloids. *Frontiers in Plant Science*, 7: 1–14.

Mukkata, K., Kantachote, D., Wittayaweerasak, B., Megharaj, M., & Naidu, R. 2018. The potential of mercury resistant purple nonsulfur bacteria as effective biosorbents to remove mercury from contaminated areas. *Biocatalysis and Agricultural Biotechnology*, 17: 93–103.

Naguib, M. M., El-Gendy, A. O., & Khairalla, A. S. 2018. Microbial diversity of operon genes and their potential rules in mercury bioremediation and resistance. *The Open Biotechnology Journal*, 12(1): 56–77.

Perpetuo, E. A., Souza, C. B., & Nascimento, C. A. O. 2011. Engineering bacteria for bioremediation, progress in molecular and environmental bioengineering. IntechOpen. doi: 10.5772/19546.

Puranik, P. R., & Paknikar, K. M. 1999. Biosorption of lead, cadmium, and zinc by *Citrobacter strain* MCM B-181: Characterization studies. *Biotechnology Progress*, 15: 228–237.

Rathore, E., Pal, P., & Biswas, K. 2017. Layered metal chalcophosphate (K-MPS-1) for efficient, selective, and ppb level sequestration of Pb from water. *Journal of Physical Chemistry C*, 121: 7959–7966.

Saiano, F., Ciofalo, M., Cacciola, S. O., & Ramirez, S. 2005. Metal ion adsorption by Phomopsis sp. biomaterial in laboratory experiments and real wastewater treatments. *Water Research*, 39: 2273–2280.

Shamim, S. 2018. Biosorption of heavy metals. IntechOpen. doi:10.5772/intechopen.72099.

Shanmugam, S., Karthik, K., Veerabagu, U., Hari, A., Swaminathan, K., Al-Kheraif, A. A., & Whangchai, K. (2021). Bi-model cationic dye adsorption by native and surface-modified *Trichoderma asperellum* BPL MBT1 biomass: From fermentation waste to value-added biosorbent. *Chemosphere*, 277: 130311.

Silva, R. M. P., Rodriquez, A. A., De Oca, J. M. G. M., & Moreno, D. C. 2009. Biosorption of chromium, copper, manganese and zinc by *Pseudomonas aeruginosa* AT18 isolated from a site contaminated with petroleum. *Bioresource Technology*, 100: 1533–1538.

Singh, S., Lee, W., DaSilva, N. A., Mulchandani, A., & Chen, W. 2008. Enhanced arsenic accumulation by engineered yeast cells expressing Arabidopsis thaliana phytochelatin synthase. *Biotechnology & Bioengineering*, 99: 333–340.

Srivastava, S., Agrawal, S., & Mondal, M. 2015. A review on progress of heavy metal removal using adsorbents of microbial and plant origin. *Environmental Science and Pollution Research*, 22: 15386–15415.

Sultan, S., Mubashar, K., & Faisal, M. 2012. Uptake of toxic Cr (VI) by biomass of exopolysaccharides producing bacterial strains. *African Journal of Microbiology Research*, 6(13): 3329–3336. doi:10.5897/AJMR12.226.

Tan, L., Cui, H., Xiao, Y., Xu, H., Xu, M., Wu, H., Dong, H., Qiu, G., Liu, X., & Xie, J. 2019. Enhancement of platinum biosorption by surface-displaying EC20 on *Escherichia coli*. *Ecotoxicology and Environmental Safety*, 169: 103–111.

Torres, E. 2020. Biosorption: A review of the latest advances. *Processes*, 8(12): 1584.

Ueda, M. 2016. Establishment of cell surface engineering and its development. *Bioscience, Biotechnology, and Biochemistry*, 80: 1243–1253. doi:10.1080/09168451.2016.1153953.

Velmurugan, P., Shim, J., You, Y., Choi, S., Kamala-Kannan, S., Lee, K. J., . . . Oh, B. T. 2010. Removal of zinc by live, dead, and dried biomass of Fusarium spp. isolated from the abandoned-metal mine in South Korea and its perspective of producing nanocrystals. *Journal of Hazardous Materials*, 182(1–3): 317–324.

Vinopal, S., Ruml, T., & Kotrba, P. 2007. Biosorption of Cd^{2+} and Zn^{2+} by cell surface-engineered *Saccharomyces cerevisiae*. *International Biodeterioration & Biodegradation*, 60(2): 96–102.

Volesky, B. 2007. Biosorption and me. *Water Research*, 41(18): 4017–4029.

Vullo, D. L., Ceretti, H. M., Daniel, M. A., Ramirez, S. A. M., & Zalts, A. 2008. Cadmium, zinc and copper biosorption mediated by *Pseudomonas veronii* 2E. *Bioresource Technology*, 99: 5574–5558.

Wang, J., Zhao, S., Ling, Z., Zhou, T., Liu, P., & Li, X. 2021. Enhanced removal of trivalent chromium from leather wastewater using engineered bacteria immobilized on magnetic pellets. *Science of the Total Environment*, 775: 145647.

Wang, N., Qiu, Y., Xiao, T., Wang, J., Chen, Y., Xu, X., Kang, Z., Fan, L., & Yu, H. 2019. Comparative studies on Pb(II) biosorption with three spongy microbe-based biosorbents: High performance, selectivity and application. *Journal of Hazardous Materials*, 373: 39–49.

Wei, W., Liu, X., Sun, P., Wang, X., Zhu, H., Hong, M., . . . Zhao, J. 2014. Simple whole-cell biodetection and bioremediation of heavy metals based on an engineered lead-specific operon. *Environmental Science & Technology*, 48(6): 3363–3371.

Yan, G., & Viraraghavan, T. 2003. Heavy-metal removal from aqueous solution by fungus *Mucor rouxii*. *Water Research*, 37: 4486–4496. doi:10.1016/S0043-1354(03)00409-3.

Yan, G., & Viraraghavan, T. 2008. Mechanism of biosorption of heavy metals by *Mucor rouxii*. *Engineering in Life Sciences*, 8: 363–371.

Yang, C. E., Chu, I. M., Wei, Y. H., & Tsai, S. L. 2017. Surface display of synthetic phytochelatins on *Saccharomyces cerevisiae* for enhanced ethanol production in heavy metal-contaminated substrates. *Bioresource Technology*, 245: 1455–1460.

Yang, T., Chen, M.-L., & Wang, J.-H. 2015. Genetic and chemical modification of cells for selective separation and analysis of heavy metals of biological or environmental significance. *TrAC Trends in Analytical Chemistry*, 66: 90–102.

Yin, K., Wang, Q., Lv, M., & Chen, L. 2019. Microorganism remediation strategies towards heavy metals. *Chemical Engineering Journal*, 360: 1553–1563.

Zhang, J., Chen, X., Zhou, J., & Luo, X. 2020. Uranium biosorption mechanism model of protonated *Saccharomyces cerevisiae*. *Journal of Hazardous Materials*, 385: 121588.

Zhang, R., & Yi, H. 2017. Enhanced Cr^{6+} biosorption from aqueous solutions using genetically engineered *Saccharomyces cerevisiae*. *Desalination and Water Treatment*, 72: 290–299.

Zhou, Y., Zhu, N., Kang, N., Cao, Y., Shi, C., Wu, P., . . . Qin, B. (2017). Layer-by-layer assembly surface modified microbial biomass for enhancing biorecovery of secondary gold. *Waste Management*, 60: 552–560.

6 Application of Genetically Modified Microorganisms for Remediation of Petrol Discharges and Related Polluted Sites

*Noreen Sajjad, Ayesha Sultan, Gulzar Muhammad,
Aiza Azam, Muhammad Arshad Raza,
Muhammad Ajaz Hussain and Liaqat Ali*

PETROLEUM

Petroleum, also called rock oil, is an oily combustible fluid that comprises hydrocarbons and different components (oxygen, nitrogen, and sulfur) in underground pockets called reservoirs. Oil refers to raw petroleum or crude oil in the industrial refining process. Raw petroleum is formed during the natural decay of animals and plants at high temperatures and pressure. Raw petroleum was first discovered, extracted, and used as fuel by the Chinese in the 4th century. Although the local use of oil dates back many centuries, the current petroleum industry has its roots in the kerosene industry of the late 19th century. The US petroleum industry was introduced by Edwin Drake's oil well drilling (21 m in depth) on Oil Creek near Titusville, Pennsylvania, in 1859 (Graul 2013).

Unrefined/raw petroleum consists of fluid hydrocarbons having at least four or more carbon atoms. Aliphatic saturated, alicyclic saturated, and aromatic hydrocarbons are the three significant constituents of oil with traces of alkenes and alkynes. Crude oil is a combination of long-chain hydrocarbons, inorganic compounds, and metals. After refining crude oil, many petrochemical products are obtained (Figure 6.1).

REASONS FOR PETROLEUM SPILLS

Petroleum products are a necessary part of our daily lives as they are a significant energy source. The demand for petroleum is increasing with the increased human population. During the last three decades, global oil consumption has risen steadily, with a total of 4.01 billion metric tons in 2020, compared to 3.6 billion metric tons of consumption in the last century. In contrast, the decline was only observed during the 2008/2009 financial crisis. The exploration, extraction, transportation, refining, and storage of petrochemicals are associated with occasional leakage and spillage of the petrochemicals, which are helpful for the energy industry but pose a severe threat to subjects' general health (Das and Chandran 2011). The release of petroleum into water directly affects marine life. Kvenvolden and Cooper (2003) reported the total amount of natural raw petroleum spillage into the marine environment at 600,000 metric tons, with an unreliability of 200,000 metric tons per year. Petrochemical emissions pollute soil and water and negatively affects the health of all inhabitants of the earth.

DOI: 10.1201/9781003188568-6

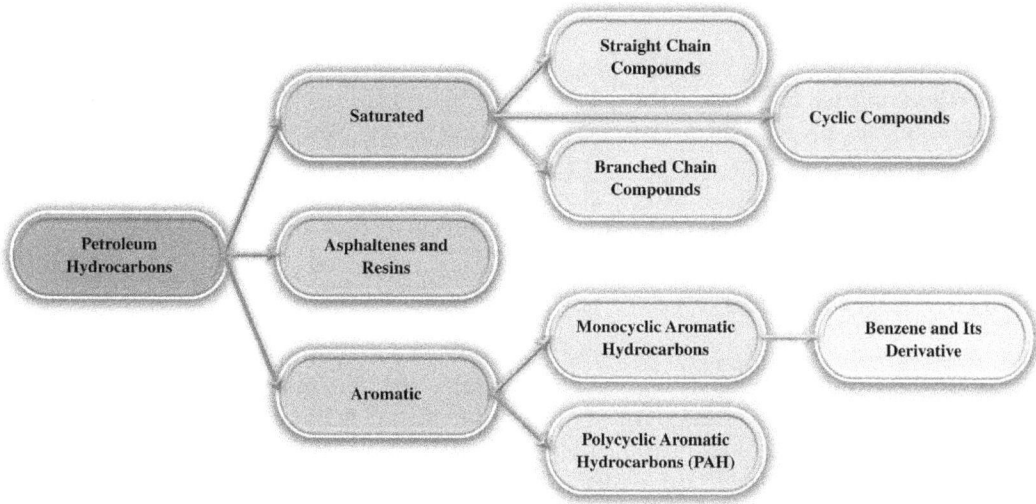

FIGURE 6.1   Various petrochemicals obtained from raw petroleum.

With the improvement of sea transportation and constant utilization of offshore oil, oil contamination has heavily influenced marine ecosystems and is currently one of the most significant worldwide concerns. Oil contamination due to spillage/leakage additionally affects the environment by setting off physical and natural changes (Ndimele 2018).

EFFECTS OF OIL SPILLAGE

ON SOIL

Oil pollution influences the nature of surface soil as it reduces soil fertility. Oil brings soil particles together and reduces soil porosity. Unrefined petroleum impurities mutilate the germination rate of the soil, destroy farmland, kill trees and plants, permeate into groundwater, and cause severe threats to marine life. Furthermore, the formation of an oily coating on the soil surface is attributed to oil spillage, which reduces the accessibility of carbon dioxide to plants (Ezeji, Anyadoh and Ibekwe 2007).

ON MARINE LIFE

Water is nature's gift, so if contaminated, will have harmful effects on life. Oil can cause severe problems in aquatic animals, including alteration in behaviors (reproduction and feeding) and causing taint (when organisms intake hydrocarbons, causing an "off" seafood flavor, making consumption by humans impossible until the disappearance of taint), and loss of habitat. Oil affects sea larvae and damages the wetland's value and structure, destroying the insulating power of mammals with fur. Thus, animals face some of the worst effects. Oil leaks disturb ecosystem biodiversity, aquatic food chains, and nursery grounds, and further cause marine life poisoning (Kvenvolden and Cooper 2003).

ON SOCIETY

Oil spillages have severe impacts on the environment, which further affects oil-producing human communities. The spillages demolish the scenic beauty of water bodies and other water qualities like swimming, drinking, fishing, recreation, and domestic use. Eventually, aquatic lives are lost, impacting fisherfolk households dependent on fishing.

Webler and Lord (2010) reported that oil spills affect humans in three significant ways:

1. Oil spills affect ecological processes, causing health damage by ingesting seafood with bio-accumulated oil toxins.
2. Oil leak stressors alter intermediary processes, e.g., economic effects on fishers due to oil spillage.
3. Stressors harm humans directly, e.g., health effects caused by inhalation of oil vapors. Therefore, ecosystem recovery entails conservation and human health protection.

REMEDIATION PROCESS

Spillage/leakage of crude oil has always negatively affected nature. The most considerable spillage incidence of oil spillage that occurred in the history of humankind was the Persian Gulf oil spill in 1991, which caused the spillage of millions of gallons of crude oil into the water and the neighboring area of 49 square km (Balba, Awadhi and Daher 1998). Another more recent incidence of oil spillage occurred in 2017—the Keystone pipeline accident, leading to 210,000 gallons of oil spreading on grass and agricultural area at southeast Amherst in northeast South Dakota (Smith 2017).

Such incidents are expected to happen due to the increased consumption of petroleum products. Considering the negative impacts of oil spillage on health as well as the environment, different methods are introduced for the remediation of oil spillage. Three remediation methods that are being used for oil spill removal are:

1. Physical method
2. Chemical method
3. Biological method

PHYSICAL METHODS

Since oil is less dense than water, it floats on the water's surface when it spills or leaks (into saltwater or freshwater). Therefore, oil spillage cleanup by physical means is easy. Physical methods commonly control oil spills in water lands and are fundamentally utilized as barriers to oil spreading without alteration in physical and chemical qualities. Different physical ways make use of obstructions to take care of oil spills, such as: (Yakubu 2007; Rodrigues and Tótola 2015).

   i. Booms
  ii. Adsorbent materials
 iii. Skimmers

Remediation using physical means is expensive and not applicable on a large scale despite the soil being unaffected or unmodified.

CHEMICAL METHODS

Chemical procedures are used with physical processes for remediation of oil spills, as they confine oil spreading and secure coastlines and sensitive marine living spaces. Different synthetic compounds are utilized to treat oil spills because they can change the physiochemical characteristics of oils. The synthetic compounds used to treat oil leaks are:

  i. Dispersants
 ii. Solidifiers

Physical and synthetic procedures are typically used to remediate polluted regions, and these methods are incredibly significant. However, these procedures require large equipment, and the natural outcomes of eliminating contaminations with these methods might bring about massive air contamination (Rodrigues and Tótola 2015).

Therefore, there is a need to explore alternative methods for effective remediation of oil spills that do not contribute to environmental pollution.

BIOREMEDIATION METHODS

Different microorganisms with different enzymatic potentials are used for the degradation of oil compounds. Some enzymes derived from microbes target branched, cyclic, or linear alkanes, while some affect aromatics, i.e., mono- or poly-nuclear, and others may treat both alkanes and aromatics mutually (Yakubu 2007; Joutey et al. 2013).

Bioremediation is a technique used to treat environmental contamination with the metabolic power of microorganisms. Bioremediation transforms pollutants without changing from one medium to another, degrades many different contaminants, and has various activities back-to-back and some degree of impurity resistance. Bacteria have essential roles in bioremediation, whereas other microbes (fungi, protozoa, and algal) may also affect the process. The microorganism is selected with an enhanced investigation for biotechnological application productions and oil degradations (Yakubu 2007). Bioremediation procedures are more efficient and economical for restoring oil-polluted regions than traditional cleanup methods (Okoh 2003).

The two main methods used for oil leak treatment are bioaugmentation and biostimulation.

Bioaugmentation

Bioaugmentation (also known as microbial seeding) used potential microbes specific for pollutants and suspended microbes using a stabilizing agent. The microbes are not helpful until restored in solution and used with micronutrients and stimulants (Yakubu 2007; Venosa et al. 1996). Bioaugmentation is of special importance when the degradation of petroleum substrates by primordial microbes may not be effective (Leahy and Colwell 1990) or stressed due to current subjection to oil leak (Yakubu 2007; Forsyth, Tsao and Bleam 1995). Under such conditions, oil-degrading microbes, used to augment primordial populations, have been introduced for bioremediation of oil-contaminated sites. The bioaugmentation efficiency is measured by the seed microbe's ability to demolish petroleum components and maintain genetic viability, stability during storage, and endurance in harsh surroundings. It can efficiently cope with initial microbes and degraded contaminants (Goldstein, Mallory and Alexander 1985).

Biostimulation

In biostimulation, environmental modification is done for the stimulation of bacteria to perform efficient remediation (Adams, Tawari-Fufeyin and Igelenyah 2014) and introduce limited nutrients and electron acceptor atoms, such as N, O, and P (Ndimele et al., 2015; Rhykerd et al., 1999). Perfumo et al. (2007) introduced biostimulation at a site filled excessively with nutrients and electron acceptors to elevate the number and action of primordial microbes. Margesin and Schinner (2001) introduced biostimulation as biological remediation with enhanced conditions—i.e., nutrient addition, temperature, aeration, and pH control—to enhance impurity degradation. The added nutrients boosted the microorganism's development but not inherent hydrocarbons degraders and set up a competition between resident bacteria (Ndimele et al., 2015; Adams, Tawari-Fufeyin and Igelenyah 2014). The method is beneficial because bioremediation involves indigenous microbes spread properly within the subsurface and are highly compatible with their surroundings.

Raskin et al. (1996) introduced phytoremediation as a technique that potentially used plants to eliminate pollutants from the surroundings or cause harm. Phytoremediation could be categorized as a unique technique of bioremediation or biodegradation. Frick, Farrel and Germida (1999) used plants to degrade polluted areas by plants. Some plant species can grow on contaminated sites and extract contaminants from growing sites. Some plants assemble trace metals in tissues (Ndimele 2013) while others can volatilize the pollutants (Brooks 1998; Terry and Zayad 1994). The pollutants are also filtered by the roots of plants in water (Brooks and Robinson 1998).

GENETICALLY MODIFIED MICROORGANISMS (GMMS)

The ecosystem requires large quantities of quality water to support life. Urbanization, growing population, and emerging economies are posing serious threats to the aquatic ecosystem via discharging biological, chemical, and physical pollutants such as heavy metals, dyes, pigments, and organic compounds. The treatment of pollutants is a growing concern and different physical, chemical, and biological methods have been introduced in this case (Ashraf et al. 2021; Altaf et al. 2021; Muhammad et al. 2020; Saif et al. 2022; Iqbal et al. 2022). Among these methods, the method of using genetically modified organisms is cost-effective and efficient. Naturally occurring mild microbial strains can slowly degrade pollutants; hence, bioaugmentation using genetically designed microbial strains can be employed for quick and effective remediation in polluted locales. Genetically modified organisms (GMOs) include animals, plants, or microorganisms. However, genetically engineered microorganisms (GEMs) or genetically modified microbes (GMMs) (bacteria and fungi including yeasts) were modified by humans using the latest biotechnology. Based on gene insertion in microbes, GEMs give characteristics of various microorganisms. Since bioremediation uses microbes or enzymes to destroy and eliminate contaminants from the environment, GEMs can also be used efficiently for bioremediation. The bacteria have a high capacity and can eliminate environmental pollutants; therefore, bacteria are of central importance as GEMs (Pant et al. 2021).

Use of GEMs for Remediation of Petrochemicals and Associated Pollutants in Oil Spills

Genetically engineered microbes and plants are used for *in-situ* bioremediation of many organic pollutants such as toluene, trichloroethylene, and polychlorinated biphenyls (Tanja et al. 2004; Reichenauer et al. 2008). The fall in the survival rate of GE bacteria is assumed to be due to stresses such as environmental conditions, introduced foreign genes, and competition with other microbes for nutrients and resources. Biotic (predation, antagonism, and competition) and abiotic (adsorption, temperature, moisture, and pH) environmental factors affect GE bacteria's survival and make it challenging to introduce a properly working modeling scheme. For the production purposes of engineered bacteria, indigenous microbial flora is advantageous over the exotic strains of GE bacteria.

The inserted genes depend on the rate of bioremediation, innate capabilities, and a total population of bacteria to bear stressful surrounding conditions. Hence, selecting and engineering the correct bacterial cell with fast growth, large population, significant nutrient responses, and effective pollutant removal without environmental risk is essential for obtaining a clean and eco-friendly environment, as in (Figure 6.2).

The ability of plant species to accumulate toxic metals has been reported for seven decades (Stomp et al. 1993) and some can enhance the degradation of polycyclic aromatic hydrocarbons (Aprill and Sims 1990) and 2, 5-dichloro-benzoate (Crowley et al. 1996). Radwan, Sorkhoh and El-Nemr (1995) have shown that many crop species can grow (up to 10% according to weight) in soil with crude oil, and can clean the crude oil rhizosphere. Rhizoremediation occurs naturally because flavonoids and other compounds released by roots can stimulate the growth of PAHs and chlorinated aliphatic compound-degrading bacteria (Pilon-Smits 2005). The rate of exudation depends on the age of a plant, the availability of mineral nutrients, and the nature of the contaminants. Organic compounds present in root exudates may provide carbon and nitrogen for the growth and long-term survival of microorganisms (Anderson and Walton 1995; Yee, Maynard and Wood 1998; Kuiper et al. 2004) that are capable of degrading organic pollutants and chlorinated aliphatic compounds (Pilon-Smits 2005).

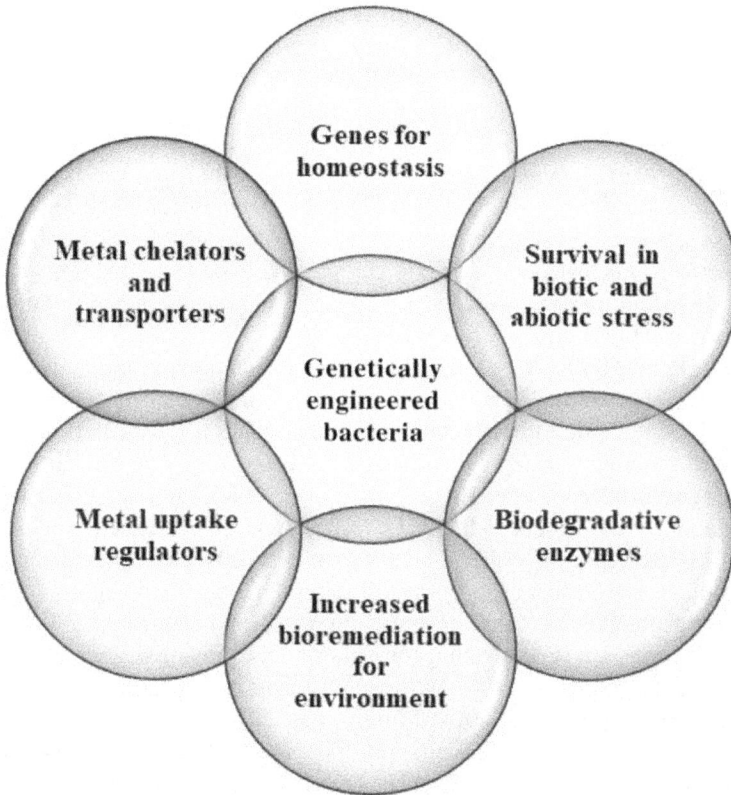

FIGURE 6.2   Potential applications of GEM for remediation of pollutants.

Trichloroethylene is considered a carcinogenic and groundwater pollutant at contaminated waste sites (McCarty 1997). Anderson and Walton (1995) have reported that plants and wild-type bacteria can promote the degradation of trichloroethylene in the rhizosphere; a noteworthy example is the conversion of 30% of [14C] TCE to $^{14}CO_2$ by legume *Lespedeza cuneata*. The trees belonging to the genus *Populus*, which includes cottonwoods, poplars, and aspens, is proved effective for rhizoremediation. Poplars grow fast within 3–5 months/year and contain long roots reaching the water table and curing the saturated zone. A 5-year-old tree could process 53 gal of water per day, causing the spreading of trichloroethylene in water treated by engineered bacteria. The field trials resulted in genetically engineered bacteria failing to give desired results due to a lower survival rate in the absence of plants (Shields and Francesconi 1996). The rhizosphere favors the symbiotic growth of engineered strains under controlled conditions adopted for roots. The roots provide steady redox conditions, optimal attachment locations, and a stable food supply of exudates comprising enzymes, amino acids, organic acids, phenolic compounds, and complex carbohydrates (Aprill and Sims 1990; Walton and Anderson 1990). The bacterial cells in the rhizosphere are 2–3 times larger than those in outlying soil (Aprill and Sims 1990).

BIOREMEDIATION OF HALOGENATED HYDROCARBONS

Organic compounds such as 1,2-dichloroethane and 1,2,3-trichloropropane are anthropogenic, do not show bio-effects, and can only be found at contaminated sites. Both the compounds are utilized or formed extensively and are found in polluted areas based on sudden leaks or improper local waste disposal. Due to greater solubility in water, both can contaminate groundwater after local discharge. The bacterial degradation of 1,2-DCE under toxic circumstances was observed by (Janssen et al. 1985) using a pure *Xanthobacter autotrophicus* culture. The primary reaction path involves hydrolysis of the C–Cl bond by haloalkane dehalogenase to 2-chloroethanol, oxidized to chloroacetic acid in two steps by dehydrogenase enzyme and glycolic acid by another enzyme, hydrolytic dehalogenase.

The plasmid with the dehalogenase gene was present in organisms that can grow on 1,2-DCE. Identical HA dehalogenases present in 1,2-DCE-degrading *Xanthobacter* and *Ancylobacter* cells act on 1,2-DCE, and strains are separated from various geographic locations such as South Korea, Africa, Australia, and Germany. Furthermore, *XA* GJ10 can readily degrade 1,2-DCE under different redox conditions in the absence of oxygen while nitrate captures electrons (Dinglasan-Panlilio et al. 2006).

Hage and Hartmans (1999) separated *Pseudomonas* sp. strain from 1,2-DCE biofilm. The investigators observed that this strain uses 1,2-DCE as a carbon and energy source. Interestingly, it does not require vitamins and other nutrients for appropriate growth. The strain bonding with 1,2-DCE is relatively high. It involves oxidation as the first step of degradation of 1,2-DCE instead of hydrolytic dehalogenation, a common initial step with other 1,2-DCE utilizers. The mechanism of degradation is presented in Scheme 6.1.

The genetic adaption was introduced in *X. autotrophicus* GJ10 and genetic and biochemical studies of related strains, and 1,2-DCE synthesis started in the 1920s. Genes having catabolic enzymes for artificial compounds are present on plasmids. They are linked with transposons (a DNA sequence whose position in a genome can create or reverse mutations and change cells' genetic identity and genome size). Transposons are very advantageous to alter DNA inside a living organism; otherwise, element insertion enables gene activation and transfer. Insertion of elements can also flank regions having HA dehalogenase (Song et al. 2004; Munro et al. 2016). Janssen et al. (1985) reported the gene expression for the HA dehalogenase enzyme. The substrate needs a second protein to recognize 1,2-DCE and HA dehalogenase, with which recent mutations in the protein are also indicated.

Pries et al. (1994) and Pikkemaat (2002) conducted laboratory evolution experiments to induce gene 1,2-DCE dehalogenase expression in the *Pseudomonas* strain. The strains were then subjected to selection pressure to accept 1-chlorohexane as substrate. The mutated bacterial strains

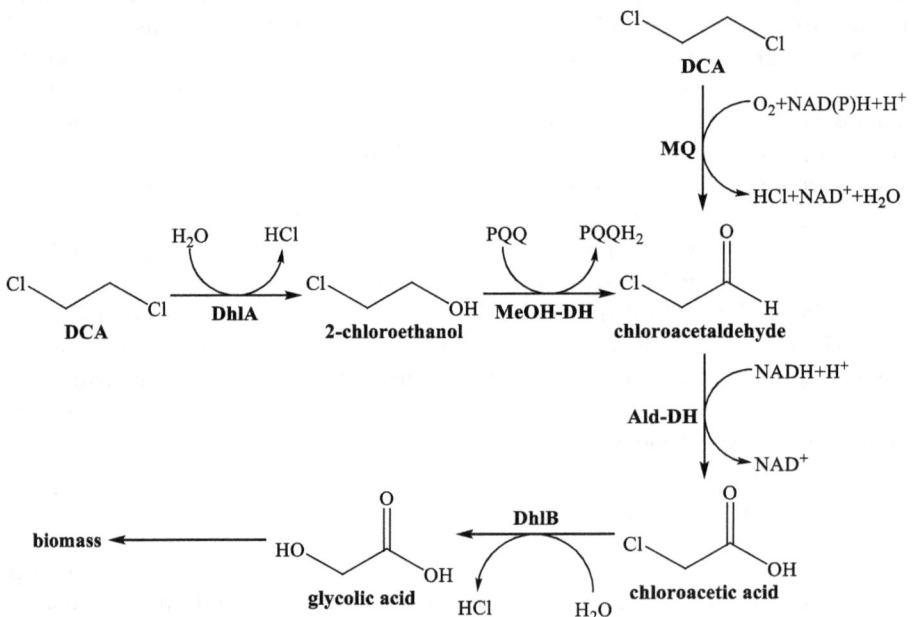

SCHEME 6.1   Mechanism of degradation of DCE.

were separated and could reproduce well in oxygen, vitamins, and yeast extract using 1,2-DCE as the only carbon source. The strains can grow via cross-feeding even if vitamins are not present. The growth-linked 1,2-DCE treatment process was found to be easy. Stucki, Thüer, and Bentz (1992) explained that 1,2-DCE degrading cells (*Pseudomonas* DE 1,5 *X. autotrophicus* GJ10) injected in fixed-bed bioreactors (laboratory scale) might be able to eliminate 1,2-DCE from groundwater.

In another study, *X. autotrophicus* GJ10 was used to treat gaseous waste and 1,2-DCE polluted wastewater in a bioreactor (Freitas dos Santos and Livingston 1994, 1995), which was designed so that biofilms degraded 1,2-DCE on membrane tubes surface by avoiding contact between 1,2-DCE and aerifying gas. The method resulted in over 99% removal of 1,2-DCE from polluted water having 1600 mg.L^{-1} of 1,2-DCE at 0.75h residence times. Biodegradation was confirmed via measurement of Cl- and CO_2 evolution in a bioreactor. The 1,2-DCE was not analyzed in bio medium over the operating process. The authors tried to develop a mathematical model for the phenomenon occurring in the biofilm; the model provided solutions that were consistent with experimental findings. However, empirically 1,2-DCE degradation rate in biofilm was not dependent on O_2 concentration, while model predictions demanded O_2 to be limiting.

FULL-SCALE PROCESSES FOR REMOVAL OF 1,2-DCE FROM GROUNDWATER: A CASE STUDY

The 1,2-DCE was removed using bacteria by degradation. The process was initiated by two companies "Ciba Specialty Chemicals, Inc." and "successor Novartis AG, Germany" using contaminated groundwater from the pharmaceutical production plant. 1,2-DCE replaced petroleum ether, i.e., non-flammable solvent, to obtain pancreatic from grained and dried calf's stomach tissue. The 1,2-DCE contamination (1–200 mg L^{-1}) was detected during site deconstruction and necessary rapid action. The sources were removed, and an extraction wells gallery was installed in the pump to treat groundwater.

Traditional sand filtration and charcoal adsorption methods were designed to treat groundwater (20 m^{3}h^{-1}). Initially, the process introduced was expensive due to poor carbon adsorption of 1,2-DCE demanding a replacement. The remediation technologies, i.e., 1,2-DCE air stripping and UV-ozone aeration, were considered inefficient and costly. Due to the efficiency and low cost of 1,2-DCE degrading bacteria, the sand and carbon filters of the plant were inoculated with 1,2-DCE-degrading bacteria and equipped with dosing stations for ammonium phosphate and H_2O_2. Later, the rise in feed concentrations of 1,2-DCE increases biological step capacity by installing a rotating disk biological contactor also inoculated with 1,2-DCE-degrading bacteria (Janssen et al. 1985).

Removal of Crude Oil Using Genetically Engineered Microorganisms

The availability of oil pollutants to microorganisms limits the removal of oil pollutants like petroleum hydrocarbon removal. It is challenging because hydrocarbons bind with soil components and vary in vulnerability to microbial attack. The susceptibility of hydrocarbons to microbial degradation in decreasing order is:

Linear alkanes > branched alkanes > small aromatics cyclic alkanes

Some compounds, such as polycyclic aromatic hydrocarbons (PAHs), having high molecular weight, may not be remediated at all (Perry 1984). The potential to bioremediate polluted water or soil is affected by physicochemical conditions. Using microorganisms, the biodegradation of hydrocarbons depends on genetic complement, genetic expression, and hydrocarbon properties. Thus, the parameters affecting bioremediation could be categorized into three categories (Figure 6.3) (Singh and Ward 2004).

The organic pollutant, i.e., petroleum hydrocarbon, is of significant concern due to its persistence, toxicity, complexity, and worldwide availability. The most common petroleum hydrocarbons are branched, aliphatic, cyclo-aliphatic alkanes, and aromatics (monocyclic and polycyclic) such as naphthalene, phenanthrene, fluorene, fluoranthene, anthracene, benzo-anthracene, and pyrene.

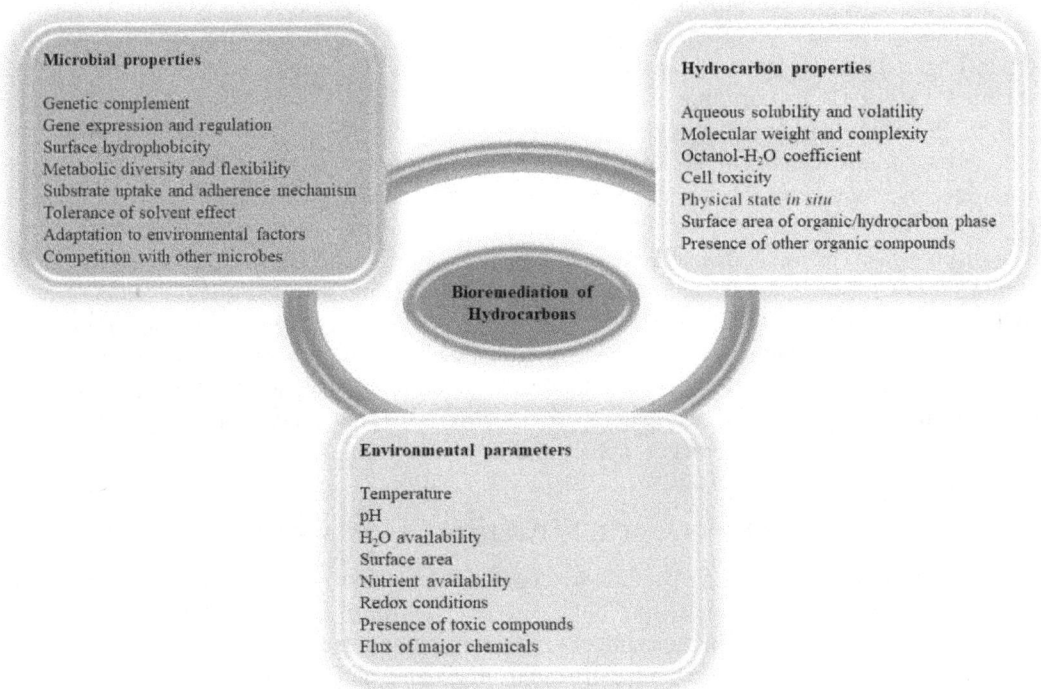

FIGURE 6.3 Schematic diagram of main parameters affecting hydrocarbon biodegradation in the environment to treat polluted water.

Crude oil may contain combined cyclo-aliphatic-aromatic structures. Each fraction of petroleum is made of hundred hydrocarbons except defined composition; that's why fractions are dissimilar in volatility, toxicity, bioavailability, persistence, and degradability. The complex order of compounds presents enormous challenges in developing efficient bioremediation processes and could be explained by different contamination effects (Atlas and Hazen 2011).

The microbes used to degrade hydrocarbons are bacteria, yeasts, and filamentous fungi (van Beilen and Funhoff 2007; Wentzel et al. 2007). Although, hydrocarbons are insoluble in water and less bioavailable to bacteria, bacteria could degrade when surface tension is lowered (Wentzel et al. 2007) by surfactants (Fritsche and Hofrichter 2000). Biosurfactants may be phospholipids, glycolipids, fatty acids, lipoproteins, lipopeptides, lipid polymers, biopolymers, and neutral lipids that can be linked to hydrocarbon droplets by elevating cell surface hydrophobic character via adhesion structure syntheses like mycolic acids, lipopolysaccharides, proteins, and exopolymers (Abbasnezhad, Gray, and Foght 2011; Christina and Tony 2021).

The hydrophobic fimbriae and emulsan (known as attachment pilus used by bacteria to adhere to one another) synthesized by GE bacterium, *Acinetobacter venetianus* RAG-1, elevates droplet attachment. Another GEM *Rhodococcus erythropolis*-20S-E1-c has an outer hydrophobic coating of mycolic acids, therefore, hydrophobic exterior structures enhance the access to hydrophobic surfaces and have various mechanical and dynamic qualities for droplet attachment (Dorobantu et al. 2008; Abbasnezhad, Gray and Foght 2011). The lateral diffusion process is responsible for uptake, i.e., diffusion of hydrocarbon molecules laterally into the external membrane, where hydrocarbons are obtained from the transporter's lumen. When hydrocarbons are in proximity to bacteria, they can enter a bacterial cell in which breakdown is achieved by catabolic machinery.

Microorganisms can activate hydrocarbons by advanced mechanisms that further produce metabolic intermediates, funneling to central metabolic pathways. In some nutrient-limited niches, the microorganisms may take the upper hand by oxidizing substrates. During aerobic catabolism, the

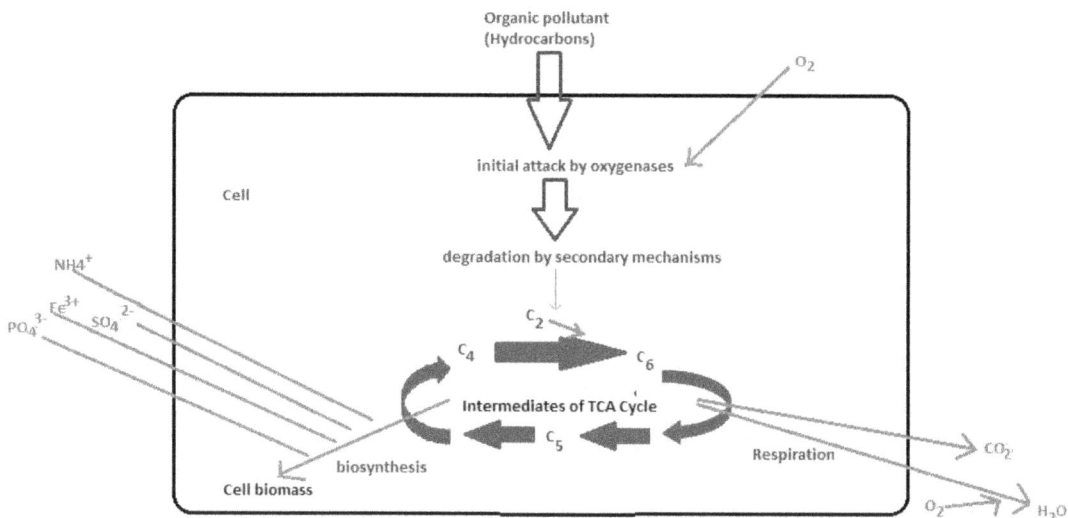

FIGURE 6.4 Mechanism of aerobic degradation of hydrocarbons by microorganisms.

first step is to add hydroxyl groups in the hydrocarbon skeleton. The addition of two hydroxyl groups is catalyzed by dioxygenases, whereas monooxygenases introduce oxygen atoms into hydrocarbon. During anaerobic degradation, activation is obtained when fumarate or CO_2 are coupled with SO_4^{2-}, hydrocarbons, and NO_3^{-1}, which are fatal electron acceptors (Callaghan 2013). However, alkane's anaerobic degradation occurrence rate is low as compared to aerobic (Wentzel et al. 2007).

For the majority of the organic pollutants, complete degradation occurs under aerobic conditions. Figure 6.4 represents the main aerobic steps involved in the degradation of hydrocarbons (Fritsche and Hofrichter 2000). The initial step is the enzyme-catalyzed oxidative process mediated by oxygenase and peroxidases. The secondary degradation mechanism converts the pollutants into other intermediate byproducts, which serve as precursors for other processes such as the tricarboxylic acid cycle. These intermediates serve as precursors for the biosynthesis of cell biomass, i.e., pyruvate, succinate, and acetyl-CoA.

Alkane Hydrocarbon Degradation by GEMs

Straight-chain hydrocarbons having 5–11 carbons are the chief components of petrochemicals. These alkanes are degraded via alkane oxygenase-mediated oxidation. These catalytic processes can be executed by multimeric *monooxygenase* or cytochrome P450 *monooxygenase*. The monooxygenase is a member of a large family of proteins with Fe-S, Fe-Fe, heme, and Cu in their active site. These enzymes are specific for inserting a single O-atom into a non-activated hydrocarbon under mild conditions (van Beilen and Funhoff 2007; Wang and Shao 2012).

In aerobic oxidation, the first primary alcohol is formed by the oxidation of alkane, as shown in Scheme 6.2, which is later further oxidized to an aldehyde; the subsequent conversion of an aldehyde into fatty acid and later β-oxidation of fatty acid leads to the formation of CO_2 and H_2O. Although alkanes are majorly converted into primary alcohols, secondary and tertiary alcohol formation can also occur (Rojo 1987).

Jones, Knight, and Byrom (1970) reported the presence of biodegradable aromatic hydrocarbons in marine deposits; these hydrocarbons had their origin in petroleum. This finding served as a starting point since it indicated that aromatic hydrocarbons of petroleum origin, which are very difficult to degrade, can be degraded in the natural setting of an aqueous environment.

According to van Beilen and Funhoff (2007), the mechanism of degradation of alkane involves *integral-membrane non-heme di-iron monooxygenase* catalyzed hydroxylation of the alkane at

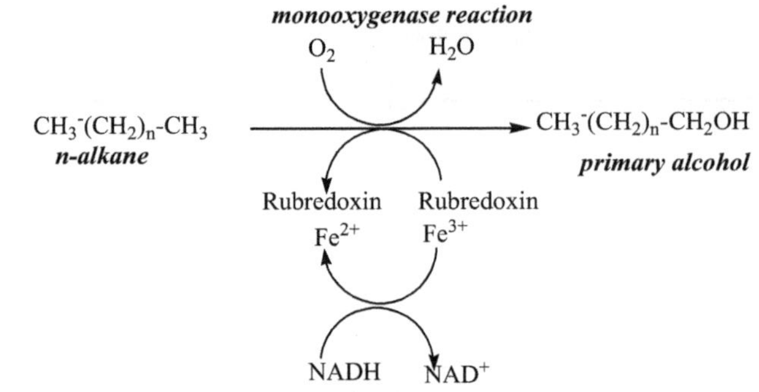

SCHEME 6.2   Aerobic oxidation of alkanes to alcohols.

TABLE 6.1

Enzymes Useful for the Breakdown of Straight-Chain Hydrocarbons Found in Different GEMs

Enzyme (Operon)	Oxidizable substrate/ Oxidation reaction	GEM as enzyme source	Reference
Alkane monooxygenase LadA	degrades up to C36 long-chain alkanes	*Geobacillus thermodenitrificans* NG80–2	Throne-Holst et al. 2007
alkBFGHJKL	*n*-alkanes into fatty acids	GPo1 OCT plasmid from *P. putida*	Rojo (1987)
pAH (AlkB)	C3–C13 *n*-alkanes	*Pseudomonas putida GPo1*	van Beilen et al. (2006)
AlkB-like alkane monooxygenase AlkM	Hydroxylation of C12–C36 *n*-alkanes	*Acinetobacter baylyi ADP1*	Rojo 1987
AlkR-mediated alkM	C7–C11 *n*-alkanes	*Acinetobacter baylyi ADP1*	Rojo 1987
alkMa	>C22 n-alkanes	*Acinetobacter* sp. *M1*	Rojo 1987
alkMb	C16–C22 n-alkanes	*Acinetobacter* sp. *M1*	Rojo 1987
Alkane hydroxylase (AlmA)	degrades <C32 n-alkanes	*Acinetobacter* sp. *DSM17874*	Rojo 1987

---

terminal carbon. The enzyme is a complex that comprises rubredoxin, rubredoxin reductase, and the enzyme hydroxylase, bound to the membrane. The reduction occurs via electron transfer sequentially from NADH *via* FAD (its cofactor) to soluble rubredoxin by rubredoxin reductase, then to PAHs, and finally into paraffin.

Rojo (1987) discovered that the substrate selectivity and specificity of PAHs isolated from different bacterial taxa (e.g., Acinetobacter, Alcanivorax, Burkholderia, Mycobacterium, Nocardia, Oleiphilus, Prauserella, Pseudomonas, and Rhodococcus) differed significantly. It is also unworthy that bacteria with monooxygenases similar to AlkB may digest intermediate (C5–C11) and long (C12) chain alkanes. Table 6.1 lists some enzymes useful for the breakdown of straight-chain hydrocarbons found in different GEMs.

Three significant hydroxylation systems are the methanotrophic bacteria enzyme, methane monooxygenases, and cytochrome P450. *Methylocella, Methylocystis, Methylococcus, Methylomicrobium, Methylomonas*, and *Methylosinus* all have cytoplasmic soluble forms of methane monooxygenase, which may oxidize straight-chain alkanes, alkenes, halo-alkanes, and saturated alicyclic hydrocarbons using a [2Fe-2S] catalytic core with FAD and NADH as cofactors. On the other hand, particulate methane monooxygenases contain mono or di Cu core, having hydroxylated substrates with 1–5 carbons, such as alkanes, alkenes, and halogenated alkanes.

Cytochrome P450 is made up of several oxidase systems found in both prokaryotes (Archaea, bacteria) and eukaryotes (eukaryotes) (fungi, plants, and animals). *Proteobacteria* and *Actinobacteria* contain soluble cytochromes P450, useful in the oxidation of alkanes (van Beilen et al. 2006). Seven *R. erythropolis* strains contain cytochromes (CYP153 and 3–5 AlkB-like pAHs), while *A. borkumensis* strains AP1 and SK2 have cytochromes, CYP153, and two AlkB homologs each.

Sulfate-reducing *Deltaproteobacteria* oxidizes alkanes anaerobically. *Desulfococcus oleovorans* (strain DSM 6200/Hxd3) is a gram-negative, sulfate-reducing bacterium that can degrade alkanes. This strain was found for the first time in a northern German oil field's saline water phase of an oil-water separator. Hxd3 is a deltaproteobacterium that can grow anaerobically on C_{12}-C_{20} alkanes as a carbon source and oxidize them to CO_2. Hxd3 activates the C-3 of n-alkanes, forming a carboxylic acid with fewer carbons than the parent substrate (So, Phelps and Young 2003). Desulfatibacillum alkenivorans AK-01 takes carbon from C13–C18 n-alkanes, 1-hexadecene, and 1-pentadecene and converts them to CO_2 (Callaghan et al. 2006; Callaghan 2013).

BACTERIAL-FACILITATED DECOMPOSITION OF AROMATIC HYDROCARBONS

The breakdown of aromatic hydrocarbons also involves oxidation as the initial step. Other mechanisms (may) involve attaching microbes to substrates and producing biosurfactants by the microbe and uptake mechanisms. Scheme 6.3 generalizes the degradation of aromatic hydrocarbons via an oxidation mechanism.

The aerobic oxidation of aromatic hydrocarbons (biphenyl, phenanthrene, pyrene, and naphthalene) occurs via initial deoxygenation mediated with a ring-hydroxylating dioxygenase (RHD) when *P. putida* mt-2 is used as an enzyme source (Peng et al. 2008; Pagnout et al. 2007). However, using the same bacterium, the oxidative pathway for methyl-substituted aromatic hydrocarbons (i.e., xylene/toluene) involves PAH-RHD monooxygenation of a methyl group with assistance.

P. putida mt-2's PAH-RHD complex is less complicated than the oxygenases of alkanes (Kweon et al. 2010; Iwai et al. 2011). The Rieske non-heme Fe catalytic core of the PAH-RHD complex is responsible for alkane oxidation; creating trans-diol intermediate is the important step in this process; however, if a cis-diol is generated, it is re-aromatized by dihydrodiol dehydrogenase. The aromatic diol is then cleaved at ortho or meta location, depending on whether the catabolic pathway prefers this position (Peng et al. 2008). The oxidation of a fused aromatic system with at least two rings produces various carboxylated aromatic intermediates, transformed into aromatic

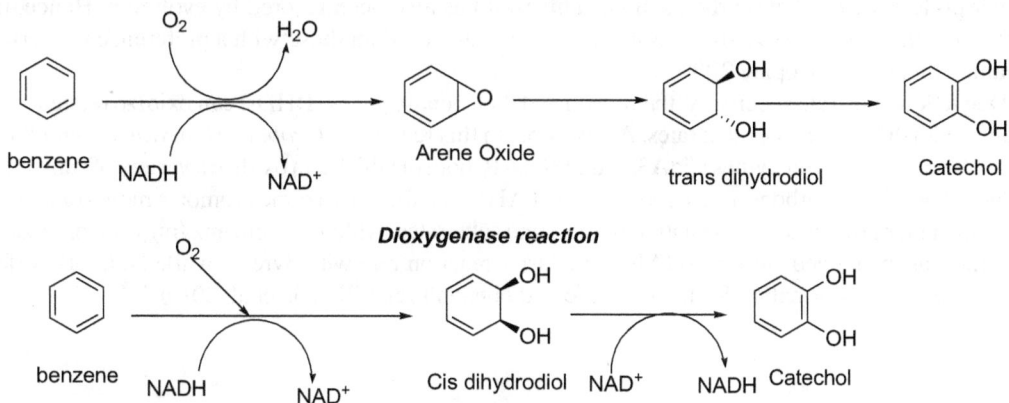

SCHEME 6.3 Degradation of aromatic hydrocarbons by oxidation.

TABLE 6.2

The Effects of Different Enzymes from Various GEMs on Aromatic Substrates

GEM	Enzyme involved	Substrate	Product	Reference
P. putida mt-2	xylene monooxygenase	toluene	benzaldehyde	Domínguez-Cuevas et al. 2006
P. putida F1	toluene dioxygenase TodC1C2BA	toluene	cis-toluene dihydrodiol.	Domínguez-Cuevas et al. 2006
Burkholderia Cepacia G4	specific monooxygenases	toluene	o-cresol	Domínguez-Cuevas et al. 2006
Ralstonia Pickettii PKO1	specific monooxygenases	toluene	m-cresol	Domínguez-Cuevas et al. 2006
Pseudomonas mendocina KR1	specific monooxygenases	toluene	p-cresol	Domínguez-Cuevas et al. 2006
Thauera aromatica K172	benzylsuccinate synthase	toluene	Addition of fumerate group to methyl group of toluene	Domínguez-Cuevas et al. 2006
Azoarcus sp. *T*	benzylsuccinate synthase	toluene	Addition of fumerate group to methyl group of toluene	Domínguez-Cuevas et al. 2006
P. putida mt-2	xyl genes from pWW0 plasmid	toluene	(alkyl)benzoate	Domínguez-Cuevas et al. 2006

metabolites. The primary difficulty with aromatic hydrocarbon degradation is that the parent substrate produces more deadly compounds, implying that aromatic degradation is ineffective (Cámara et al. 2004; Pieper and Seeger 2008).

The volatile organic portion of crude oil contains benzene, toluene, ethylbenzene, and xylenes (BTEX). Shinoda et al. (2004) identified one anaerobic and five aerobic toluene breakdown pathways. Table 6.2 lists numerous enzymes from various GEMs that can be used to oxidize BTEX.

The nah-genes taken from the plasmid NAH7 control naphthalene breakdown in *P. putida* G7, which has many similarities to the xyl catabolic route. Table 6.3 highlights the role of several GEM operons in polycyclic aromatic hydrocarbon degradation. The bacterial-mediated degradation is not substrate-specific, and for the same reason, one bacterial strain can degrade several PAHs. Bacteria using carbon from organic compounds for their development have an advantage over bacteria with a higher preference for their carbon source. This trait has also been favored by evolution. Hence, the bacteria with more diverse carbon usage are more abundant than those with a preference for certain organic compounds (Copley 2000).

Due to low substrate specificity, the reductive dehalogenase genes (RHDs) can oxidize related compounds with different conversion rates. An example in this regard is *Mycobacterium vanbaalenii* PYR-1, which possesses two enzymes NidA3B3 and NidAB, both of which can oxidize a variety of substrates, including non-hydrocarbons, hydrocarbons, and PAHs. The former enzyme promotes mono-oxidation, while the latter promotes dioxygenation of the same substrates. Although both enzymes can oxidize all substrates as mentioned earlier, NidAB has a faster reaction rate with pyrene, while NidA3B3 offers better rates of transformation for fluoranthene and phenanthrene (Kweon et al. 2010).

BIPHENYLS

Biphenyls are common and important constituents of crude oil and are of special importance since these serve as precursors for polychlorobiphenyls. Both biphenyl and polychlorobiphenyls can be

TABLE 6.3

Operons from Different GEMs for degradation of PAHs

GEM	Operon	Substrate	Product	Reference
P. putida G7 (NAH7 plasmid)	nahAaAbAcAdBFCED	Naphthalene	salicylate	Peng et al. 2008
	nahGTHINLOMKJ	Salicylate	pyruvate and acetaldehyde	
	Nah	Naphthalene	2,3-dihydoxyhihydronaphthalene	
Pseudomonas fluorescens 5R (plasmid pKA1)	Nah	Anthracene	hydroxynaphthoic acid	
		Phenanthrene		
Mycobacterium vanbaalenii PYR-1	NidAB	Non-hydrocarbons carbazole	Mono-oxygenation	Kweon et al. 2010
		Toluene	Mono-oxygenation	
		Naphthalene	Mono-oxygenation	
		m-Xylene	Mono-oxygenation	
		Anthracene	Mono-oxygenation	
		Phenanthrene	Mono-oxygenation	
		Fluoranthene	Mono-oxygenation	
		Pyrene	Mono-oxygenation	
		Benz[a]anthracene	Mono-oxygenation	
		Benzo[a]pyrene	Mono-oxygenation	
		Dibenzothiophene	Mono-oxygenation	
Mycobacterium vanbaalenii PYR-1	NidA3B3	Non-hydrocarbons carbazole	di-oxygenation	Kweon et al. 2010
		Toluene	di-oxygenation	
		m-Xylene	di-oxygenation	
		Naphthalene	di-oxygenation	
		Phenanthrene	di-oxygenation	
		Anthracene	di-oxygenation	
		Fluoranthene	di-oxygenation	
		Pyrene	di-oxygenation	
		Benz[a]anthracene	di-oxygenation	
		Benzo[a]pyrene	di-oxygenation	
		Dibenzothiophene	di-oxygenation	

degraded by PBH genes coded enzymes coded in Achromobacter, Burkholderia, Pseudomonas, Rhodococcus, and Sphingomonas strains of bacteria (Pieper and Seeger 2008). Biphenyl-2,3-dioxygenase is another PBH-coded enzyme that facilitates the cis-dihydroxylation of one of the biphenyl rings as the initial stage of the enzymatic process. The BDO of *Burkholderia xenovorans* LB400 oxidizes monochlorobiphenyl to hexachlorobiphenyl and has a low substrate specificity (Seeger and Pieper 2009). Natural and synthesized isoflavonoids and biphenyls containing F, Br, NO_2, OH, dibenzofuran, and dibenzodioxin moieties are acceptable substrates for LB400 BDO (Seeger, Cámara, and Hofer 2001; Overwin et al. 2012).

ENZYMATIC DEFENSE SYSTEM OF BACTERIA AGAINST AROMATIC HYDROCARBONS

Bacteria use aromatic hydrocarbons as a carbon source for their growth. However, these substrates can also prove to be lethal for the bacteria. To overcome the toxic influence of aromatic compounds, the bacteria possess a highly intricate system of genes that involve regulating processes for effectively responding to stress and promoting the degradation of toxic chemicals. It has been observed that bacteria isolated from environments polluted by crude oil exhibited more pronounced *catalase* and *peroxidase* activities. These enzymes are defense mechanisms of these bacteria responsible for the degradation of pollutants and neutralizing/reducing the harmful impact of reactive oxygen species (Bučková et al. 2010).

One of the most abundant bacteria in polluted environments is *Pseudomonas putida*. The *P. putida* strain DOT-T1E is solvent-resistant and flourishes in high concentrations of monoaromatic hydrocarbons. DOT-T1E has unique mechanisms for regulating lipid fluidity, responding to stress stimuli, enhancing energy generation, and stimulating specific efflux pumps for extruding solvents to the medium (Domínguez-Cuevas et al. 2006).

Similar mechanisms exist in *B. xenovorans LB400*, whereby the presence of toxic aromatic compounds stimulates the mediation of stress responses by different proteins. The *OxyR regulon* comprises genes responsible for peroxide metabolism & protection and redox balance. *Alkyl hydroperoxide reductase* and *catalase* are antioxidant enzymes associated with OxyR regulon and play a major role in the cellular response to oxidants. Upregulation of OxyR regulon has been observed in *P. putida KT2440* (pWW0) during toluene degradation (Agulló et al. 2007; Domínguez-Cuevas et al. 2006).

DESIGNING BIOREMEDIATION PROCESSES

When a petroleum spillage occurs, the choice of a suitable remediation strategy depends on the physicochemical aspects of the polluted site, the extent of the spill, and on the time that passed since the spillage event occurred. Bioremediation can either be carried out at the pollution site (*in situ*) or transported to a different appropriate place (*ex situ*) for treatment. The choice of methodology is based on the nature of the pollutant and other rate-related controlling parameters. In the case of petroleum spillage, four types of situations are usually encountered:

1. The spillage of petrochemicals results in an excess of carbon sources at the site, which results in the limitation of other nutrients responsible for microbial growth. It has been found that higher spillage on soil results in the binding of nutrients, thus making them less available. Therefore, under such circumstances, an additional amount of phosphorus and nitrogen must be added to restore the balance of nutrients and promote microbial growth responsible for bioremediation.

2. Hydrocarbon spillage results in poor aeration in the affected site, whether soil or H_2O. Insufficiency of oxygen retards the biodegradation rate. Under such a scenario, air injection or stirring of the matrix promotes the aeration necessary for the aerobic degradation of petrochemicals.

3. Once bound to the soil matrix, the hydrocarbons cannot be degraded easily, reducing their bioavailability. This issue can be resolved by using surfactants, which can improve the solubility and hence the bioavailability of the hydrocarbons. An effective, efficient, and economical alternative is to use biosurfactants that improve the bioavailability of aromatic hydrocarbons and have the added advantage of being non-toxic and biodegradable.

4. No availability of microbes capable of bioremediation at the polluted site is also a common scenario. Under such conditions, the addition of microbes either as pure culture or consortia can promote bioremediation.

The economic aspects of bioremediation promoted by microbes are determined by the amount of soil sample, extent and depth of pollution, and soil type. Usually, sands are easier to remediate than clay/silt soil. In general, if the polluted site is small, petrochemicals tend to seep deeper, and soil particles are smaller. The smaller particle size of soil leads to stronger binding of petroleum, and hence the cost of treatment will be higher in clay/silt, which has a smaller size than sand (US Federal Remediation Technologies Roundtable 2014).

For bioremediation processes, genetic engineering is widely used to boost the normal capacity of microbes (Pant et al. 2021). The knowledge of biotechnology, ecology, genetic engineering, and biochemical procedures is emerging and crucial for effective *in-situ* treatment of contaminants by GMOs (Liu et al. 2019). Genetically modified bacteria are effective alternatives, as they have increased capacities to remediate and degrade toxic compounds in crude oil. Because they are modified with oil-degrading enzymes, bacteria like *Pseudomonas putida* and *P. aeruginosa* degrade crude oil (Nagata et al. 2014; Aybey and Demirkan 2016). *Pseudomonas fluorescens HK44* and *pseudomonas putida KT2442* degrade naphthalene (Kawasaki, Watson, and Kertesz 2012; Boronin and Kosheleva 2014), and *Rhodococcus RHA1* bioremediates 4- chlorobenzoate (Li et al. 2016).

CONCLUSION

Accidental spillage of petroleum products is of grave concern to the environment. Petrochemicals, especially their hydrocarbon components, are well known for their toxicity. Oil spills usually occur during processing, refining, and transportation. Spillages cause damage to aquatic life in waters and reduced soil fertility on land. Currently, prevalent methods for remediation are very expensive, especially when the amounts of contaminants are very large.

Several microbes present in aqueous or terrestrial environments can degrade petrochemicals and their related contaminants. Bioremediation results in the complete conversion of the organic compound into CO_2, H_2O, inorganic compounds, or the transformation of complex organic compounds into simpler ones. Bioremediation using microbes is a promising method that is cost-effective and efficient. Engineered microbes, especially bacteria, open a frontier for research with wide implications. Tailoring bacteria's genomes can help develop GEMs with enzyme systems capable of degrading any specific type of organic pollutant.

REFERENCES

Abbasnezhad H., M. Gray, and J. M. Foght. 2011. Influence of adhesion on aerobic biodegradation and bioremediation of liquid hydrocarbons. *Appl. Microbiol. Biotechnol.* 92: 653–675.

Adams G. O., P. Tawari-Fufeyin, and E. Igelenyah. 2014. Bioremediation of spent oil contaminated soils using poultry litter. *Res. J. Eng. Appl. Sci.* 3: 124–130.

Agulló L., B. Cámara, P. Martínez, V. Latorre, and M. Seeger. 2007. Response to (chloro) biphenyls of the polychlorobiphenyl-degrader *Burkholderia xenovorans* LB400 involves stress proteins also induced by heat shock and oxidative stress. *FEMS Microbiol Lett.* 267(2): 167–175.

Altaf M., N. Yamin, G. Muhammad, M. A. Raza, M. Shahid, and R. S. Ashraf. 2021. Electroanalytical techniques for the remediation of heavy metals from wastewater. In *Water Pollution and Remediation: Heavy Metals*, pp. 471–511. Springer, Cham.

Anderson T. A., and B. T. Walton. 1995. Comparative fate of [14C]trichloroethylene in the root zone of plants from a former solvent disposal site. *Environ. Toxicol. Chem.* 14: 2041–2047.

Aprill W., and R. C. Sims. 1990. Evaluation of the use of prairie grasses for stimulating polycyclic aromatic hydrocarbon treatment in soil. *Chemosphere.* 20: 253–265.

Ashraf R. S., Z. Abid, M. Shahid, Z. U. Rehman, G. Muhammad, M. Altaf, and M. A. Raza. 2021. Methods for the treatment of wastewaters containing dyes and pigments. In *Water Pollution and Remediation: Organic Pollutants*, pp. 597–661. Springer, Cham.

Atlas R. M., and T. C. Hazen. 2011.Oil biodegradation and bioremediation: A tale of the two worst spills in U.S. history. *Environ. Sci. Tech.* 45: 6709–6715.

Aybey A., and E. Demirkan. 2016. Inhibition of quorum sensing-controlled virulence factors in *Pseudomonas aeruginosa* by human serum paraoxonase. *J Med Microbial.* 65: 105–113.

Balba M. T., N. Al Awadhi, and R. Al Daher. 1998. Bioremediation of oil-contaminated soil: Microbiological methods for feasibility assessment and field evaluation. *J Microbiol Methods.* 32: 155–164.

Beilen V., B. Jan, and E. G. Funhoff. 2007. Alkane hydroxylases involved in microbial alkane degradation. *Appl. Microbiol. Biotechnol.* 74: 13–21.

Beilen V., B. Jan, E. G. Funhoff, A. van Loon, A. Just, L. Kaysser, M. Bouza, R. Holtackers, M. Röthlisberger, Z. Li, and B. Witholt. 2006. Cytochrome P450 alkane hydroxylases of the CYP153 family are common in alkane-degrading eubacteria lacking integral membrane alkane hydroxylases. *Appl. Environ. Microbiol.* 72: 59–65.

Boronin A. M., and I. A. Kosheleva. 2014. The role of catabolic plasmids in biodegradation of petroleum hydrocarbons. In *Current Environmental Issues and Challenges*, pp. 159–168. Springer.

Brooks R. R., and B. H. Robinson. Ed. 1998. *Aquatic Phytoremediation by Accumulator Plants, Plants That Hyperaccumulate Heavy Metals: Their Roles in Phytoremediation, Microbiology, Archeology, Mineral Exploration and Phytomining*, R. R. Brooks, Ed., pp. 203–226. CAB International, Oxon.

Bučková M., J. Godočíková, M. Zámocký, and B. Polek. 2010. Screening of bacterial isolates from polluted soils exhibiting catalase and peroxidase activity and diversity of their responses to oxidative stress. *Current Microbiol.* 61(4): 241–247.

Callaghan A. V. 2013. Enzymes involved in the anaerobic oxidation of n-alkanes: From methane to long-chain paraffins. *Frontiers Microbiol.* 4: 89.

Callaghan A. V., M. G. Lisa, G. K. Kropp, M. S. Joseph, and Y. Y. Lily. 2006. Comparison of mechanisms of alkane metabolism under sulfate-reducing conditions among two bacterial isolates and a bacterial consortium. *Appl. Environ. Microbiol.*72: 4274–4282.

Cámara B., C. Herrera, M. González, E. Couve, B. Hofer, and M. Seeger. 2004. From PCBs to highly toxic metabolites by the biphenyl pathway. *Environ. Microbiol.* 6: 842–850.

Christina N., and G. Tony. 2021. Biosurfactants and their applications in the oil and gas industry: Current state of knowledge and future perspectives. *Front. Bioeng. Biotechnol.* 9: 1–19.

Christophe P., G. Frache, P. Poupin, B. Maunit, J.-F. Muller, and J.-F. Férard. 2007. Isolation and characterization of a gene cluster involved in PAH degradation in Mycobacterium sp. strain SNP11: Expression in *Mycobacterium smegmatis* mc2155. *Res. J. Microbiol.* 158: 175–186.

Copley S. D. 2000. Evolution of a metabolic pathway for degradation of a toxic xenobiotic: The patchwork approach. *Trends Biochem Sci.* 25(6): 261–265.

Crowley D. E., M. V. Brennerova, C. Irwin, V. Brenner, and D. D. Focht. 1996. Rhizosphere effects on biodegradation of 2,5-dichlorobenzoate by a bioluminescent strain of root-colonizing *Pseudomonas fluorescens. FEMS Microbiol. Ecol.* 20: 79–89.

Das N., and P. Chandran. 2011. Microbial degradation of petroleum hydrocarbon contaminants: An overview. *Biotech Res Int.* 2011: 1–13.

Dietmar P. H., and M. Seeger. 2008. Bacterial metabolism of polychlorinated biphenyls. *J. Mol. Microbiol. Biotec.* 15: 121–138.

Dinglasan-Panlilio M. J., S. Dworatzek, S. Mabury, and E. Edwards. 2006. Microbial oxidation of 1,2-DCE under anoxic conditions with nitrate as electron acceptor in mixed and pure cultures. *FEMS Microbiol. Ecol.* 56: 355–364.

Domínguez-Cuevas P., J.-E. González-Pastor, S. Marqués, J.-L. Ramos, and V. de Lorenzo. 2006. Transcriptional tradeoff between metabolic and stress-response programs in Pseudomonas putida KT2440 cells exposed to toluene. *J. Biol. Chem.* 281: 11981–11991.

Dorobantu L. S., S. Bhattacharjee, J. M. Foght, and M. R. Gray. 2008. Atomic force microscopy measurement of heterogeneity in bacterial surface hydrophobicity. *Langmuir.* 24: 4944–4951.

Ezeji U. E., S. O. Anyadoh, and V. I. Ibekwe. 2007. Clean up of crude oil-contaminated soil. *TAET.* 1: 54–59.

Forsyth J. V., Y. M. Tsao, and R. D. Bleam. 1995. Bioremediation: When is augmentation needed? In *Bioaugmentation for Site Remediation*, R. E. Hinchee, J. Fredrickson, and B. C. Alleman, Eds., pp. 1–14. Battelle Press, Columbus, OH.

Freitas dos Santos L. M., and A. G. Livingston. 1994. Extraction and biodegradation of a toxic volatile organic compound (1,2-DCE) from waste-waters in a membrane bioreactor. *Appl. Microbiol. Biotechnol.* 42: 421–431.

Freitas dos Santos L. M., and A. G. Livingston. 1995. Novel membrane bioreactor for detoxification of VOC wastewaters: Biodegradation of 1,2-DCE. *Water Res.* 29: 179–194.

Frick C. M., R. E. Farrel, and J. J. Germida. Ed. 1999. Assessment of phytoremediation as an in situ technique for cleaning oil contaminated sites. PTAC petroleum Technology Alliance Canada Calgary.

Fritsche W., and M. Hofrichter. Ed. 2000. *Aerobic degradation by microorganisms, in Environmental Processes-Soil Decontamination*. Wiley-VCH.

Goldstein R. M., M. Mallory, and M. Alexander. 1985. Reasons for possible failure of inoculation to enhance biodegradation. *Appl. Environ. Microbiol.* 50: 977–983.

Graul M. 2013.*Oil and Natural Gas*. Dorling Kindersley Ltd. Digital edition. www.rgn.unizg.hr.

Hage, J. C., and S. Hartmans. 1999. Monooxygenase-mediated 1,2-dichloroethane degradation by Pseudomonas sp. strain DCA1. *Appl. Environ. Microbiol.* 65: 246–2470.

Iqbal M. M., G. Muhammad, M. A. Raza, M. S. Aslam, M. A. Hussain, and Z. Shafiq. 2022. Fibrous membranes for water purification: Focusing on dye removal. In *Membrane Based Methods for Dye Containing Wastewater*, pp. 79–120. Springer, Singapore.

Iwai S., T. A. Johnson, B. Chai, S. A Hashsham, and J. M. Tiedje. 2011. Comparison of the specificities and efficacies of primers for aromatic dioxygenase gene analysis of environmental samples. *Appl. Environ. Microbiol.* 77: 3551–3557.

Janssen D. B., A. Scheper, L. Dijkhuizen, and B. Witholt. 1985. Degradation of halogenated aliphatic compounds by Xanthobacter autotrophicus GJ10. *Appl. Environ. Microbiol.* 49: 673–677.

Jones J. G., M. Knight, and J. A. Byrom. 1970. Effect of gross pollution by kerosine hydrocarbons on the microflora of a moorland soil. *Nature.* 227: 1166–1166.

Joutey N. T., W. Bahafid, H. Sayel, and N. El Ghachtouli. 2013. Biodegradation: Involved microorganisms and genetically engineered microorganisms: 289–319. DOI: 10.5772/56194.

Kawasaki A., E. R. Watson, and M. A. Kertesz. 2012. Indirect effects of polycyclic aromatic hydrocarbon contamination on microbial communities in legume and grass rhizospheres. *Plant Soil.* 358: 169–182.

Kuiper I., E. L. Lagendijk, G. V. Bloemberg, and B. Lugtenberg. 2004. Rhizoremediation: A beneficial plant-microbe interaction. *J. Mol. Plant Microbe Interact.* 17: 6–15.

Kvenvolden K. A., and C. K. Cooper. 2003. Natural seepage of crude oil into the marine environment. *Geo-Mar. Lett.* 23: 140–146.

Kweon O., S-J. Kim, J. P. Freeman, J. Song, S. Baek, and C. E. Cerniglia. 2010. Substrate specificity and structural characteristics of the novel Rieske nonheme iron aromatic ring-hydroxylating oxygenases NidAB and NidA3B3 from *Mycobacterium vanbaalenii* PYR-1. *MBio.* 1: e00135–10.

Leahy J. G., and R. R. Colwell. 1990. Microbial degradation of petroleum in the environment. *Microbiol. Rev.* 53: 305–315.

Li C., C. Zhang, G. Song, H. Liu, G. Sheng, and Z. Ding. 2016. Characterization of a protocatechuate catabolic gene cluster in *Rhodococcus ruber* OA1 involved in naphthalene degradation. *Ann Microbiol.* 66: 469–478.

Liu L., M. Bilal, X. Duan, and H. M. N. Iqbal. 2019. Mitigation of environmental pollution by genetically engineered bacteria—Current challenges and future perspectives. *Sci Total Environ.* 667: 444–454.

Malcolm S. S., and S. C. Francesconi. 1996. Molecular techniques in bioremediation. *Biotechnol. Res. Ser.* 6: 341–390.

Margesin R., and F. Schinner. 2001. Bioremediation (natural attenuation and bio stimulation) of diesel oil contaminated soil in an alpine glacier skiing area. *Appl. Environ. Microbiol.* 67: 3127–3133.

María José R-S., V. Mendez, L. Agullo, and M. Seeger. 2013. Genomic and functional analyses of the gentisate and protocatechuate ring-cleavage pathways and related 3-hydroxybenzoate and 4-hydroxybenzoate peripheral pathways in *Burkholderia xenovorans* LB400. *PloS One.* 8: e56038.

McCarty P. L. 1997. Breathing with chlorinated solvents. *Science.* 276: 1521–1522.

Mimmi T-H., A. Wentzel, T. E. Ellingsen, H-K. Kotlar, and S. B. Zotchev. 2007. Identification of novel genes involved in long-chain n-alkane degradation by Acinetobacter sp. strain DSM 17874. *Appl. Environ. Microbiol.* 73: 3327–3332.

Muhammad G., A. Mehmood, M. Shahid, R. S. Ashraf, M. Altaf, M. A. Hussain, and M. A. Raza. 2020. Biochemical methods for water purification. In *Methods for Bioremediation of Water and Wastewater Pollution*, pp. 181–212. Springer, Cham.

Munro E., E. Liew, M. Ly, and N. Coleman. 2016. A new catabolic plasmid in Xanthobacter and Starkeya spp. from a 1,2-DCE-contaminated site. *Appl. Environ. Microbiol.* 82: 5298–5308.

Nagata Y., M. Tabata, S. Ohhata, and M. Tsuda. 2014. Appearance and evolution of γ-hexachlorocyclohexane-degrading bacteria. In *Biodegradative Bacteria*, pp. 19–41. Springer, Tokyo, Japan.

Ndimele P. E., A. Jenyo-Oni, K. S., Chukwuka, C. C. Ndimele, and I. A. Ayodele. 2015. Does Fertilizer 9N15P15K12) amendment enhance phytoremediation of petroleum polluted aquatic ecosystem in the presence of water hyacinth (Eichhornia crassipes [Mart.] Solms)? *Environ. Technol.* 36: 2502–2514.

Ndimele P. E., and C. C. Ndimele. 2013. Comparative effects of biostimulation and phytoremediation on crude oil degradation and absorption by water hyacinth (Eichhornia Carrsipes [Mart]Solms). *Int J Environ Stud.* 70: 241–258.

Okoh A. I. 2003. Biodegradation of bonny light crude oil in soil microcosm by some bacterial strains isolated from crude oil flow stations saver pits in Nigeria. *Afr. J. Biotechnol.* 2: 104–108.

Overwin H., M. González, V. Méndez, M. Seeger, V. Wray, and B. Hofer. 2012. Dioxygenation of the biphenyl dioxygenation product. *Appl. Environ. Microbiol.* 78(12): 4529–4532.

Pagnout C., G. Frache, P. Poupin, B. Maunit, J-F. Muller, and J-F. Férard. 2007. Isolation and characterization of a gene cluster involved in PAH degradation in Mycobacterium sp. strain SNP11: Expression in *Mycobacterium smegmatis* mc2155. *Res. Microbiol.* 158: 175–186.

Pant G., D. Garlapati, U. Agrawal, R. G. Prasuna, T. Mathimani, and A. Pugazhendhi. 2021. Biological approaches practised using genetically engineered microbes for a sustainable environment: A review. *J. Hazard Mater.* 405.

Peng R-H., A-S. Xiong, Y. Xue, X-Y. Fu, F. Gao, W. Zhao, Y-S. Tian, and Q-H. Yao. 2008. Microbial biodegradation of polyaromatic hydrocarbons. *FEMS Microbiol. Rev.* 32: 927–955.

Perfumo A., I. M. Banat, R. Marchant, and L. Vezzulli. 2007. Thermally enhanced approaches for bioremediation of hydrocarbon contaminated soils. *Chemosphere.* 66: 179–184.

Perry J. J. 1984. Microbial metabolism of cyclic alkanes. In *Petroleum Microbiology*, R. M. Atlas, Ed., pp. 61–98. Macmillan, New York, NY, USA.

Pieper D. H., and M. Seeger. 2008. Bacterial metabolism of polychlorinated biphenyls. *J Mol Microbiol Biotechnol.* 15: 121–138.

Pilon-Smits E. 2005. Rhizoremediation: A beneficial plant-microbe interaction. *Annu Rev Plant Biol.* 56: 15–39.

Pries F., A. J. van den Wijngaard, R. Bos, M. Pentenga, and D. B. Janssen. 1994. The role of spontaneous cap domain mutations in haloalkane dehalogenase specificity and evolution. *J. Biol. Chem.* 269: 17490–17494.

Radwan S., N. Sorkhoh, and I. El-Nemr. 1995. Oil biodegradation around roots. *Nature.* 376: 302.

Rhykerd R. L., B. Crews, K. J. Mclnnes, and R. W. Weaver. 1999. Impact of bulking agents, forces aeration and tillage on remediation of oil-contaminated soil. *Bioresour. Technol.* 6: 279–285.

Rodrigues E. M., and M. R. Tótola. 2015. Petroleum: From basic features to hydrocarbons bioremediation in oceans. *Open Access Libr.* 2: e2136. http://dx.doi.org/10.4236/oalib.1102136.

Saif S., T. Saif, M. A. Raza, G. Muhammad, M. M. Iqbal, N. Ur Rehman, and M. A. Hussain. 2022. An introduction to membrane-based systems for dye removal. In *Membrane Based Methods for Dye Containing Wastewater*, pp. 1–22. Springer, Singapore.

Seeger M., B. Cámara, and B. Hofer. 2001. Dehalogenation, denitration, dehydroxylation, and angular attack on substituted biphenyls and related compounds by a biphenyl dioxygenase. *J Bacteriol.* 183: 3548–3555.

Shinoda Y., Y. Sakai, H. Uenishi, Y. Uchihashi, A. Hiraishi, H. Yukawa, H. Yurimoto, and N. Kato. 2004. Aerobic and anaerobic toluene degradation by a newly isolated denitrifying bacterium, Thauera sp. strain DNT-1. *Appl. Environ. Microbiol.* 70(3): 1385–1392.

Singh A., and O. P. Ward. 2004. *Biodegradation and Bioremediation*, p. 4. SpringerLink.

Smith M., and J. Bosman. 2017. Keystone Pipeline leaks 210,000 gallons of oil in South Dakota. *The New York Times.*

So C. M., C. D. Phelps, and L. Y. Young. 2003. Anaerobic transformation of alkanes to fatty acids by a sulfate-reducing bacterium, strain Hxd3. *Appl. Environ. Microbiol.* 69: 3892–3900.

Song J. S., D. H. Lee, K. Lee, and C. K. Kim. 2004. Genetic organization of the dhlA gene encoding 1, 2-DCE dechlorinase from Xanthobacter flavus UE15. *J. Microbiol.* 42: 188–193.

Stomp, A. M., K. H. Han, S. Wilbert, and M. P. Gordon. 1993. Genetic improvement of tree species for remediation of hazardous wastes. *In Vitro Cell. Dev. Biol. Plant.* 29: 227–232.

Stucki G., M. Thüer, and R. Bentz. 1992. Biological degradation of 1, 2-DCE under groundwater conditions. *Water Res.* 26: 273–278.

Terry N., and A. M. Zayad, 1994. *Selenium Volatilization by Plants. Selenium in the Environment*, J. R. Frankenberger and S. Benson, Eds., pp. 343–367. Marcel Dekker, New York, NY.

Throne-Holst M., A. Wentzel, T. E. Ellingsen, H. K. Kotlar, and S. B. Zotchev. 2007. Identification of novel genes involved in long-chain n-alkane degradation by Acinetobacter sp. strain DSM 17874. *Appl. Environ. Microbiol.* 73: 3327–3332.

Venosa A. D., M. T. Suidan, B. A. Wrenn, K. L. Strohmeier, J. R. Haines, B. L. Eberhart, D. King, and E. Holder. 1996. Bioremediation of an experimental oil spill on the shoreline of Delaware Bay. *Environ. Sci. Technol.* 30: 1764–1775.

Walton B. T., and T. A. Anderson. 1990. Microbial degradation of trichloroethylene in the rhizosphere: Potential application to biological remediation of waste sites. *Appl. Environ. Microbiol.* 56: 1012–1016.

Wanpeng W., and Z. Shao. 2012. Diversity of flavin-binding monooxygenase genes (almA) in marine bacteria capable of degradation long-chain alkanes. *FEMS Microbiol. Ecol.* 80: 523–533.

Webler T., and F. Lord. 2010. Planning for the human dimensions of oil spills and spill response. *J. Environ. Manage.* 45: 723–738.

Yakubu M. B. 2007. Biological approach to oil spills remediation in the soil. *Afr. J. Biotechnol.* 6: 2735–2739.

Yee D. C., J. A. Maynard, and T. K. Wood. 1998. Rhizoremediation of trichloroethylene by a recombinant, root-colonizing *Pseudomonas fluorescens* strain expressing toluene ortho -monooxygenase constitutively. *Appl. Environ. Microbiol.* 64: 112–118.

7 Recombinant *E. coli* in the Rejuvenation of the Polluted Environment

Anirudh Pratap Singh Raman, Prashant Singh and Pallavi Jain

INTRODUCTION

Water is an extremely valuable element in this world; without it, life would be impossible. Water covers the bulk of the earth; around 97% of it is ocean water, which cannot be used for drinking, cooking or other purposes. This is due to the high salinity of the water. Although salt is present in large concentrations in ocean water, other compounds such as magnesium, sulfate, potassium, and calcium are also present, distinguishing the qualities of ocean water from potable water. There are various dissolved gases and water including nitrogen, carbon dioxide, and oxygen. Nitrogen is mostly found in fertilizers which are used to improve soil fertility and plant growth. If nitrogen is washed into the ocean, it damages aquatic life and pollutes the water, which is bad for coral (Landrigan et al. 2020). Oceans serve as crucial carbon sinks, absorbing and storing a significant amount of CO_2, which effectively mitigates the greenhouse effect. Although we generally have confidence in the purity of ocean water, it is essential to be vigilant about its overall quality, especially considering the pollution stemming from human activities on land (Urgun-Demirtas, Stark and Pagilla 2006). Industrial effluents and sewer discharge are the two main pollutants of open surface water. However, non-point sources of pollution, such as acid rain, agriculture and open drainage cannot be easily identified. The amount and quality of water are two of the most pressing issues we face in this era. These pollutants comprise extremely dangerous complex compounds that are harmful to both human health and the environment. Physical and chemical dispersants were used in traditional remedies and approaches which could wind up being harmful to the environment (Mitra et al. 2020). Modern waste management methods are based on the concept of bioremediation, which is the optimal option. The biological process of biodegradation is the biological destruction of a substance. It's a natural occurrence. Bioremediation is an engineering methodology to break down a material using biological mechanisms (algae, bacteria, fungi and others). It is a method of removing pollutants from the contaminated areas, including water, soil, sludge and residue, by utilising pathogenic organisms such as algae (*Chlorella vulgaris, Anabaena variabilis*), bacteria (*E. coli, pseudomonas* sp.) and fungi (*Rhizopus arrhizus, Aspergilus niger*, and others) that convert hazardous substances into non-toxic forms. Genetic engineering is currently a popular strategy for increasing microorganism biodegradation activity for environmental rejuvenation using modern scientific technologies (Dris et al. 2015; EFSA Journal 2011).

OVERVIEW OF GREEN TECHNOLOGY

For the restoration of polluted surface water bodies, several green solutions have been intensively researched. Many processes engaged in environmental restoration use biological and eco-technological approaches that range from traditional to novel hybrid technologies.

DOI: 10.1201/9781003188568-7

ADVANCEMENT IN BIOREMEDIATION

Bioaugmentation was added into the scenario by integrating genetically modified bacteria-contaminated areas for rapid and dependable breakdown; hence, organic and non-organic organisms are significantly less important and sluggish decomposers of trash and toxins in the environment. These genetic alterations are a collection of strategies for modifying the genetic material of domestic animals and plants in order to achieve certain consequences (such as substitution, induced mutation and hybridization). Genetic engineering is the method of modifying genes that entails making a precise modification to a plant's, microbe's, or animal's gene sequence in order to accomplish a desired effect. Microorganisms (fungi or bacteria) that have been altered by humans through molecular biotechnological *in-vitro* processes are known as GEMs (genetically engineered microorganisms) (M-Ridha et al. 2020). GEMs have the qualities of a range of microorganisms due to the implantation of genes in the strains of bacteria. GEMs can be effectively used for bioremediation since bioremediation is a technology that promotes the destruction and elimination of toxins from the environment. Bacterial strains that can digest a wide range of pollutants including biphenyls, polycyclic, aromatics PCBs, nitroaromatic compounds and oil components. However, some of the most persistent and harmful xenobiotics, such as nitrated and halo-aromatic chemicals as well as a number of pesticides, are stable and chemical resistant in natural environment and are not successfully eliminated by numerous methods. These limits have prompted the development of bacterial strains with efficient catabolic pathways that outperform other microorganisms in terms of bioremediation. Composting electro-bioremediation, microbial assisted phytoremediation, and other technologies based on biostimulation and bioaugmentation are examples of bioremediation (Pant et al. 2021). Genetic engineering is becoming more popular as scientific discoveries improve the natural capability of microbes for cleanup. De Carcer et al. (2007) found that a full understanding of biotechnology and ecology, as well as biochemical processes, is required for successful bioremediation using GMOs (Jain and Bajpai 2012).

ESCHERICHIA COLI (*E. COLI*)

The ability of *E. coli* to utilize aromatic compounds as a primary generator of carbon and energy has long been studied. As seen in Figure 7.1, *E. coli* is a common bacterial species found in the guts of most warm-blooded creatures, with 500 distinct types. *E. coli*'s capacity to fill a variety of ecological niches is thought to play a role in its intestinal success. As a result of its ability to flourish both anaerobically and aerobically, *E. coli* can colonize parts of the intestine where oxygen is

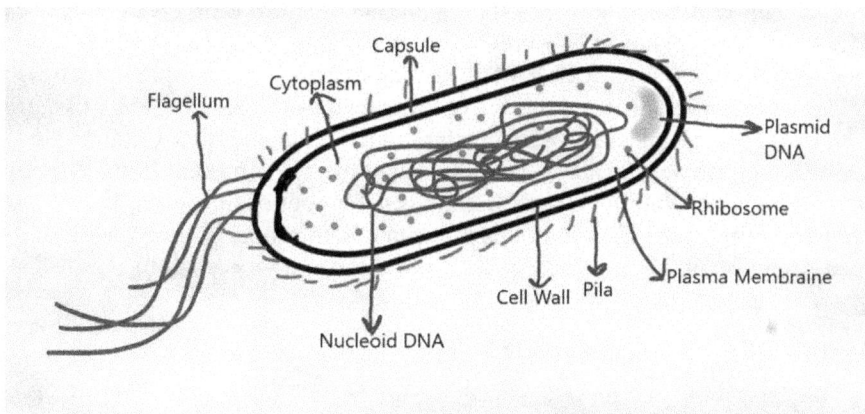

FIGURE 7.1 *E. coli* bacteria cell diagram.

beneficial to the environment (Medfu Tarekegn, Zewdu Salilih and Ishetu 2020). These settings can be found near the respiratory epithelium, where oxygen molecules from the blood can flow through the epithelium and reach the microbes clinging to it. *E. coli* may have a role in the formation and maintenance of anaerobic conditions in the large intestine by reducing the potential for oxidation-reduction, which favors stringent anaerobes while implanted in the mucous membrane overlaying intestinal epithelial cells. *E. coli* reproduces with a prompting time of 80 minutes. The *E. coli* colony in the small intestine, on the other hand, grows slowly and is excreted in the faeces. Although *E. coli* is not recognized for its endovascular survival, it does share with its soil-dwelling comparable species, such as Klebsiella species, the ability to thrive under a variety of physicochemical conditions, such as adaptability over extended periods of time. *E. coli* has the potential to reside in different environments such as soil, sediments, water and food, and it is likely to spend similar periods of time in each of these main habitats. The bacteria were largely studied in equatorial seas but remained physically active and thrived when tested using nutrient analysis techniques, implying that it could be an organic habitant of these habitats and hence a component of previously created culture. In protozoa, *E. coli* can proliferate, replicate and survive (Díaz et al. 2001; Rodríguez-Lázaro et al. 2007).

INTRASPECIES IN *E. COLI*

E. coli strains can be categorized into several phylogenetic groups, including A, B1, B2, C, D, E, F and clade I. Non-pathogenic *E. coli*, which are found, among other places, on the mucosa of the gastrointestinal tract, are most commonly associated with groups A or B1. Pathogenic *E. coli* strains responsible for intestinal infections are linked to phylogenetic groups A, B1 or D. *E. coli* strains responsible for extraintestinal infections are primarily associated with groups B2 and D. Furthermore, group E is closely related to group D (Dale and Woodford 2015; Köhler and Dobrindt 2011). The K-12 strain which has been examined extensively and represents the best-understood organisms at the genetic level is classified as part of that group (Selinger et al. 2000). The wild type of K-12 strain was isolated from the faeces of a recovering cholera patient, with subgroups and variants following in 1944. Because they are difficult to remove from the human intestine, K-12 strains are regarded as the standard for physiologically secure gene cloning and transcription procedures (Gray and Tatum 1944). K-12 strains were not found in any of the 226 human and environmental semen samples examined, implying that the existence of *E. coli* isolates is extremely rare. Some bacterial strains such as *E. coli*, B & W are *E. coli* pathogens found in research labs and are not related to K-12. *E. coli* C is a prototrophic strain of *E coli* F that was one of the first to show that it could recombine (act as a receiver) with K-12 in the laboratory investigation; *E. coli* B/R, a radiation-resistant mutation of strain B, has been widely used (Dale and Woodford 2015). S.A. Waksman is credited with isolating the ATCC 9637 wild-type *E. coli* strain from graveyard dirt (Perna et al. 2001). The majority of people have the genetics that define the quality shared by all members of species often known as the basic combination of genes for an organism. Pathogenic organisms and secondary metabolite genes, for example, are dispersed across each strain's genome (flexible gene pool), dictating traits present in many but not all *E. coli* individuals, and contributing to the dynamic genetic pool of the genus. In this perspective, intraspecies variability in the *E. coli* strains that can create primary carbon and energy via different aromatic acids is important (Díaz et al. 2001; Riley and Serres 2000).

RECOMBINANT *E. COLI*

The host cell has been widely employed in the design and manufacture of hybrid pathways based on the integration of enzymes from various sources. In *E. coli*, the recombinant process is useful for research in a variety of fields including protein, medicine, agriculture and the environment. This is because of the simplicity, economic viability, quick high-density growth, well-characterized genetics and a large number of viable expression systems that are available. Recombinant biocatalysts, for example, have been developed that can use benzene, one of the most hazardous components of petroleum, as a low-cost index ingredient in the production of a high-cost product called L-Dopa. *P. putida* and F1 toluene dioxygenase and toluene cis-glycol dioxygenase, as well as *Citrobacter freundii* throsine phenol-lyase were used to establish hybrid pathways in *E. coli* cells (Ortega-González et al. 2013). In the presence of pyruvate and ammonia, toluene dioxygenase and toluene dehydrogenase create catechol from benzene. Although benzene is toxic to recombinant *E. coli*, the procedure's efficacy can be increased by using solvent-resistant *E. coli*. As demonstrated in Figure 7.2, the bulk of techniques for generating microbial strains for aromatic component overproduction have focused on improving the metabolic processes within the targeted product-specific pathways. Many genetic illnesses such as phenylketonuria, have been successfully treated with recombinant *E. coli* bacteria that have been enhanced to convert aromatic compounds. The prototype human hereditary ailment is hyperphenylalaninemia, which is caused by decreased phenylalanine hydroxylase activity, the enzyme in the cell responsible for catabolism at the oxidation of the majority of phenylalanine dietary intake. The phenylalanine ammonia-lyase gene from *Rhodoporidium toruloides* was cloned and expressed in *E. coli*, allowing recombinant cells to be exposed to physiologically insignificant amounts of ammonia and CI. Hippurate is a harmless substance that once converted to benzoic acid is easily excreted through the urine. These findings are only the beginning of an intriguing and potentially useful disease prevention strategy based on the use of genetically engineered bacteria (Adetunji and Panpatte 2021; Dangi et al. 2019; Fraiture et al. 2020).

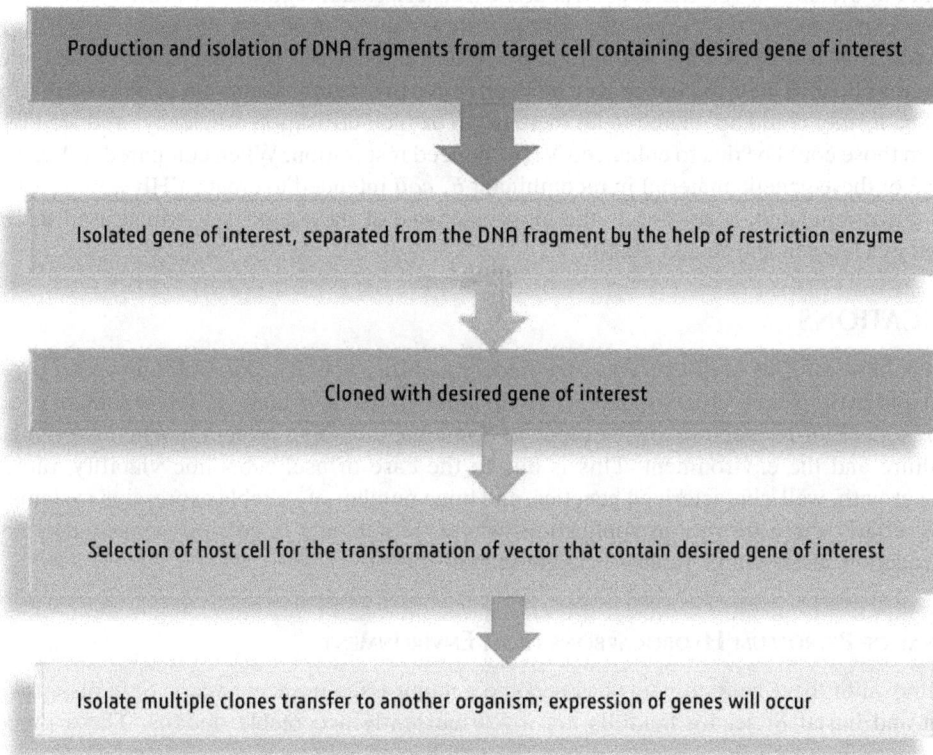

FIGURE 7.2 Schematic representation of the mechanism of GMOs.

APPROACHES THAT UPREGULATE UNMODIFIED CODING REGIONS

Fusion with Unique Advantages

Organophosphorus hydrolase breaks down organophosphate insecticides (OPH). *E. coli* were genetically modified to create OPH fusion proteins with domains that permitted them to be produced on the cell surface. In a model bioreactor, the generated recombinant cells successfully decomposed organophosphorus insecticides having an advantage because of the lack of cell membrane to substance diffusion. Recombinant *E. coli* was designed to have a heterologous bacterial OPH gene coupled with a gene encoding that promoted its secretion to the phagosome in a similar way (Sonter et al. 2021).

Bacterial Haemoglobin Technology

As previously stated, oxygenases, both natural and synthetic, are required for the degradation of a variety of organic pollutants, particularly aromatics. They do, however, require a sufficient amount of oxygen to function properly. The availability of oxygen in low-oxygen environments such as soil and aquatic sediments can be an issue for bioremediating such chemicals. For example, adding oxygen to a site is a costly technical solution. The only solution to this problem could be to cultivate aerobic bacteria that have the ability to reproduce and break down biological contaminants under stressful environments. One of the approaches being employed is engineering oxygenated bacteria to produce the bacterial haemoglobin gene (vgb). The generation of bacterial haemoglobin (VHb) may deliver more oxygen to oxygenases responsible for increasing their bioactivity as well as respiratory chain and growth (Pathania et al. 2002). In microaerobic circumstances, the haemoglobin concentration of bacteria carrying vgb increases around 10-fold. Positive transcriptional done by the Arc and Fnr systems in vgb exerts control at the transcriptional regulation *E. coli* with bearing vgb. Vgb was extracted from a Vitreoscilla genome library and cloned into the bacterial strain *E. coli* vector pUC8 to form pUC8:16. Then, to create pSC160, pUC8:16 was placed into a wide range of host vector pKT230. PSC160 has proven to be effective in the transformation of bioremediation bacteria. The former could be related to an increase in oxygen utilisation efficiency produced by VHb, although those could be due to enhanced VHb-induced respiration. When compared to VHb-free *E. coli*, 9% of these genetic material in recombinant *E. coli* intended to create VHb were substantially up- or down-regulated, according to the most extensive of these analyses, which used microarray technology (Janssen and Stucki 2020).

APPLICATIONS

E. coli is known for its various forms and biodegradability activity. It is well known for producing carbon and hydrogen by utilising organic compounds in the environment. Recombinant process in *E. coli* is ideal for research in a wide range of domains, including protein biochemistry, medicine, agriculture and the environment. This is due to the ease of use, economic viability, rapid high-density growth, well-characterized genetics, and huge number of suitable expression systems accessible. Similarly, there are various applications where recombinant *E. coli* are the most useful strains in degradation, as shown in Table 7.1 (Dighton et al. 1997).

Removal of Petroleum Hydrocarbons from Environment

The removal of these contaminants has become a major issue in recent years. Safe disposal, incineration and burial in secure landfills are a few currently acceptable options. These procedures can become prohibitively expensive when the number of pollutants is considerable. Some methods such as landfills, incinerators, pyrolysis, as well as gasification, are effective, but contaminants produce byproducts that have negative environmental and public health consequences after

TABLE 7.1

Various Applications of Using Recombinant *E. coli*

Bacteria	Genetically modified organism	Purpose	Reference
E. coli	*E. coli* JM10	Decolorisation of CI Direct Blue 71	Liu et al. 2009
E. coli SE5000 strain	Express nix A gene	Nickel from water bodies	Farnham and Dube 2015
E. coli	pL-Ds Red–pL-OPH and Pds plasmid insertion	organo-phosphorous pesticides	L. Li et al. 2014
E. coli	atrazine chlorohydrolase gene obtained from *Pseudomonas* sp. strain ADP	atrazine pesticide removal	Popova, Matafonova and Batoev 2019
E. coli	recombinant alkaline catalase in *E. coli*	Removal of H_2O_2	Song et al. 2016
E. coli BL21 (DE3)	gene cluster (phtBAabcdCR) isolated from *Gordonia* sp. and cloned with *E. coli*	catabolism of phthalate acid	D. Li et al. 2016
E. coli	Gene (cphC-I and cphB) obtained from *Arthrobacter chlorophenolicus*, encoding the flavin monooxygenase complex were cloned in *E. coli*	4-chlorophenol degradation	Kang et al. 2017
E. coli	*Escherichia coli* (*E. coli*) Nissle (EcN)	Probiotic	Scaldaferri et al. 2016
E. coli strain BL21(DE3).	Halohydrin dehalogenase (HheC)51 and epoxide hydrolase (EchA) 60 obtained from that organism, just like the plasmid-based route studied in *E. coli*	toxic persistent pollutant,	Kurumbang et al. 2014
E. coli	*E. coli* used as host and expression of *Comamonas* sp. strain CNB-1 genes	4-chloronitrobenzene and nitrobenzene	Lacombe et al. 2010
E. coli	MerR protein from *Shigella flexneri* cloning and expressed in *E. coli*	mercury detoxification	Qin et al. 2006
E. coli	Recombinant *E. coli* via PCR technique with *Staphylococcus aureus*	arsenate detoxification	Crameri et al. 1997

treatment. Thermal and chemical methods commonly extracted hydrocarbons from polluted sites have significant limitations and can be costly, and secondary pollutants can cause recontamination. The most often utilised technologies for soil remediation are mechanical, burying, evaporation, dispersion and washing. However, these methods are costly and can result in partial pollutant breakdown. The biological remediation procedure is based on the use of petroleum. The biological rejuvenation approach works on the idea of using petroleum hydrocarbons as the primary carbon and energy source. Organic compounds such as paraffin, grease, creosote, and polycyclic aromatic hydrocarbons are cultured and genetically modified microbes (PAHs). Acclimatization of organisms occurs after continuous exposure to the polluting substance, with the pollutant gradually degrading as a result. Plants contain biological processes that may be involved in heavy metal and petroleum hydrocarbon detoxification. The process is known as phytoremediation. As a result, we can use the recombinant *E. coli* to clean up contaminations. The biodegradation of petroleum and other hydrocarbon pollutants by natural populations of microorganisms is one of the most important methods for removing these pollutants from the ecosystem (Ambaye et al. 2022; Salari et al. 2022).

REMOVAL OF HYDROGEN PEROXIDE (H_2O_2)

In emerging countries, the textile industry is traditionally a significant sector. Despite accounting for a significant share of total industrial production value, pollution and usage of water (100 L/kg cloth) have severely hampered its expansion. In the textile industry, H_2O_2 is commonly used to accomplish the bleaching process in an alkaline environment. The key impediment to be overcome is residual H_2O_2 in both the water purification and dying process. Biodegradation of H_2O_2 by the process of alkaline catalase is a viable, cost-effective method and environmentally benign technology in the textile industry in recent times (Cuerda-Correa et al. 2020). Catalase (CAT) is an antioxidant enzyme containing heme-containing and homotetrameric that catalyzes the conversion of H_2O_2 to water and oxygen. The process is generally used in textile and paper industries to degrade and remove leftovers of H_2O_2 following peroxide bleaching due to its great efficiency (Hubmann, Jammer and Monschein 2022). Numerous alkaline catalases have been successfully produced and described from biological populations. Despite this, issues on an industrial scale like expensive production costs, low productivity and complex fermentation bioprocess still hinder catalase manufacture. The high alkali resilience of alkaline catalases was successfully expressed in bacterial strain *Bacillus* sp. cultures, suggesting that they could be useful in the textile sector. The desired genes of *Bacillus* sp. cultures were successfully cloned and expressed in *E. coli* with extracellular enzyme production. Zhenxiao Yu et al. in their work refined the induction technique and paired it with ethanol to obtain high extracellular production of enzymes in recombinant *E. coli*. The total activity of catalase was 78,462 U/ml, giving the highest value with an extracellular ratio of 92.5% (Yu et al. 2016). The utilisation of Kat A for H_2O_2 removal was evaluated due to its high stability at a temperature of 50 °C and alkaline conditions (pH 8–10). The bio-catalysis process employing Kat A has a greater rate of removal, is less expensive, and uses less water than the old process (Ku et al. 2020).

DYES

Dyes and pigments serve a crucial function in various industries, including paper, pulp, textile and others, owing to their cost-effectiveness and remarkable stability, which significantly enhance the appearance properties of the products. The dyes, on the other hand, can reduce gas solubility in the aquatic habitat, limiting the growth of aquatic plants and wildlife. When these contaminants penetrate directly or indirectly into humans, they cause acute, mild and chronic diseases. Furthermore, injuries to essential organs of the digestive system, central nervous system and reproductive system are prevalent. Furthermore, certain colors are naturally carcinogenic. Dyes are now considered significant environmental contaminants worldwide due to their excessive emission in industrial effluent and their harmful impacts on the environment and health system. Physical and biochemical (oxidation, reduction, complex metric techniques, neutralization and so on) approaches are used. Mohanad et al. in their work used *E. coli* and *Bacillus* sp. bacterial strains for the biodegradation of dyes RB49 and RR195 from aqueous solutions (M-Ridha et al. 2020). The RSM model based on optimal design was used to extract experimental variables. The desirability function was used to test the second order equation for p-values, adequacy, F-value and future replication. The best conditions for removal of these dyes were 99.6 mg/L at a temperature of 39.9° C at an initial concentration of 14.9 $V_{Biomass}/V_{solution}$ for 1 day incubation; pH should be around 6–7. Endogenous bacteria have proved their ability to bioremediate a number of dyes. GMOs have various applications opening up several new ways for the creation of highly sensitive methods (Ge et al. 2006).

XENOBIOTICS

Xenobiotics are artificial and manmade compounds that, due to their complicated organic structure, can stay in the ecosystem for decades. They are very well recognized as the principal class of pollutants involved with chemically produced toxicity in humans; additionally, they induce direct threats to the ecosystem within a short exposure. The use of high-specificity, high-efficiency microbial techniques for xenobiotic treatment has expanded the field of bioremediation beyond traditional ways. However, the widespread application led to the release of these chemical compounds in the ecosystem, which contaminates soil and water bodies, resulting in toxicity to biota and humans. These xenobiotics are easily exposed to humans by inhalation, absorption through the skin or swallowing. As a result, various health dangers have emerged, including adverse reproductive effects, neurodegeneration, cardiac issues and central nervous system problems. Therefore, removing these chemicals from the environment is a huge challenge. To eliminate the contaminants, numerous physical and chemical procedures have been used, including burning, solvent extraction and UV irradiation, but these have several downsides, including incomplete combustion, the introduction of additional chemicals, high cost and complexity (Wang et al. 2022). To rejuvenate the environment, eco-friendly techniques such as bioremediation have been established. These approaches are reliable and cheap, and they entail using microbes to eliminate these substances by using them as a carbon source or bio-transforming them into benign, more water-soluble molecules. For pollutant degradation, many bioremediation strategies have been used, including biostimulation, natural attenuation, bioaugmentation and composting. Microbial degradation is classified as the utilisation of microorganisms to break down molecules such as xenobiotics into simple molecules. The bacteria use them and their metabolites as their only supply of carbon or nitrogen, thus eliminating them. In the present, genetic engineering techniques for creating microorganisms with improved degradative capabilities have paved the path for the environmental detoxification of such persistent contaminants. *In-vitro* and *in-vivo* techniques, including gene cloning and strain-to-strain transfer of a complete plasmid, can be employed to create such customized bacteria. Microbes use a variety of methods, including oxidation, reduction and dehalogenation to bioremediate organic xenobiotics from contaminated sites. One of the most extensively used triazine-based herbicides is atrazine (2-chloro-4-ethylamino-6-isopropylamino-1,3,5-triazine). In mammals, acute poisoning causes skin irritation, disorientation and nausea. It has been linked to teratogenic and endocrine disruptive effects, as well as sex reversal in frogs. Numerous bacterial species have been found to convert atrazine to ammonia and carbon dioxide in various ways. One of the methods to use recombinant DNA in Atrazine bioremediation was carried out using *E. coli* cells carrying the atrazine chlorohydrolase gene obtained from *Pseudomonas* sp. strain ADP. When these cells were put in the contaminated location, the concentration of atrazine decreased, indicating that the remediation was efficient. However, there are also significant environmental and ecological problems with using GMMs in the field (Agarwal et al. 2022; Hussain et al. 2018; Nagata 2020).

TOXIC METALS

Within the past few decades, concentrations of anthropogenic pollutants have increased to a larger extent. The availability of heavy metals in marine biota is a major concern both in terms of prediction of effects of metals on aquatic life and possible human health risks. For inorganic pollutants like heavy metals (Cr, MN, Fe, Hg, Cd, Pb) and other metalloids present in water, the main challenge is in assessing the environmental risks and health hazards. These elements are not prone to degradation like other organic pollutants. The effectiveness and bioavailability of inorganic salts is easily traced by some reactions like reduction/oxidation, complexation, adsorption and precipitation reactions. In the presence of oxygen the most redox-sensitive metals form oxide compounds, which strongly trapped heavy metals and metalloids. Under depletion of oxygen, these compounds undergo reduction and released toxic elements like iron and manganese are generally formed in the water system which are devoid of oxygen and cause color problems in water if present in concentrations greater than 0.3 mg/l. Sometimes they act as a source of energy for certain types of microorganisms, and this makes it a challenging task to trace the exact effects of these metals. Advancements in spectroscopy technology give a new direction for identifying the presence of heavy metals and in tracing microbial activities. The large classes of different mineral phases and their possible interaction between solutes, which are related to the adsorption process, make the evaluation of metal pollution difficult for the environmentalist; for example, Sodium in water makes the metal surface corrosive and proves to be toxic for plants as well as for the kidney and lungs of humans. Nitrates in excess affect infants and result in diseases such as Methemoglobinemia or Blue baby disease. The key disadvantage of this area is not having significant information on predicting the level of effects of inorganic pollutants on human health and the ecosystem (Kumar Gupta 2014; Liu et al. 2022).

Bacterial Remediation Capacity of Heavy Metal

Microorganisms intake heavy metals either through bioaccumulation or adsorption. Several microorganisms such as fungi, bacteria and algae are being used to eliminate these heavy metals from the environment. The application of metal-resistant strains in consortium, single and immobilised forms for metal degradation has given successful results, while immobilised forms may have high adsorption rates to adsorb heavy metals (Huang et al. 2017).

RADIOACTIVE POLLUTANTS

Radioactive wastes are produced by a variety of nuclear industries, research organizations, hospitals, and other institutions, which aggregate and cause radioactivity exposure. The radioactive substance combined with heavy metals makes the process of bioremediation difficult as natural microorganisms are unable to remediate pollutants, providing an opportunity for GMOs. The biotransformation of radioactive contaminants into an insoluble form that precipitates is part of the bioremediation of radioactive substances. When compared to wild microorganisms, GMOs have high resistance and potential of survival in highly radioactive environments for bioremediation of radioactive substances. There are no publications that suggest wild-type microorganisms' ability to break down or transform radioactive elements into safer forms. The bacteria *Deinococcus radiodurans*, which is now being genetically engineered to ingest and metabolize organic compounds and ionic mercury from highly radioactive nuclear wastes and radioactive iodine, has been discovered to be known as the most radioresistant microorganism. As a result, using genetically modified bacteria to convert or destroy radioactive materials might be the safest and easiest option (Groudev, Bratcova and Komnitsas 1999; Macías et al. 2012).

ACHIEVEMENTS OF BIOREMEDIATION TECHNOLOGY

Bioremediation solutions are being developed by a number of research institutions and corporations. These businesses run on razor-thin profit margins, and just a few have a clear focus. Only very few research labs and firms are commercially utilising genetic engineering for bioremediation. Nonetheless, considerable progress in the molecular genetics of biotransformation systems in bioremediation continues to be made. In India, bioremediation of petroleum oil-contaminated sites is now conceivable. TERI'S "Oilzapper" was recently developed by the Tata Energy Research Institute (TERI, New Delhi). It is a bacterial system that degrades crude oil and has been designed to reclaim damaged lands. TERI employs microorganisms that have been isolated from the natural environment and possess the unique ability to consume hazardous substances commonly found in oil spill locations. These bacteria are grown and nurtured under controlled laboratory conditions.

The mixture is then combined with a specified carrier material and packaged in polybags for easy delivery to petroleum contamination areas. TERI has given these carrier-based microorganisms the name "Oilzapper". The Oilzapper was tested in a field at the Mathura refinery, and encouraging results were found. A similar operation has been carried out at the Barauni oil refinery. As commercial bioremediation agents for oil spill cleaning, only a few nutrient additives have been produced and are available on the market (Chitrakoot et al. 2014).

CONCLUSION

A number of branches including geology, ecology, engineering, microbiology and chemistry work together in the development of bioremediation processes. In the bioremediation of polluted material, microbes are the key stimulus. However, present knowledge of biological contributions to bioremediation and their impacts on the ecosystem and microbial community are still often considered a "Pandora's box". The molecular microbiological techniques using *E. coli* and its application in the removal of contaminants from the environment are outlined in this chapter. The recombinant techniques in *E. coli* enhance its ability to degrade harmful chemicals and will provide new insights into processes of optimization and assessing ecosystem impact, making bioremediation a more dependable and safer approach.

REFERENCES

Adetunji, Charles Oluwaseun, and Deepak Panpatte. 2021. "Microbial Rejuvenation of Polluted Environment." 3. doi: 10.1007/978-981-15-7459-7.

Agarwal, Neha, Vijendra Singh Solanki, Amel Gacem, Mohd Abul Hasan, Brijesh Pare, Amrita Srivastava, Anupama Singh, Virendra Kumar Yadav, Krishna Kumar Yadav, Chaigoo Lee, Wonjae Lee, Sumate Chaiprapat, and Byong-Hun Jeon. 2022. "Bacterial Laccases as Biocatalysts for the Remediation of Environmental Toxic Pollutants: A Green and Eco-Friendly Approach—A Review." *Water* 14 (24): 4068. doi:10.3390/w14244068.

Ambaye, Gebregiorgis Teklit, Alif Chebbi, Francesca Formicola, Shiv Prasad, Franco Hernan Gomez, Andrea Franzetti, and Mentore Vaccari. 2022. "Remediation of Soil Polluted with Petroleum Hydrocarbons and Its Reuse for Agriculture: Recent Progress, Challenges, and Perspectives." *Chemosphere* 293: 133572. doi:10.1016/j.chemosphere.2022.133572.

Chitrakoot, Gandhi, Gramoday Vishwavidyalaya, Pushpendra Singh, Rajesh Sharma, and Ravindra Singh. 2014. "Microorganism as a Tool of Bioremediation Technology for Cleaning Environment: A Review Biodeterioration on Khajuraho Monuments View Project Studies on Alpha Amylase Production by Free and Immobilized Cells View Project Ravindra Singh Microorganism as a Tool of Bioremediation Technology for Cleaning Environment: A Review." *Proceedings of the International Academy of Ecology and Environmental Sciences* 4. www.iaees.org.

Crameri, Andreas, Glenn Dawes, Emilio Rodriguez, Simon Silver, and Willem P. C. Stemmer. 1997. *Molecular Evolution of an Arsenate Detoxification Pathway by DNA Shuffling.* www.maxygen.com.

Cuerda-Correa, Eduardo Manuel, María F. Alexandre-Franco, and Carmen Fernández-González. 2020. "Advanced Oxidation Processes for the Removal of Antibiotics from Water. An Overview." *Water* 12 (1): 102. doi:10.3390/w12010102.

Dale, Adam, and Neil Woodford. 2015. "Extra-Intestinal Pathogenic *Escherichia Coli* (ExPEC): Disease, Carriage and Clones." *Journal of Infection* 71 (6): 615–626. doi:10.1016/j.jinf.2015.09.009.

Dangi, Arun Kumar, Babita Sharma, Russell T. Hill, and Pratyoosh Shukla. 2019. "Bioremediation Through Microbes: Systems Biology and Metabolic Engineering Approach." *Critical Reviews in Biotechnology.* Taylor and Francis Ltd. doi:10.1080/07388551.2018.1500997.

De Cárcer, Daniel Aguirre, Marta Martín, Martina MacKova, Thomas MacEk, Ulrich Karlson, and Rafael Rivilla. 2007. "The Introduction of Genetically Modified Microorganisms Designed for Rhizoremediation Induces Changes on Native Bacteria in the Rhizosphere but Not in the Surrounding Soil." *ISME Journal* 1 (3): 215–223. doi:10.1038/ismej.2007.27.

Díaz, Eduardo, Abel Ferrández, María A. Prieto, and José L. García. 2001. "Biodegradation of Aromatic Compounds by Escherichia Coli." *Microbiology and Molecular Biology Reviews* 65 (4). American Society for Microbiology: 523–569. doi:10.1128/mmbr.65.4.523-569.2001.

Dighton, John, Helen E. Jones, Clare H. Robinson, and John Beckett. 1997. "The Role of Abiotic Factors, Cultivation Practices and Soil Fauna in the Dispersal of Genetically Modified Microorganisms in Soils." *Applied Soil Ecology* 5 (2): 109–131. doi: 10.1016/S0929-1393(96)00137-0.

Dris, Rachid, Hannes Imhof, Wilfried Sanchez, Johnny Gasperi, François Galgani, Bruno Tassin, and Christian Laforsch. 2015. "Beyond the Ocean: Contamination of Freshwater Ecosystems with (micro-)plastic Particles." *Environmental Chemistry* 12 (5): 539–550. doi:10.1071/EN14172.

EFSA Journal. 2011. "Guidance on the Risk Assessment of Genetically Modified Microorganisms and Their Products Intended for Food and Feed Use" 9 (6). Wiley-Blackwell Publishing Ltd. doi:10.2903/j. efsa.2011.2193.

Farnham, Kate R., and Danielle H. Dube. 2015. "A Semester-Long Project-Oriented Biochemistry Laboratory Based on Helicobacter Pylori Urease." *Biochemistry and Molecular Biology Education* 43 (5): 333–340. John Wiley and Sons Inc. doi:10.1002/bmb.20884.

Fraiture, Marie Alice, Marie Deckers, Nina Papazova, and Nancy H. C. Roosens. 2020. "Are Antimicrobial Resistance Genes Key Targets to Detect Genetically Modified Microorganisms in Fermentation Products?" *International Journal of Food Microbiology* 331 (October). Elsevier B.V. doi:10.1016/j. ijfoodmicro.2020.108749.

Ge, Baosheng, Song Qin, Lu Han, Fan Lin, and Yuhong Ren. 2006. "Antioxidant Properties of Recombinant Allophycocyanin Expressed in Escherichia Coli." *Journal of Photochemistry and Photobiology B: Biology* 84 (3): 175–180. doi:10.1016/j.jphotobiol.2006.02.008.

Gray, C. H., and E. L. Tatum. 1944. "X-Ray Induced Growth Factor Requirements in Bacteria." *Proceedings of the National Academy of Sciences* 30 (12): 404–410. doi:10.1073/pnas.30.12.4044.

Groudev, S. N., S. G. Bratcova, and K. Komnitsas. 1999. "Treatment of Waters Polluted with Radioactive Elements and Heavy Metals by Means of a Laboratory Passive System." *Minerals Engineering* 12 (3): 261–270. doi:10.1016/S0892-6875(99)00004-7.

Gupta, Mahendra Kumar, Kiran Kumari, Amita Shrivastava, and Shikha Gauri. 2014. "Bioremediation of Heavy Metals Polluted Environment by Using Resistance Bacteria." *Journal of Environmental Research and Development* 8 (4): 883–889. www.researchgate.net/publication/264004661.

Hubmann, Anna M., Alexandra Jammer, and Stephan Monschein. 2022. "Activities of H_2O_2-Converting Enzymes in Apple Leaf Buds During Dormancy Release in Consideration of the Ratio Change Between Bud Scales and Physiologically Active Tissues." *Horticulturae* 8 (11): 982. doi:10.3390/horticulturae8110982.

Huang, Xianhui, Linfeng Yu, Xiaojie Chen, Chanping Zhi, Xu Yao, Yiyun Liu, Shengjun Wu et al. 2017. "High Prevalence of Colistin Resistance and Mcr-1 Gene in Escherichia Coli Isolated from Food Animals in China." *Frontiers in Microbiology* 8 (APR). Frontiers Research Foundation. doi:10.3389/ fmicb.2017.00562.

Hussain, Imran, Gajender Aleti, Ravi Naidu, Markus Puschenreiter, Qaisar Mahmood, Mohammad Mahmudur Rahman, Fang Wang, Shahida Shaheen, Jabir Hussain Syed, and Thomas G. Reichenauer. 2018. "Microbe and Plant Assisted-Remediation of Organic Xenobiotics and Its Enhancement by Genetically Modified Organisms and Recombinant Technology: A Review." *Science of the Total Environment*. Elsevier B.V. doi:10.1016/j.scitotenv.2018.02.037.

Jain, Pankaj Kumar, and Vivek Bajpai. "Biotechnology of Bioremediation-A Review." doi:10.6088/ ijes.20120301310533.

Janssen, Dick B., and Gerhard Stucki. 2020. "Perspectives of Genetically Engineered Microbes for Groundwater Bioremediation." *Environmental Science: Processes and Impacts*. Royal Society of Chemistry. doi:10.1039/c9em00601j.

Kang, Christina, Jun Won Yang, Wooyoun Cho, Seonyeong Kwak, Sungyoon Park, Yejee Lim, Jae Wan Choe, and Han S. Kim. 2017. "Oxidative Biodegradation of 4-Chlorophenol by Using Recombinant Monooxygenase Cloned and Overexpressed from Arthrobacter Chlorophenolicus A6." *Bioresource Technology* 240. Elsevier Ltd: 123–129. doi:10.1016/j.biortech.2017.03.078.

Köhler, Christian-Daniel, and Ulrich Dobrindt. 2011. "What Defines Extraintestinal Pathogenic Escherichia Coli?" *International Journal of Medical Microbiology* 301 (8): 642–647. doi:10.1016/j.ijmm.2011.09.006.

Ku, Seockmo, Suyoung Yang, Hyun Ha Lee, Deokyeong Choe, Tony V. Johnston, Geun Eog Ji, and Myeong Soo Park. 2020. "Biosafety Assessment of Bifidobacterium Animalis Subsp. Lactis AD011 Used for Human Consumption as a Probiotic Microorganism." *Food Control* 117 (November). Elsevier Ltd. doi:10.1016/j.foodcont.2019.106985.

Kurumbang, Nagendra Prasad, Pavel Dvorak, Jaroslav Bendl, Jan Brezovsky, Zbynek Prokop, and Jiri Damborsky. 2014. "Computer-Assisted Engineering of the Synthetic Pathway for Biodegradation of a Toxic Persistent Pollutant." *ACS Synthetic Biology* 3 (3). American Chemical Society: 172–181. doi:10.1021/sb400147n.

Lacombe, Alison, Vivian C. H. Wu, Seth Tyler, and Kelly Edwards. 2010. "Antimicrobial Action of the American Cranberry Constituents; Phenolics, Anthocyanins, and Organic Acids, Against Escherichia Coli O157:H7." *International Journal of Food Microbiology* 139 (1–2): 102–107. doi:10.1016/j. ijfoodmicro.2010.01.035.

Landrigan, Philip, John J. Stegeman, Lora E. Fleming, Denis Allemand, Donald M. Anderson, Lorraine C. Backer, Françoise Brucker-Davis, Nicolas Chevalier, Lilian Corra, Dorota Czerucka, Marie-Yasmine Dechraoui Bottein, Barbara Demeneix, Michael Depledge, Dimitri D. Deheyn, Charles J. Dorman, Patrick Fénichel, Samantha Fisher, Françoise Gaill, François Galgani, William H. Gaze, Laura Giuliano, Philippe Grandjean, Mark E. Hahn, Amro Hamdoun, Philipp Hess, Bret Judson, Amalia Laborde, Jacqueline McGlade, Jenna Mu, Adetoun Mustapha, Maria Neira, Rachel T. Noble, Maria Luiza Pedrotti, Christopher Reddy, Joacim Rocklöv, Ursula M. Scharler, Hariharan Shanmugam, Gabriella Taghian, Jeroen A. J. M. van de Water, Luigi Vezzulli, Pál Weihe, Ariana Zeka, Hervé Raps, and Patrick Rampal. 2020. "Human Health and Ocean Pollution." *Annual Global Health* 86 (1): 151. doi:10.5334/ aogh.2831.

Li, Dandan, Jiali Yan, Li Wang, Yunze Zhang, Deli Liu, Hui Geng, and Li Xiong. 2016. "Characterization of the Phthalate Acid Catabolic Gene Cluster in Phthalate Acid Esters Transforming Bacterium-Gordonia Sp. Strain HS-NH1." *International Biodeterioration and Biodegradation* 106 (January). Elsevier Ltd: 34–40. doi:10.1016/j.ibiod.2015.09.019.

Li, Lin, Binting Wang, Shuai Feng, Jinnian Li, Congming Wu, Ying Wang, Xiangchun Ruan, and Minghua Zeng. 2014. "Prevalence and Characteristics of Extended-Spectrum β-Lactamase and Plasmid-Mediated Fluoroquinolone Resistance Genes in Escherichia Coli Isolated from Chickens in Anhui Province, China." *PLoS One* 9 (8). doi:10.1371/journal.pone.0104356.

Liu, Guangfei, Jiti Zhou, Jing Wang, Mi Zhou, Hong Lu, and Ruofei Jin. 2009. "Acceleration of Azo Dye Decolorization by Using Quinone Reductase Activity of Azoreductase and Quinone Redox Mediator." *Bioresource Technology* 100 (11): 2791–2795. doi:10.1016/j.biortech.2008.12.040.

Liu, Yan, Xiaoliang Wang, Sujin Nong, Zehui Bai, Nanyu Han, Qian Wu, Zunxi Huang, and Junmei Ding. 2022. "Display of a Novel Carboxylesterase CarCby on Escherichia Coli Cell Surface for Carbaryl Pesticide Bioremediation." *Microbial. Cell Factory* 21 (97): 1–14. doi:10.1186/s12934-022-01821-5.

M-Ridha, Mohanad J., Sahar I. Hussein, Ziad T. Alismaeel, Mohammed A. Atiya, and Ghazi M. Aziz. 2020. "Biodegradation of Reactive Dyes by Some Bacteria Using Response Surface Methodology as an Optimization Technique." *Alexandria Engineering Journal* 59 (5). Elsevier B.V.: 3551–3563. doi:10.1016/j.aej.2020.06.001.

Macías, Francisco, Manuel A. Caraballo, Tobias S. Rötting, Rafael Pérez-López, José Miguel Nieto, and Carlos Ayora. 2012. "From Highly Polluted Zn-Rich Acid Mine Drainage to Non-Metallic Waters: Implementation of a Multi-Step Alkaline Passive Treatment System to Remediate Metal Pollution." *Science of the Total Environment* 433 (September): 323–330. doi:10.1016/j.scitotenv.2012.06.084.

Medfu Tarekegn, Molalign, Fikirte Zewdu Salilih, and Alemitu Iniyehu Ishetu. 2020. "Microbes Used as a Tool for Bioremediation of Heavy Metal from the Environment." *Cogent Food and Agriculture*. Informa Healthcare. doi:10.1080/23311932.2020.1783174.

Mitra, Anindita, Soumya Chatterjee, Sampriti Kataki, Rajesh P. Rastogi, and Dharmendra K. Gupta. 2020. "Bacterial Tolerance Strategies Against Lead Toxicity and Their Relevance in Bioremediation Application." *Environment Science and Pollution Research International* 28 (12): 14271–14284. doi:10.1007/s11356-021-12583-9.

Nagata, Yuji. 2020. "Special Issue: Microbial Degradation of Xenobiotics." *Microorganisms*. MDPI AG. doi:10.3390/microorganisms8040487.

Ortega-González, Diana Katherine, Diego Zaragoza, José Aguirre-Garrido, Hugo Ramírez-Saad, César Hernández-Rodríguez, and Janet Jan-Roblero. 2013. "Degradation of Benzene, Toluene, and Xylene Isomers by a Bacterial Consortium Obtained from Rhizosphere Soil of Cyperus Sp. Grown in a Petroleum-Contaminated Area." *Folia Microbiolica (Praha)* 58 (6): 569–577. doi:10.1007/s12223-013-0248-4.

Pant, Gaurav, Deviram Garlapati, Urvashi Agrawal, R. Gyana Prasuna, Thangavel Mathimani, and Arivalagan Pugazhendhi. 2021. "Biological Approaches Practised Using Genetically Engineered Microbes for a Sustainable Environment: A Review." *Journal of Hazardous Materials*. Elsevier B.V. doi:10.1016/j. jhazmat.2020.124631.

Pathania, Ranjana, Naveen K. Navani, Anne M. Gardner, Paul R. Gardner, and Kanak L. Dikshit. 2002. "Nitric Oxide Scavenging and Detoxification by the Mycobacterium Tuberculosis Haemoglobin, HbN in Escherichia Coli." *Molecular Microbiology* 45 (5): 1303–1314. doi:10.1046/j.1365-2958.2002. 03095.x.

Perna, N. T., G. Plunkett 3rd, V. Burland, B. Mau, J. D. Glasner, D. J. Rose, G. F. Mayhew, P. S. Evans, J. Gregor, H. A. Kirkpatrick, G. Pósfai, J. Hackett, S. Klink, A. Boutin, Y. Shao, L. Miller, E. J. Grotbeck, N. W. Davis, A. Lim, E. T. Dimalanta, K. D. Potamousis, J. Apodaca, T. S. Anantharaman, J. Lin, G. Yen, D. C. Schwartz, R. A. Welch, and F. R. Blattner. 2001. "Genome Sequence of Enterohaemorrhagic Escherichia Coli O157:H7." *Nature* 409 (6819): 529–533. doi:10.1038/35054089.

Popova, Svetlana, Galina Matafonova, and Valeriy Batoev. 2019. "Simultaneous Atrazine Degradation and E. Coli Inactivation by UV/S2O82-/Fe2+ Process Under KrCl Excilamp (222 nm) Irradiation." *Ecotoxicology and Environmental Safety* 169 (March). Academic Press: 169–177. doi:10.1016/j.ecoenv.2018.11.014.

Qin, Jie, Lingyun Song, Hassan Brim, Michael J. Daly, and Anne O. Summers. 2006. "Hg(II) Sequestration and Protection by the MerR Metal-Binding Domain (MBD)." *Microbiology* 152 (3): 709–719. doi:10.1099/mic.0.28474-0.

Rodríguez-Lázaro, David, Bertrand Lombard, Huw Smith, Artur Rzezutka, Martin D'Agostino, Reiner Helmuth, Andreas Schroeter et al. 2007. "Trends in Analytical Methodology in Food Safety and Quality: Monitoring Microorganisms and Genetically Modified Organisms." *Trends in Food Science and Technology* 18 (6): 306–319. doi:10.1016/j.tifs.2007.01.009.

Riley, M., and M. H. Serres Interim. 2000. "Report on Genomics of Escherichia Coli." *Annual Review of Microbiology* 54: 341–411. doi:10.1146/annurev.micro.54.1.341.

Salari, Marjan, Vahid Rahmanian, Seyyed Alireza Hashemi, Wei-Hung Chiang, Chin Wei Lai, Seyyed Mojtaba Mousavi, and Ahmad Gholami. 2022. "Bioremediation Treatment of Polyaromatic Hydrocarbons for Environmental Sustainability." *Water* 14 (23): 3980. doi:10.3390/w14233980.

Scaldaferri, Franco, Viviana Gerardi, Francesca Mangiola, Loris Riccardo Lopetuso, Marco Pizzoferrato, Valentina Petito, Alfredo Papa et al. 2016. "Role and Mechanisms of Action of Escherichia Coli Nissle 1917 in the Maintenance of Remission in Ulcerative Colitis Patients: An Update." *World Journal of Gastroenterology*. Baishideng Publishing Group Co. doi:10.3748/wjg.v22.i24.5505.

Selinger, D. W., K. J. Cheung, R. Mei, E. M. Johansson, C. S. Richmond, F. R. Blattner, D. J. Lockhart, and G. M. Church. 2000. "RNA Expression Analysis Using a 30 Base Pair Resolution Escherichia Coli Genome Array." *Nature Biotechnolgy* 18 (12): 1262–1268.

Song, Chengjie, Liping Wang, Jie Ren, Bo Lv, Zhonghao Sun, Jing Yan, Xinying Li, and Jingjing Liu. 2016. "Comparative Study of Diethyl Phthalate Degradation by UV/H2O2 and UV/TiO2: Kinetics, Mechanism, and Effects of Operational Parameters." *Environmental Science and Pollution Research* 23 (3). Springer Verlag: 2640–2650. doi:10.1007/s11356-015-5481-8.

Sonter, Shruti, Shringika Mishra, Manish Kumar Dwivedi, and Prashant Kumar Singh. 2021. "Chemical Profiling, in Vitro Antioxidant, Membrane Stabilizing and Antimicrobial Properties of Wild Growing Murraya Paniculata from Amarkantak (M.P.)." *Scientific Reports* 11 (1). Nature Research. doi:10.1038/s41598-021-87404-7.

Urgun-Demirtas, Meltem, Benjamin Stark, and Krishna Pagilla. 2006. "Use of Genetically Engineered Microorganisms (GEMs) for the Bioremediation of Contaminants." *Critical Reviews in Biotechnology*. doi:10.1080/07388550600842794.

Wang, Xingwen, Muhammad Umair Sial, Muhammad Amjad Bashir, Muhammad Bilal, Qurat-Ul-Ain Raza, Hafiz Muhammad Ali Raza, Abdur Rehim, and Yucong Geng. 2022. "Pesticides Xenobiotics in Soil Ecosystem and Their Remediation Approaches Pesticides Xenobiotics in Soil Ecosystem and Their Remediation Approaches." *Sustainability* 14 (6): 3353. doi:10.3390/su14063353.

Yu Z, H. Zheng, X. Zhao, S. Li, J. Xu, and H. Song. (2016). "High Level Extracellular Production of a Recombinant Alkaline Catalase in E. Coli BL21 Under Ethanol Stress and Its Application in Hydrogen Peroxide Removal After Cotton Fabrics Bleaching." *Bioresources Technology* 214: 303–310. doi:10.1016/j.biortech.2016.04.110.

8 Molecular Cloning for the Production of Genetically Modified Organisms for Bioremediation

Dhaval Patel, Jyoti Solanki, Aharon Azagury and R.B. Subramanian

INTRODUCTION

'Molecular gene cloning' is not a new word in science today; it is an evergreen area of research in the field of molecular biology. It deals with the manipulation of gene sequences thereby altering the characteristic of an organism. It is considered the pioneer of molecular cloning techniques. Molecular cloning is not only an approach for gene manipulation, but it also opens the door to understanding the functions of a number of genes. The root of this technique was found in November 1973 for the first time in the research article published by A.C.Y. Chang, H.W. Boyer, R.B. Helling and S.N. Cohen under the title "Construction of Biologically Functional Bacterial Plasmids *In Vitro*" (Cohen et al., 1973). Cloning experiments have become achievable after the discovery of the restriction endonuclease enzyme that cuts the DNA precisely at the desired sequence. The cloning technique is not only restricted to transfer the gene from prokaryotic cell to prokaryotic cell, but also to eukaryotic cell and vice versa. This technique has given tremendous hope not only to the biological sciences but also to the chemical sciences along with considerable controversial opinions regarding public and environmental health concerns due to apparent laboratory hazards and ethical issues (Cohen, 2013). However, due to the specificity, sensitivity and effectiveness of this technique, it has become popular and is considered as one of the essential techniques in standard laboratory practice.

The molecular cloning technique has the capacity to change the genetic makeup and thereby phenotypic character of an organism. In other words, by transferring a desired fragment of the DNA, one can add or delete a particular characteristic in the host cell. This phenomenon of making genetically engineered organisms is exceptionally helpful in environmental cleanup (Lovely, 2003).

Anthropological activities have caused detrimental effects on the environment in several ways, which are of foremost concern across the globe. Environmental pollution cleanup using a biological entity is known as "bioremediation". It is a relatively safe and eco-friendly solution to problems related to environmental pollution. Bioremediation is a biotechnological process of employing bacteria, whole plant or plant enzymes for the removal of contaminants, toxins and pollutants from soil, water and any other media. Among them, bacteria are the most utilized for bioremediation because of their smaller generation time and diverse metabolic activity (Widada et al., 2002). A range of approaches are utilized in bioremediation; for instance, the optimization of enzymes and metabolic pathways in an organism (Pieper and Reineke, 2000). The use of metabolic latency for the removal of environmental pollution is the safest and most economical alternative (Pieper and Reineke, 2000). The success of bioremediation processes depends on a combination of biological processes along with physical and chemical treatments (Singh et al., 2020). Bioremediation processes have several

DOI: 10.1201/9781003188568-8

limitations; for instance, microorganisms have restricted capability of degrading high molecular weight compounds like polyhydroxyalkanoate (PHA) or the degradation process might generate toxic intermediates during microbial action such as vinyl chloride from dehalogenation of trichloroethylene (Megharaj et al., 2014). Therefore, to make the bioremediation process more effective, genetically engineered bacteria can be used.

The present chapter focuses on the molecular techniques used to make engineered bacteria for bioremediation purposes. The chapter includes the basics of the cloning technique, the sourcing of genes and the screening of the recombinants, as well as recent advances and ethics related to the use of genetically engineered organisms for bioremediation processes.

BASIC MOLECULAR TECHNIQUE

The molecular cloning technique includes the following steps:

i. Isolation of desired DNA fragment or gene of interest called the "insert".
ii. Construction of a recombinant molecule by ligation of the insert with cloning or expression vector.
iii. Insertion or transformation of the recombinant vector into a suitable host (now a genetically engineered organism).
iv. Screening and selection of the recombinant host.

The journey of molecular cloning starts with the search for a desired gene (or gene of interest) and an appropriate cloning vector, which is followed by construction of recombinant molecules and a search for a suitable host for the propagation of the recombinant construct.

SOURCE OF "INSERT"

An insert is a short DNA sequence or gene (i.e., gene of interest) that is responsible for a particular phenotype or trait. The screening and selection of such a gene are central to molecular cloning. More than 300 inserts have been reported and cloned in cultivable bacteria to date (Widada et al., 2002). The best source for a gene that encourages bioremediation is the contaminated/polluted soil, water or environmental site itself because the condition at that site generates a selective pressure on microbiota to change their metabolic activities (Widada et al., 2002) and enforces the bacteria to degrade complex molecules (xenobiotics) for its own survival. Figure 8.1 indicates the possible flow diagram for the selection of a suitable gene from polluted soil. Bacteria are mainly involved in mobilization, transformation and detoxification processes during bioremediation (Malik et al., 2021). Bacteria possess vast arrays of genetic makeup in carrying out such complex biological processes, and therefore, it becomes difficult to focus on a specific gene for cloning. Bacteria have the most valuable and diverse genes and/or operons, but the search for specific genes is a monotonous job for cloning. Molecular biology techniques help in search of such genes of interest and/ or microbes in two ways. The first is the *culture-dependent method* in which cultivable bacteria are isolated from environmental samples. The second is the *culture-independent method* in which extraction of nucleic acid is done directly from an environmental sample, followed by amplification of DNA or cDNA by polymerase chain reactions (PCR) (Widada et al., 2002) (Figure 8.1). Some of these techniques are as follows:

1. *DNA-based methods* for the detection of a gene from environmental samples. In this technique, a labeled DNA probe is used for hybridization to identify a specific genome in the environmental sample. The major advantage of this technique is that it can monitor a specific genome in the environment in cultivable and/or non-cultivable bacteria and can be detected (colony hybridization technique) with higher sensitivity (with slot blot and

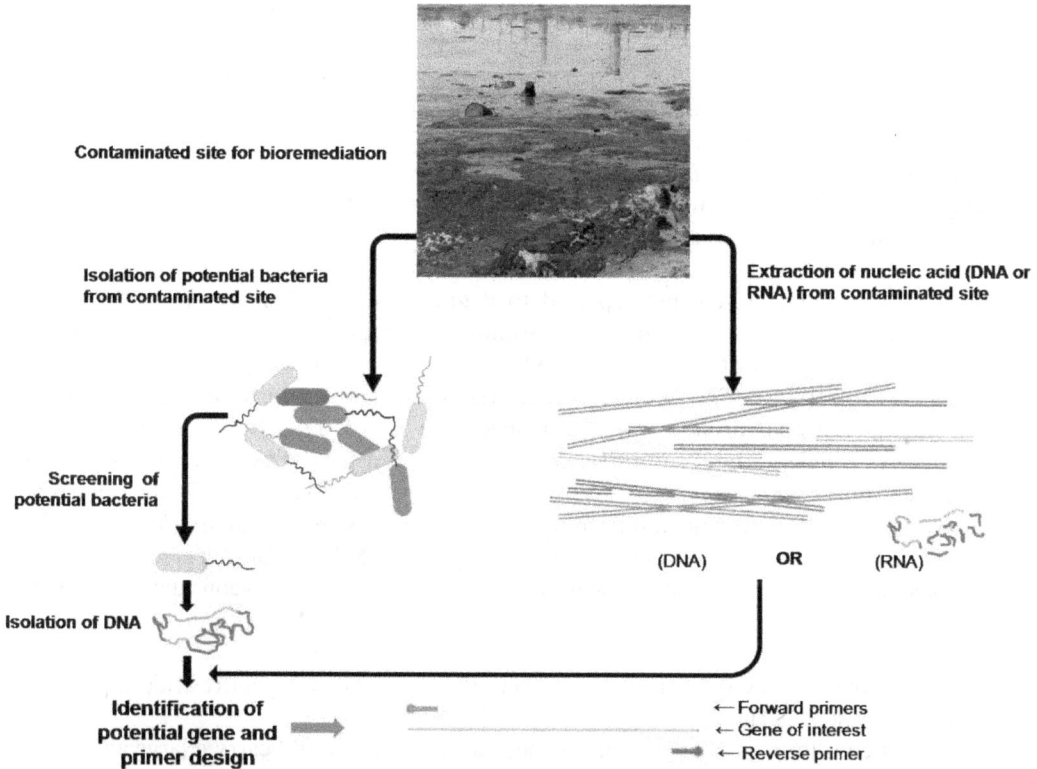

FIGURE 8.1 Isolation of desired DNA fragments or insert from contaminated sample either by isolation and purification of bacteria or by extracting nucleic acid (DNA or RNA) from environmental sample (the metagenomic approach).

southern blot) (Widada et al., 2002; Atlas, 1992; Sayler and Layton, 1990). One major drawback of the technique is that it does not differentiate between living and dead organisms.

2. *RNA-based methods*, which provide important information on gene expression as well as the viability of live cells under the given environmental condition. Therefore, RNA-based techniques are more reliable than DNA-based methods. On the top, isolated RNA (more specifically, mRNA) from an organism show direct relation between the environmental condition and *in-situ* metabolic activity of the organism of interest (Widada et al., 2002; Wilson et al., 1999). Briefly, the technique includes extraction of mRNA from live bacteria from polluted environments, followed by a reverse-transcription polymerase chain reaction (RT-PCR) to generate cDNA. The cDNA obtained in this way provides an actual picture of metabolically active microorganisms. Using RT-PCR, the detection and quantification of individual gene structures and expressions are made achievable. For eukaryotic gene expression, the differential display (DD) technique is used, which is also an RNA-based technique. (Our major focus here is only the bacterial cell, therefore DD technique is not discussed here.)

3. The *genetic fingerprinting technique* is applied to detect the pattern and profile of genetic diversity among the microbial community present at the polluted site (Widada et al., 2002). This technique detects small nucleotide changes in DNA sequences and is also helpful in separating them. It provides important information regarding the prevalence of critical genes in a bacterial community that actively engage in biodegradation. The sensitivity of the technique is increased using polymerase chain reaction (PCR), which helps in the

amplification of the less abundant genes (Widada et al., 2002; Scheegurt-Mark and Kulpa-Chaler, 1998).

4. *PCR-based direct amplification*, which employs a set of primers, prepared using available consensus DNA sequence, amplifies and detects the novel catabolic gene sequence directly from environmental DNA (Widada et al., 2002).

5. *Next Generation Sequencing* is used to discover insight into biodegradation pathways among the microbial community, the structure of the microbial community and their interaction (Malik et al., 2021; Lovely, 2003).

The first enzyme characterized and reported to degrade hydrophobic aromatic compounds is "multi-component-dioxygenase" from *Sphingomonas* sp. strain RW1. The gene responsible for the production of α-subunit of enzyme is *dxn*A1, which gives substrate specificity to the enzyme (Armengaud et al., 1998). The second characterized enzyme is carbazole 1,9a-dioxygenase (CARDO), composed of two larger subunits named as α and β. The gene that encodes for this enzyme is *car*Aa (single protein) from *Pseudomonas* sp. strain CA10. Apart from this gene other reported genes from *Pseudomonas* sp. are *car*BC and *car*Ad (Sato et al., 1997). Another gene or gene family includes the *nah*-gene family from the *Pseudomonas* and *Burkholderia* strains (Pieper and Reineke, 2000) and *bhp* gene from *Rhodococcus* sp. strain M15–3 (Bosma et al., 1999).

To engineer a host bacterium, the following strategies must be taken into consideration (Keasling and Bang, 1998):

1. The recombinant gene (which is inserted into the host cell) must not restrict the normal growth of host bacteria.

2. The genetic vector should be designed in such a way that it would carry a large amount of DNA but not put a burden on the cell (preferably smaller in size).

3. Vector must remain stable within the host cell.

4. If possible, use the field application vectors (FAVs) to propagate a host cell for a selective temporary niche during field application (Lajoie et al., 1993).

5. If possible, include killer genes or suicide cassettes in the vector to eliminate the genetically engineered organisms once their role is over.

Bioremediation processes through genetically engineered bacteria are studied extensively (Table 8.1). Some of the successfully engineered bacteria along with the purpose for which they were designed are listed in Table 8.2. Besides the bacteria, many fungi are also isolated and screened for their biodegradation potential for a long time. Some of them are listed in Table 8.3 along with the molecule upon which they act.

Vector and Recombinant DNA Construction

A vector is a small, self-replicated, double-stranded DNA molecule that acts as a vehicle to carry and transfer the gene of interest into the host cell. It must have an origin for replication, multiple cloning sites (MCS) and marker genes (selectable and reporter gene) for detection (in most cases, an antibiotic resistance gene). In general, two types of vectors are used in genetic engineering. The first is a cloning vector and another is an expression vector. Cloning vectors are used to propagate a gene of interest into host cells while expression vectors allow transcription and translation of a gene of interest to produce proteins or enzymes. An expression vector carries a promoter and terminator sequence for expression, while a cloning vector does not carry such sequences (Carson et al., 2012). The general features of cloning vectors and expression vectors are depicted in Figure 8.2(a) and Figure 8.2(b), respectively. The most suitable molecules to use as vectors are plasmids, transposons and some viruses (phages). Among them, plasmids are most utilized compared to transposons and phages. Plasmids are small extracellular, circular strands of DNA that are generally less needed for the normal growth of microbes, but

TABLE 8.1

Genetically Engineered Bacteria Used for Bioremediation

Genetically engineered bacteria	Gene cloned	Purpose and site of bioremediation	Reference
Sphingobium sp. JQL4–5	Methyl parathion hydrolase gene (*mpd*) gene	Methyl parathion and fenpropathrin degradation from wastewater treatment sludge of an insecticide factory.	Yuanfan et al., 2010
Sphingomonas desiccabilis and *Bacillus idriensis*	*ars*M	Arsenic removal by methylation by biovolatilization (over 30 days).	Liu et al., 2011
Cupriavidus metallidurans MSR33	*mer* genes (*mer*-B, *mer*-G and other)	Removal of mercury from polluted water.	Rojas et al., 2011
Pseudomonas putida SB32 and *Pseudomonas monteilli* SB35	*czc* gene	Cadmium removal from contaminated soil by performing an *in-situ* experiment using the soybean plant.	Jain and Bhatt, 2013
Pseudomonas putida MC4.	Dehalogenase gene (*dha*A31)	Aerobic bioremediation of a recalcitrant chlorinated hydrocarbon 1,2,3-Trichloropropane (TCP).	Samin et al., 2014
Pseudomonas putida KT2440	Organophosphate degrading gene (*mpd* gene) and pyrethroid-hydrolyzing carboxylesterase gene (*pyt*H gene)	Degradation of mixed pesticides including methyl parathion, fenitrothion, chlorpyrifos, permethrin, fenpropathrin, and cypermethrin (0.2 mM each) completely within 15 days from fumigated and non-fumigated soils.	Zuo et al., 2015
E. coli BL21 with pUCP-P$_{la}$c-*gfp*^{mut2}	*ars*R gene	Inexpensive alternative solution for As (III) removal from contaminated environment.	Salvi et al., 2017

they give a selective advantage to the bacteria to survive in adverse conditions. The plasmids associated with the metabolism of natural substances are of more interest as they are more useful for designing a vector (Sayler, 1991). Presently, an expression vector is the preferred choice for most researchers (Wasilkowski et al., 2012). A large number of vectors is constructed from naturally occurring plasmids. The *field application vectors* (FAVs) are also exploited to propagate in host cells, which help them to survive in temporary environmental conditions during field experiments (Lajoie et al., 1993).

Cloning vectors are of two types, based on their ability to propagate in a range of diverse host species. The first type is *broad-host-range plasmid* and the second is *narrow-host-range plasmid*. The advantage of a broad-host-range plasmid is that it can be cloned in various bacteria belonging to different species. Moreover, for the construction of such a recombinant plasmid, it is necessary to make out that the plasmid should have a medium to high copy number so that the expression of a character is never compromised. For instance, plasmid RK2 and RSF1010 which are used for the construction of a mini-plasmid cloning vector in many of the experiments, have copy numbers 4–7 per host chromosome and 20 per host chromosome, respectively. They both share the same origin of replication and are found inherited in stable form until the 200th generation in *Pseudomonas putida* under less disturbing (environmental) conditions. In addition, both plasmids are naturally transmitted through conjugation in other bacteria, which give an additional advantage of sustainability (of cloned-recombinant plasmids) in a microbial community under *in-situ* conditions (Keasling and Bang, 1998; Popowska and Krawczyk-Balska, 2013).

TABLE 8.2

Genetically Modified Bacteria Produced through Gene of Interest with their Functions and Respective Donor Cells

Gene	Donor	(Genetically engineered) Host	Application	Reference
Toluene mono oxygenase (TM0) gene	—	*Escherichia coli*	Trichloroethene degradation	Sayler, 1991
PCB-degradative genes (bphABC)	*Pseudomonas* sp. strain ENV307	*Pseudomonas paucimobilis* IIGP4	Polychlorinated biphenyls degradation	Lajoie et al., 1993
pUTK21 plasmid	—	*Pseudomonas fluorescens* HK44	Naphthalene degradation	Sayler and Ripp, 2000
Atrazine chlorohydrolase gene	—	*Escherichia coli* AtzA	Atrazine degradation	Strong et al., 2000
pRO103 Plasmid	—	*Ralstonia eutropha* and *Escherichia coli* HB101	2,4-dichlorophenoxyacetic acid/2-oksoglutaric dioxygenasedegradation	Lipthay et al., 2001
merT-merP and MT genes	—	*Escherichia coli*	Removal of Hg^{2+} from electrolytic wastewater	Deng and Wilson, 2001
Cytochrome CYP1A gene	—	*Saccharomyces cerevisiae*	Dibenzo-p-dioxin, 1-monochlorodibenzo-p-dioxin degradation	Sakaki et al., 2002
pTOM plasmid	*Burkholderia cepacia* G4	*Burkholderia cepacia* L.S.2.4	Toluene degradation	Barac et al., 2004
operon *bph, gfp*	*E. coli rrn*BP1	*Pseudomonas. fluorescens* F113wt	Chlorinated biphenyl degradation	Boldt et al., 2004
-	*Burkholderia* sp.	*Pseudomonas fluorescens* ATCC 17400	2,4-dinitrotoluene degradation	Monti et al., 2005
Plasmid pNF142::TnMod-OTc	—	*Pseudomonas putida* KT2442	Naphthalene degradation	Filonov et al., 2005
p-TOM-Bu61 plasmid	—	*Burkholderia cepacia* VM1468	Toluene degradation	Taghavi et al., 2005
merT -merP protein and metallothionein (MT)	—	*Escherichia coli* JM109	Bioremediation of contaminated wastes containing mercury	Zhao et al., 2005
fbABC operon	*Burkholderi axenovorans* strain LB400	*Rhodococcus* sp. RHA1 (pRHD34:fcb)	2(4)-chlorobenzoate 2(4)-chlorobiphenyl polychlorinated biphenyl (PCB) biodegradation	Rodrigues et al., 2006
(mpd) gene	*Plesiomonas* sp. M6	*Stenotrophomonas* YC-1	Methyl parathion degradation	Yang et al., 2006
tod and xyl genes	*Pseudomonas putida*	*Deinococcus radiodurans*	Degradation of organic contaminants	Brim et al., 2006
Metal-binding peptide EC20 gene	*Pseudomonas* strain Pb2–1	*Rhizobacteria*	Trichloroethene degradation	Lee et al., 2006
pWW0 plasmid	—	*Pseudomonas putida* PaW85	Petroleum degradation	Jussila et al., 2007

TABLE 8.2 *(Continued)*
Genetically Modified Bacteria Produced through Gene of Interest with their Functions and Respective Donor Cells

Gene	Donor	(Genetically engineered) Host	Application	Reference
Nitroreductase	—	*Pseudomonas putida* pnrA	2,4,6- trinitrotoluene degradation	Van Dillewijn et al., 2008
pNB:2dsRed plasmid	*P. putida* pNB2	*Comamonas testosteroni* SB3	3-chloroaniline (3CA) degradation	Bathe et al., 2009
(pGEX-AZR) Azoreductase gene	*Rhodobacter sphaeroides*AS1.1737	*Escherichia coli* JM109	Decolorize azo dyes, C.I. Direct Blue 71	Jin et al., 2009
(pDH5) pDH5 plasmid	*Arthrobacter* sp. FG1	*Pseudomonas putida* PaW340	4-chlorobenzoic acid degradation	Massa et al., 2009
bph genes	—	*Cupriavidus necator* JMP134-X3.	Polychlorobipheny bioremediation	Saavedra et al., 2010
Haloalkane dehalogenase gene *(dhaA31)*	—	*Pseudomonas putida* MC4–5222	1,2,3-Trichloropropane (TCP) degradation	Samin et al., 2014
laccase gene	*Phanerochaete flavido-alba*	*Aspergillus niger*	Acid red and Brilliant blue R degradation	Benghazi et al., 2014
onpABC gene cluster (ortho-nitrophenol operon)	—	*Cupriavidus necator* strain JMP134-ONP	Ortho-nitrophenol and meta-nitro phenol degradation	Hu et al., 2014
Arsenite S-adenosylmethionine methyltransferase gene	*Cyanidioschyzon merolae*	*B. subtilis* 168	Oxide, and volatilize large quantities of dimethylarsine and trimethylarsine	Huang et al., 2015
Yeast laccase gene *(YlLac)*	*Yarrowia lipolytica*	*Pichia pastoris* fungus	Phenolic compounds degradation	Kalyani et al., 2015
ohb-operon and *fcb*- operon	*P. aeruginosa*	*Comamonas testosterone* VP44	Monochlorobiphenyls mineralization	Kumar et al., 2018
Pesticide degrading gene	—	*Pseudomonas putida* KTUe	Remediation of multiple pesticides contaminated soil degradation of pyrethroids, organophosphates, and carbamates	Gong et al., 2018

Plasmids are constructed for specific purposes and for specific molecules to target degradation. Some such plasmids are the pDLB101 plasmid reported by Beaudoin et al. (2000), Inc-P1 (*incompatibility group-P1*) plasmids documented by Popowska and Krawczyk-Balska (2013), pARS1 and pAIO1 studied by Drewniak et al. (2015) and pWB6 plasmid reported by Zhang et al. (2016). Most of the source plasmids used are isolated from *Pseudomonas* species (Popowska and Krawczyk-Balska, 2013; Drewniak et al., 2015).

Narrow-host-range plasmids survive in limited species of host bacteria. Applications of this plasmid are very limited because such plasmids are not transmitted horizontally in other species and the stability of them is also questionable. The only advantage of narrow-host-range plasmids is structural improvement and segregation stability in a particular strain of host (Boivin et al.,

TABLE 8.3

Fungi Used for Bioremediation

Fungi	Biodegraded molecule	Reference
Cerrena sp.	Olive mill wastewater phenolics	Mann et al., 2009
Aspergillus terreus, Trichoderma longibranchiatum, Aspergillus niger,	Pb_3, Cr_8, and Ni_{27} respectively	Joshi et al., 2011
Aspergillus terreus, Penicillium sp., *Alternaria* sp. and *Acromonium* sp.	Petroleum pollution	Mohsenzadeh et al., 2012
Pleurotus flabellatus, P. florida, P. ostreatus and *P. sajor-caju*	Direct blue 14 (DB14)	Singh et al., 2013
White-rot fungi, namely *Gymnopilus luteofolius, Kuehneromyces mutabilis,* and *Phanerochaete velutina,*	2,4,6-trinitrotoluene (TNT)	Anasonye et al., 2015
Arbuscula mycorrhizal fungi	Cadmium	Hashem et al., 2016
Rhizopus sp. CUC23, *Aspergillus fumigates* ML43 and *Penicillium radicum* PL17	Chromium	Bibi et al., 2018
Trichoderma hamatum FBL 587 and *Rhizopus arrhizus* FBL 578	Dichlorodiphenyltrichloroethane (DDT)	Russo et al., 2019
Penicillium sp. and *Trichoderma viride*	Chromium (VI)	Zapana-Huarache et al., 2020
Penicillium crysogenum FMS2	Cadmium	Din et al., 2021
Trichoderma lixii CR700	Cupper (Cu^{+2})	Kumar and Dwivedi, 2021
Cladosporium halotolerans stain xM01	Biogenic Mn oxides and cadmium (Heavy metal biodegradation)	Wang et al., 2022
Aspergillus sydowii-FJH-1	Triphenyl phosphate	Feng et al., 2022

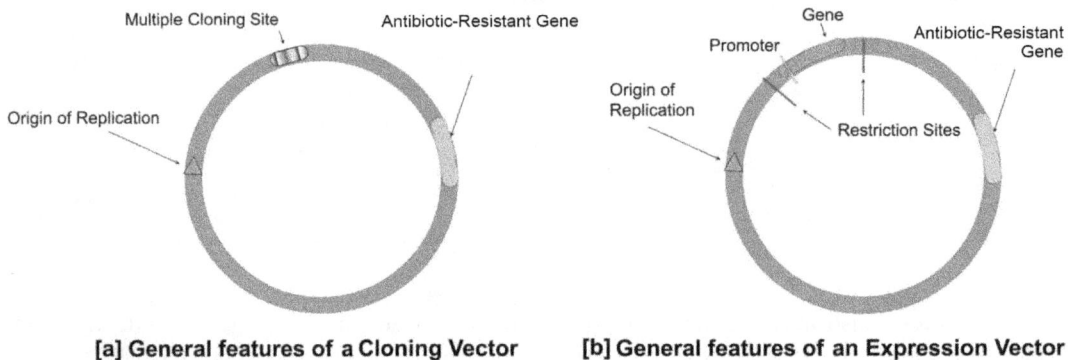

[a] General features of a Cloning Vector   **[b] General features of an Expression Vector**

FIGURE 8.2   General features of (a) Cloning vector and (b) Expression vector.

1994; Keasling and Bang, 1998). Among the reported narrow-host-range plasmid vectors, pVS1 of *P. putida* is more highly stable in *P. fluorescens* than in broad-host-range-plasmids, and also confers low metabolic burden on the host cell (Van-der Bij et al., 1996). Jones et al. (1998) reported a single-copy number narrow-host-rang vector based on F-plasmid from *Escherichia coli* with several necessary elements responsible for cell-specific-replication, segregation and antibiotic resistance gene for detection. Recently, the Bacterial Artificial Chromosome (BAC) is used to catalog an environmental genomic library and also to clone larger DNA fragments isolated from environmental samples (Lovely, 2003).

Along with the vector, restriction endonuclease (RE) enzymes are also important for molecular cloning as they are essential for the cutting of not only the vector but also the insert to generate similar ends for ligation. In most cases, RE type-II is used for such experiments over other Res (i.e., RE type-I and RE type-III). The major advantage of this RE type-II is it cut the DNA at definite 4–8 ase pair sequence specific for each RE type-II. Some examples of RE type-II are, *Eco*R1, *Bam*-III, *Ras*I, *Hae*III, *Taq*I and *Sau*3A (Lajoie et al., 1993; Paul et al., 2006; Chandra and Kumar, 2017).

INSERTION OF RECOMBINANT VECTOR INTO A HOST

Recombinant vectors are constructed by the joining of insert DNA and vectors (opened using the same RE type-II enzymes). For joining the molecules, ligase enzyme is used. Figure 8.3 (a, b) shows the general feature of the cloning vector and the process of cutting with RE type-II, which possibly produces cohesive/sticky end or blunt end in both the vector as well as the insert. Figure 8.3 (c) shows how the insert (with cohesive or blunt ends) is ligated with the vector (with similar ends produced) to create the recombinant molecule. Once the recombinant vector is constructed it is transferred into a suitable host. Figure 8.3 (d) showed the insertion and propagation of recombinant constructs in host cells. Varieties of molecular techniques for horizontal gene transfer are being employed to introduce recombinant constructs in host cells. The insertion of a recombinant construct in a host cell is carried out by various methods; for instance, electroporation, protoplast transformation, conjugation and biolistic transformation (Kumar et al., 2018). For transformation, cells are first made "competent" through various physical and chemical treatments. One such physical insertion technique for the transfer of genes into host cells is electroporation (Ike et al., 2007). Using the electroporation method, Ike et al. (2007) had transferred two expression vectors (pMpnolBMTL4nifHPCS and pMPnifHPCS) in to *Mesorhizobium huakuii* subs. *Rengei* B3 is used for bioremediation of cadmium-contaminated soil. Kumari and Das (2019) had transformed *btm*A gene in to $CaCl_2$ treated competent cells of *E. coli* BL21 (DE3) from *P. aeruginosa* N6P6. Ng et al. (2020) have used polyethylene glycol (PEG) for making the *Yarrowia lipolytica* competent for the introduction of *ws*2 and *maqu*_0168 (for heterologous wax ester syntheses) from *Marinobacter* species for fatty acid ethyl esters biosynthesis.

SCREENING OR SELECTION OF A RECOMBINANT HOST

Recombinant cells are screened and selected according to vectors used for cloning purposes. Figure 8.4 shows the general features of different vectors that can be screened by insertional inactivation (Figure 8.4a), by green fluorescent protein expression (Figure 8.4b) or by antibiotic sensitive/resistant phenomena (Figure 8.4c). In most cases, the cloning vector based on the insertional inactivation of *lac*Zα (code for β-galacosidase enzyme) by "insert" DNA is used. The transformed colony appears as white and the non-transformed colony appears blue in colour in the presence of chromogenic substance; i.e., 5-bromo-4-chloro-3-indolyl-β-D-galactopyranoside (x-gal) or 5-bromo-4-chloro-3-indolyl-β-D-glucuronide cyclohexyl ammonium salt (x-gluc). Another method that uses metabolic enzymes for the detection of transformed cells is *xyl*E, which encodes for catechol-2, 3 oxygenase (giving a yellow-coloured catabolite from catechol) and *gus*A codes for β-glucuronidase. Another screening technique includes the detection of an antibiotic-resistant gene by supplementation of the respective antibiotic in the medium. This approach also adopts an "insertional inactivation" phenomenon. Plasmids with two antibiotic-resistant genes are used for such purposes. Among them, one is used to select the plasmid receiver cell and another for insertion and detection of the cloned gene. The reporter proteins which are utilized at present are alkaline phosphatase, luciferase, β-gaucuronidase, β-lactamase and green fluorescence protein (*gfp* gene from *Aequirae Victoria* jellyfish) (Widada et al., 2002; Kumar et al., 2013).

SAFETY ASPECTS AND NOVEL ACTIVE BIOLOGICAL CONTAINMENT (ABC) STRATEGY

Genetically modified organisms are extraordinarily useful both scientifically and commercially. However, they are not naturally occurring organisms and therefore, they may disturb the existing ecological framework of the environment. This is the major obstacle in using genetically modified organisms for environmental applications such as bioremediation due to their possible risk

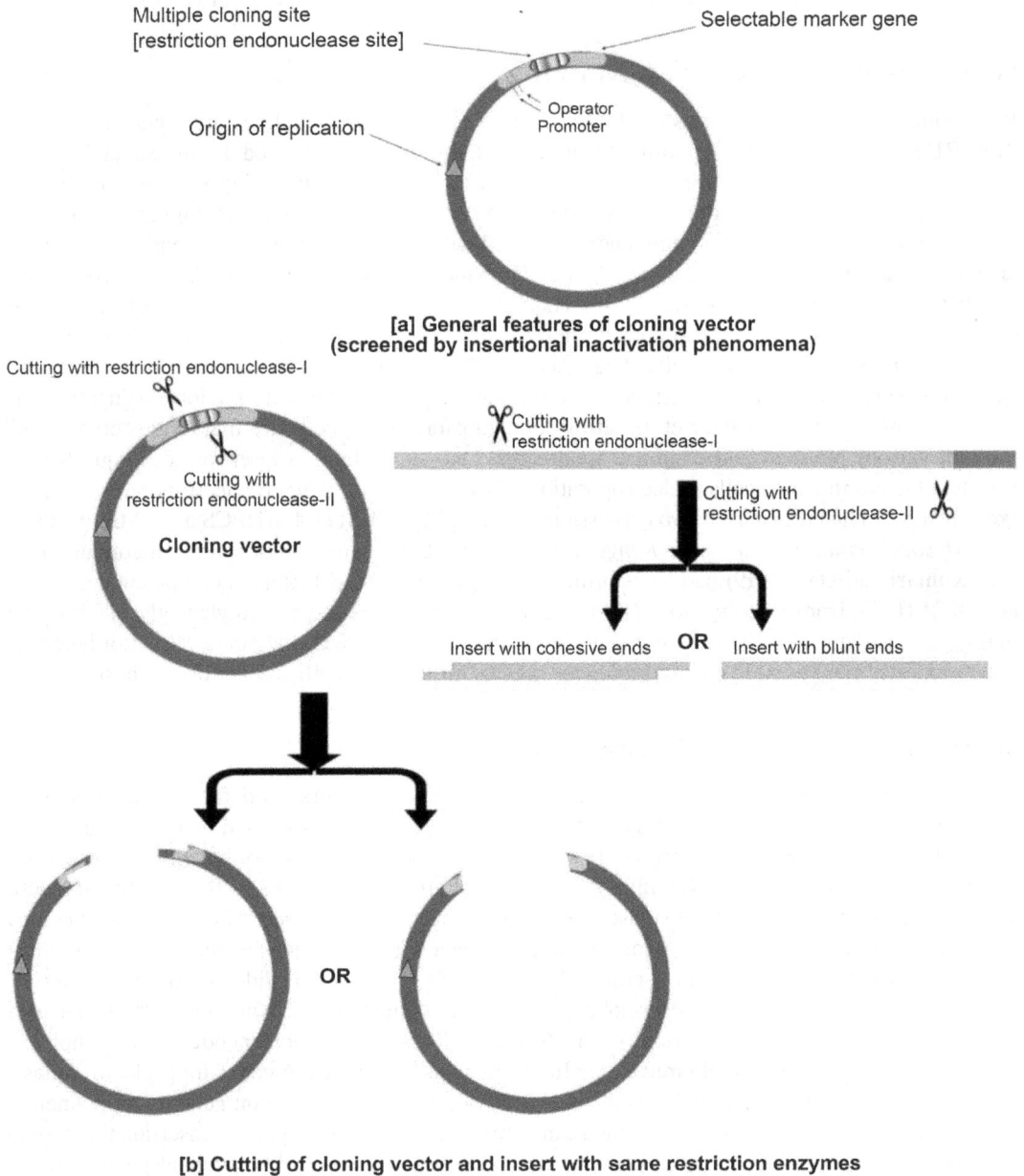

[a] General features of cloning vector
(screened by insertional inactivation phenomena)

[b] Cutting of cloning vector and insert with same restriction enzymes

FIGURE 8.3 Basic mechanism of molecular cloning to create genetically engineered bacteria.

PART-I [a] General feature of cloning vector (screened by insertional inactivation phenomena), [b] Cutting of cloning vector and insert with same restriction endonuclease (to generate similar sticky or blunt ends),

Open cloning vector (with cohesive or sticky ends) | Insert (with cohesive or sticky ends) | Recombinant DNA molecule

OR

Open cloning vector (with blunt ends) | Insert (with blunt ends) | Recombinant DNA molecule

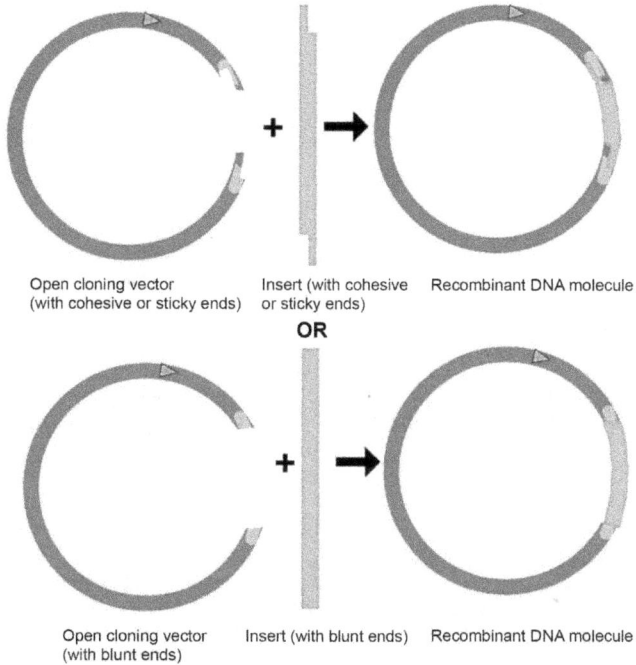

[c] Construction of recombinant DNA molecules by ligation of cloning vector (having cohesive end or blunt end) with insert (having cohesive or blunt end, respectively)

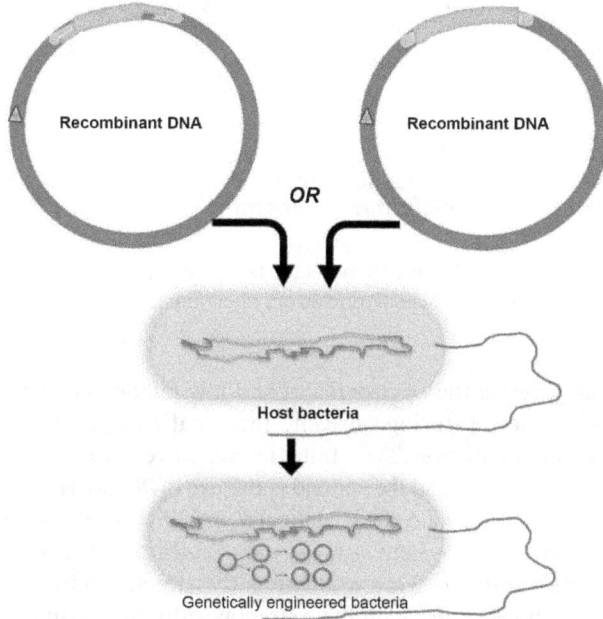

Recombinant DNA | Recombinant DNA

OR

Host bacteria

Genetically engineered bacteria

[d] Insertion of recombinant DNA in to host cell and multiplication of cloned DNA (to achieve copy number)

FIGURE 8.3 (Continued)

PART-II [c] Construction of DNA molecules by ligation of cloning vector with insert, [d] Insertion of recombinant DNA in to host cell and multiplication of cloned DNA (to achieve copy number).

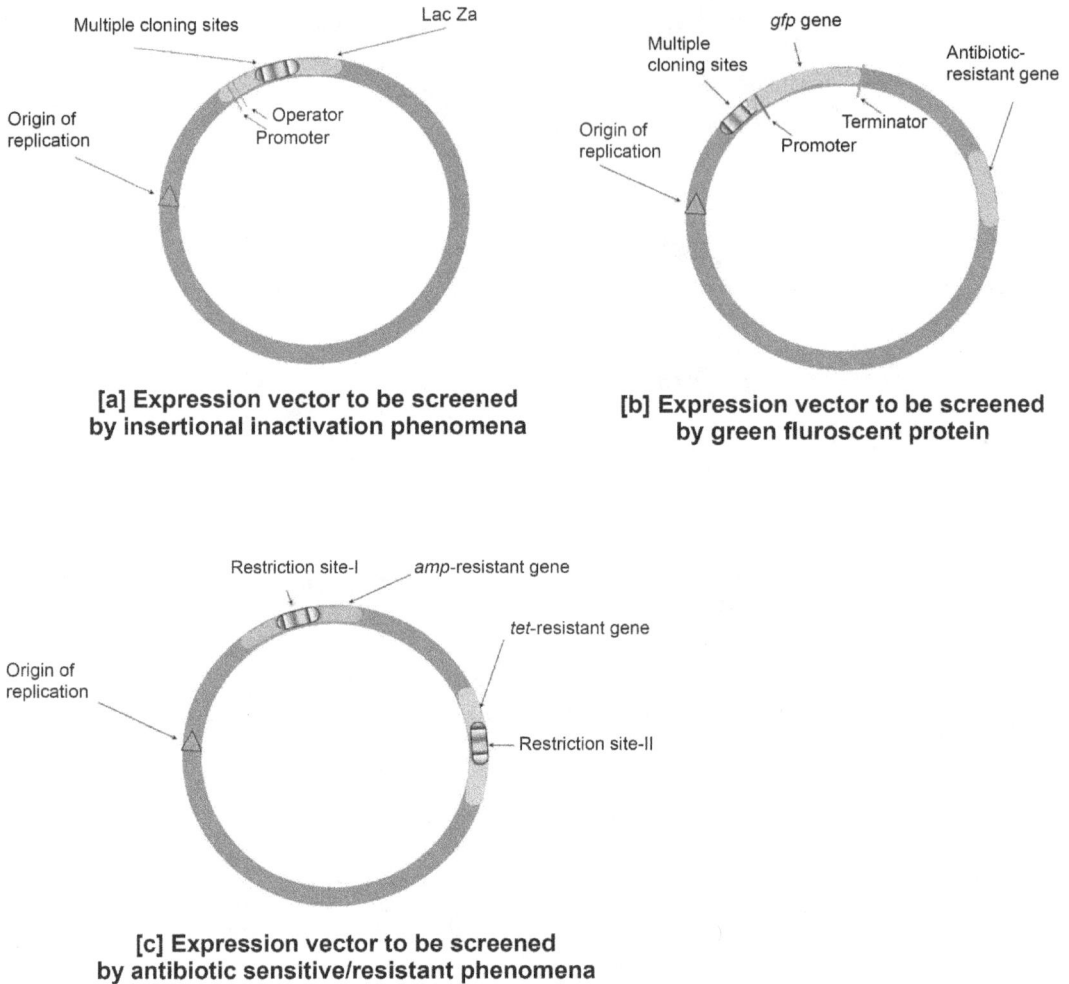

[a] Expression vector to be screened
by insertional inactivation phenomena

[b] Expression vector to be screened
by green fluroscent protein

[c] Expression vector to be screened
by antibiotic sensitive/resistant phenomena

FIGURE 8.4 General features of an expression vector to be screened by [a] Insertional inactivation, [b] Green fluorescent protein, [c] Antibiotic sensitive/resistant phenomena.

(of spreading unnatural genes in the environment) and low public acceptance (Azad et al., 2014). Therefore, researchers are now focusing on using technical safeguards in designing genetically modified organisms for bioremediation. Two strategies are currently available for this purpose. The first is the use of a "*suicide system*" and the second is the use of "*killer genes*" (Widada et al., 2002). Both systems are utilized to eliminate the genetically modified organism from the environment once their job is done. The suicide system strategy uses "suicide genes" which are activated in the absence of certain growth substrates. One example is the streptaviridin gene, which is expressed in the absence of the growth substrate to produce streptaviridin that ultimately binds to D-biotin to kill the host bacterium (Szafranski et al., 1997; Widada et al., 2002). The killer gene strategy, based on the killer product (produced due to the presence of the killer gene), kills the non-immune recipient cell after gene transfer. For instance, *gef* gene, a killer gene of *E. coli*, was incorporated into TOL plasmid (for meta-cleavage) of *Pseudomonas putida* under the control of *lac* promoter. In the presence of *m*-methylebenzoate, *gef* expression is suppressed. In the absence of *m*-methylbenzoat, *gef* (killer gene) is expressed and it will kill the host bacterium (i.e., genetically modified *P. putida*) (Molina et al., 1998; Widada et al., 2002).

The major drawbacks of these systems are as follows:

1. Bacteria designed for bioremediation are eliminated more slowly in soil than they are in liquid mediums (in the laboratory) (Ronchel and Ramos, 2001; Widada et al., 2002).
2. Mutations observed in bacteria (10^{-8}/cell and generation) which enable them to live in polluted sites for a longer time and may disturb ecological frameworks (Ronchel and Ramos, 2001; Widada et al., 2002).

The possible solution for such a problem is the use of mutant bacteria as a host which survives in the presence of specific growth substrates, for instance, *P. putida* Δ*asd* strain (mutation in *asd* gene involve in aspartate-β-semialdehyde biosynthesis), which survives only in the presence of 3-methylebenzoate—an essential metabolite (Ronchel and Ramos, 2001; Widada et al., 2002).

Conclusively, the bioremediation process is the removal of pollutants by microorganisms. The wild-type strains have limited efficiency to degrade or remove pollutants, therefore the genetically engineered organisms (with high capability to degrade pollutant characteristics added through gene transfer) are beneficial for the process. However, the following criteria are considered for the design and application of genetically modified bacteria for bioremediation purposes:

1. The selection of insert (gene of interest) should be very specific not only for pollutant removal but also for cloning purposes.
2. Choice of expression vector should be based on the size of the insert and environmental conditions (or field) for which the genetically engineered bacteria are constructed.
3. Genetically engineered bacteria must be removed from the application site due to ethical as well as ecological issues.

REFERENCES

Anasonye, F., Winquist, E., Rasanen, M., Kontro, J., Bjorklof, K., Vasilyeva, G., Jorgensen, K., Steffen, K., Tuomela, M. 2015. Bioremediation of TNT contaminated soil with fungi under laboratory and pilot scale conditions. *Int. Biodeteror. Biodegrad.* 105: 7–12. https://doi.org/10.1016/j.ibiod.2015.08.003.

Armengaud, H., Happe, B., Timmis, K.N. 1998. Genetic analysis of dioxin dioxygenease of *Sphingomonas* sp. strain RW1: Catabolic genes dispersed on the genome. *J. Bacteriol.* 180: 3954–3966. https://doi.org/10.1128/jb.180.15.3954-3966.1998.

Atlas, M. 1992. Molecular methods for environmental monitoring and containment of genetically engineered microorganisms. *Biodegradation.* 3: 137–146. https://doi.org/10.1007/bf00129079.

Azad, M.A.K., Amin, L., Sidik, N.M. 2014. Genetically engineered organisms for bioremediation of pollutants in contaminated sites. *Chin. Sci. Bull.* 58(8): 703–714. doi: 10.1007/s11434-013-0058-8.

Barac, T., Taghavi, S., Borremans, B., Provoost, A., Oeyen, L., Colpaert, Vangronsveld, J., Van der Lelie, D. 2004. Engineered endophytic bacteria improve phytoremediation of water-soluble, volatile, organic pollutants. *Nat. Biotechnol.* 22(5): 583–588. https://doi.org/10.1038/nbt960.

Bathe, S., Schwarzenbeck, N., Hausner, M. 2009. Bioaugmentation of activated sludge towards 3-chloroaniline removal with a mixed bacterial population carrying a degradative plasmid. *Bioresour. Technol.* 100(12): 2902–2909. https://doi.org/10.1016/j.biortech.2009.01.060.

Beaudoin, D.L., Bryers, J.D., Cunningham, A.B., Peretti, S.W. 2000. Mobilization of broad host range plasmid from *Pseudomonas putida* to established biofilm of *Bacillus azotoformans*I. Experiments. *Biotechnol. Bioeng.* 57(3): 272–279. https://doi.org/10.1002/(SICI)1097-0290(19980205)57:3%3C272::AID-BIT3%3E3.0.CO;2-E.

Benghazi, L., Record, E., Suárez, A., Gomez-Vidal, J.A., Martínez, J., De la Rubia, T. 2014. Production of the *Phanerochaete flavido-albalaccase* in *Aspergillus niger*for synthetic dyes decolourization and biotransformation. *World. J. Microbiol. Biotechnol.* 30(1): 201–211. https://doi. Org/10.1007/s11274–013–1440-z.

Bibi, S., Hussain, A., Hamayun, M., Rahman, H., Iqbal, A., Shah, M., Irshad, M., Qasim, M., Islam, B. 2018. Bioremediation of hexavalent chromium by endophytic fungi; safe and improved production of *Lactuca sativa* L. *Chemospere.* 211: 653–663. https://doi.org/10.1016/j.chemosphere.2018.07.197.

Boivin, R., Bellemare, G., Dion, P. 1994. Novel narrow-host-range vectors for direct cloning of foreign DNA in *Pseudomonas. Curr. Microbiol.* 28: 41–47. https://doi.org/10.1007/bf01575984.

Boldt, T.S., Sorensen, J., Karlson, U., Molin, S., Ramos, C. 2004. Combined use of different Gfp reporters for monitoring single-cell activities of a genetically modified PCB degrader in the rhizosphere of alfalfa. *FEMS Microbiol. Ecol.* 48(2): 139–148. https://doi.org/10.1016/j.femsec.2004.01.002.

Bosma, T., Kruzinga, E., Bruin, E.J.D., Poelarends, G.J., Janseen, D.B. 1999. Utilization of AD1 through heterologous expression of the haloalkane dehalogenase from *Rhodococcus* sp. strain M15–3. *Appl. Environ. Microbiol.* 65: 4575–4581. https://doi.org/10.1128/aem.65.10.4575-4581.1999.

Brim, H., Osborne, J.P., Kostandarithes, H.M., Fredrickson, J.K., Wackett, L.P., Daly, M.J. 2006. *Deinococcus radiodurans* engineered for complete toluene degradation facilitates Cr (VI) reduction. *Microbiol.* 152: 2469–2477. https://doi.org/10.1099/mic.0.29009-0.

Carson, S., Miller, H., Witherow, D. 2012. *Lab session-2: Purification and digestion of plasmid (vector) DNA. Molecular biology techniques.* 3rd ed. Elsevier. https://doi.org/10.1016/B978-0-12-385544-2.00002-8.

Chandra, R., Kumar, V. 2017. Detection of androgenic-mutagenic compounds and potential autochthonous bacterial communities during *in-situ* bioremediation of post-methanated distillery sludge. *Front. Microbiol.* 8: 887. https://doi.org/10.3389/fmicb.2017.00887.

Cohen, S.N. 2013. DNA Cloning: A personal view after 40 years. *Proc. Natl. Acad. Sci. USA.* 110(39): 15521–15529. https://doi.org/10.1073/pnas.1313397110.

Cohen, S.N., Chang, A.C.Y., Boyer, H.W., Helling, R.B. 1973. Construction of biologically functional bacterial plasmids in vitro. *Proc. Natl. Acad. Sci. USA.* 70(11): 3240–3244. doi: 10.1073/pnas.70.11.3240.

Deng, X., Wilson, D. 2001. Bioaccumulation of mercury from wastewater by genetically engineered *Escherichia coli. Appl. Microbiol. Biotechnol.* 56(1): 276–279. https://doi.org/10.1007/s002530100620.

Din, G., Dunlap, H., Ripp, S., Shah, A. 2021. Cadmium tolerance and bioremediation potential of filamentous fungus *Penicillum chrysogenum* FMS2 isolated from soil. *Int. J. Environ. Sci. Technol.* https://doi.org/10.1007/s13762-021-03211-7.

Drewniak, L., Ciezkowska, M., Radlinska, M., Sklodowska, A. 2015. Construction of recombinant broad-host range plasmids providing their bacterial host arsenic resistance and arsenite oxidation ability. *J. Bacteriol.* 196–197: 42–51. https://doi.org/10.1016/j.jbiotec.2015.01.013.

Feng, M., Zhou, J., Yu, X., Mao, W., Guo, Y., Wang, H. 2022. Insights into biodegradation mechanisms of triphenyl phosphate by a novel fungal isolate and it's potential in bioremediation of contaminated river sediment. *J. Hazard. Mater.* 424(Part-B): 127545. https://doi.org/10.1016/j.jhazmat.2021.127545.

Filonov, A.E., Akhmetov, L.I., Puntus, I.F., Esikova, T.Z., Gafarov, A.B., Izmalkova, T.Y., Boronin, A.M. 2005. The construction and monitoring of genetically tagged, plasmid-containing, naphthalene-degrading strains in soil. *Microbiol.* 74(4): 453–458.

Gong, T., Xu, X., Dang, Y., Kong, A., Wu, Y., Liang, P., Yang, C. 2018. An engineered *Pseudomonas putida* can simultaneously degrade organophosphates, pyrethroids and carbamates. *Sci. Total Environ.* 628: 1258–1265. https://doi.org/10.1016/j.scitotenv.2018.02.143.

Hashem, A., Abd-Allah, E., Alqarawi, A., Egamberdieva, D. 2016. Bioremediation of adverse impact of cadmium toxicity on *Cassia italic* Mill by arbuscula mycorrhizal fungi. *Saudi. J. Biolog. Sci.* 23(1): 39–47. https://doi.org/10.1016/j.sjbs.2015.11.007.

Hu, F., Jiang, X., Zhang, J.J., Zhou, N.Y. 2014. Construction of an engineered strain capable of degrading two isomeric nitrophenols via a sacB-and gfp-based markerless integration system. *Appl. Microbiol. Biotechnol.* 98(10): 4749–4756. https://doi.org/10.1007/s00253-014-5567-0.

Huang, K., Chen, C., Shen, Q., Rosen, B.P., Zhao, F.J. 2015. Genetically engineering *Bacillus subtilis* with a heat-resistant arsenite methyltransferase for bioremediation of arsenic-contaminated organic waste. *Appl. Environ. Microbiol.* 81(19): 6718–6724. https://doi.org/10.1128/AEM.01535-15.

Ike, A., Sriprang, R., Ono, H., Murooka, Y., Yamashita, M. 2007. Bioremediation of cadmium contaminated soil using symbiosis between leguminous plant and recombinant rhizobia with the *MTL4* and the *PCS* genes. *Chemosptere.* 66: 1670–1676. doi: 10.1016/j.chemosphere.2006.07.058.

Jain, S., Bhatt, A. 2013. Molecular and in situ characterization of cadmium resistant diversified extremophilic strains of *Pseudomonas* for their bioremediation potential. *3Biotech.* 4: 97–304. https://doi.org/10.1007/s13205-013-0155-z.

Jin, R., Yang, H., Zhang, A., Wang, J., Liu, G. 2009. Bioaugmentation on decolourization of CI Direct Blue 71 by using genetically engineered strain *Escherichia coli* JM109 (pGEX-AZR). *J. Hazard. Mater.* 163(2–3): 1123–1128. https://doi.org/10.1016/j.jhazmat.2008.07.067.

Jones, K.L., Keasling, J.D. 1998. Construction and characterization of F-plasmid-based expression vectors. *Biotechnol. Bioeng.* 59(6): 659–665.

Joshi, P., Swarup, A., Maheshwari, S., Kmar, R., Singh. N. 2011. Bioremediation of heavy metals in liquid medium through fungi isolated from contaminated sources. *Indian J. Microbiol.* 51: 482–487. https://doi.org/10.1007/s12088-011-0110-9.

Jussila, M.M., Zhao, J., Suominen, L., Lindström, K. 2007. TOL plasmid transfer during bacterial conjugation *in vitro* and rhizoremediation of oil compounds in vivo. *Environ Pollut.* 146(2): 510–524. https://doi.org/10.1016/j.envpol.2006.07.012.

Kalyani, D., Tiwari, M.K., Li, J., Kim, S.C., Kalia, V.C., Kang, Y.C., Lee, J.K. 2015. A highly efficient recombinant laccase from the yeast *Yarrowia lipolytica* and its application in the hydrolysis of biomass. *PLoS One.* 10(3): e0120156. https://doi.Org/10.1371/journal.Pone.

Keasling, J.D., Bang, S.W. 1998. Recombinant DNA techniques for bioremediation and environmentally-friendly synthesis. *COBIOT.* 9: 135–140. http://biomednet.com/elecref/0958166900900135.

Kumar, N.M., Muthukumaran, C., Sharmila, G., Gurunathan, B. 2018. Genetically modified organisms and its impact on the enhancement of bioremediation. In S. Varjani, A. Agarwal, E. Gnansounou, B. Gurunathan (eds) *Bioremediation: Applications for environmental protection and management. Energy, environment, and sustainability.* Springer, Singapore. https://doi.org/10.1007/978-981-10-7485-1_4.

Kumar, V., Dwivedi, S. 2021. Bioremediation mechanism and potential of copper by actively growing fungus Trichoderma *lixii* CR700 isolated from electroplating wastewater. *J. Environ. Manage.* 277: 11370. https://doi.org/10.1016/j.jenvman.2020.111370.

Kumar, S., Dagar, V.K., Khasa, Y.P., Kuhud, R.C. 2013. Chapter-11 Genetically modified microorganisms (GMOs) for bioremediation. In R.C. Kuhad, A. Singh (eds) *Biotechnology for environmental management and resource recovery.* Springer Press. 191–218. doi: 10.1007/978-81-322-0876-1_11.

Kumari, S., Das, S. 2019. Expression of metallothionein encoding gene *bmt*A in biofilm-forming marine bacterium *Pseudomonas aeruginosa* N6P6 and understanding its involvement in Pb (II) resistance and bioremediation. *Environ. Sci. Pollut. Res.* 26: 28763–28774. https://doi.org/10.1007/s11356-019-05916-2.

Lajoie, C.A., Zylstra, G.J., DeFlaun, M.F., Strom, P.F. 1993. Development of field application vector for bioremediation of soils contaminated with polychlorinated biophenyls. *Appl. Environ. Microbiol.* 59(6): 1735–1741. https://doi.org/10.1128/aem.59.6.1735-1741.1993.

Lee, W., Wood, T.K., Chen, W. 2006. Engineering TCE-degrading rhizobacteria for heavy metal accumulation and enhanced TCE degradation. *Biotechnol. Bioeng.* 95(3): 399–403. https://doi.org/10.1002/bit.20950.

Lipthay, J.R., Barkay, T., Sørensen, S.J. 2001.Enhanced degradation of phenoxyacetic acid in soil by horizontal transfer of the tfdA gene encoding a 2, 4-dichlorophenoxyacetic acid dioxygenase. *FEMS Microbiol. Ecol.* 35(1): 75–84. https://doi.org/10.1111/j.1574-6941.2001.tb00790.x.

Liu, S., Zhang, F., Chen, J., Sun, G. 2011. Arsenic removal from contaminated soil via biovolatilization by genetically engineered bacteria inner laboratory conditions. *J. Environ. Sci.* 23(9): 1544–1550. https://doi.org/10.1016/S1001-0742(10)60570-0.

Lovely, D.R. 2003. Cleaning up with genomics: Applying molecular to bioremediation. *Nat Rev Microbiol.* 1: 35–44. https://doi.org/10.1038/nrmicro731.

Malik, G., Arora, R., Chaturvedi, R., Paul, M. 2021. Implementation of genetic engineering and novel omics approach to enhance bioremediation: A focused review. *Bull. Environ. Contam. Toxicol.* https://doi.org/10.1007/s00128-021-03218-3.

Mann, J., Markham, J., Peiris, P., Nair, N., Spooner-Hart, R., Holford, P. 2009. Screening and selection of fungi for bioremediation of olive oil wastewater. *World J. Microbiol. Biotechnol.* 26: 567–571. https://doi.org/10.1007/s11274-009-0200-6.

Massa, V., Infantino, A., Radice, F., Orlandi, V., Tavecchio, F., Giudici, R., Barbieri, P. 2009. Efficiency of natural and engineered bacterial strains in the degradation of 4-chlorobenzoic acid in soil slurry. *Int. Biodeterior. Biodegrad.* 63(1): 112–115. https://doi.org/10.1016/j.ibiod.2008.07.006.

Megharaj, M., Venkateswarlu, K., Naidu, R. 2014. *Encyclopaedia of toxicology.* 3rd ed. Elsevier Press. 485–489. https://doi.org/10.1016/B978-0-12-386454-3.01001-0.

Mohsenzadeh, F., Rad, A., Akbari, M. 2012. Evaluation of oil removal efficiency of enzymatic activity in some fungal stain for bioremediation of petroleum polluted soils. *Iranian J. Environ. Health Sci. Eng.* 9: 26. https://doi.org/10.1186/1735-2746-9-26.

Molina, L., Ramos, C., Ronchel, M., Molin, S., Ramos, J.L. 1998. Construction of an efficient biologically contained *Pseudomonas putida* strain and its survival in outdoor assays. *Appl. Environ. Microbiol.* 64: 2072–2078. doi: 10.1128/AEM.64.6.2072-2078.1998.

Monti, M.R., Smania, A.M., Fabro, G., Alvarez, M.E., Argarana, C.E. 2005. Engineering *Pseudomonas fluorescens* for biodegradation of 2, 4-dinitrotoluene. *Appl. Environ. Microbiol.* 71(12): 8864–8872. https://doi.org/10.1128/AEM.71.12.8864-8872.2005.

Ng, T.K., Yu, A.Q., Ling, H., Juwono, N.K., Choi, W.J., Leong, S.S., Chang, M.W. 2020. Engineering *Yarrowia lipolytica* towards food waste bioremediation: Production of fatty acid ethyl ester from vegetative cooking oil. *J. Biosci. Bioeng.* 129(1): 31–40. https://doi.org/10.1016/j.jbiosc.2019.06.009.

Paul, D., Pandey, G., Meier, C., van der Meer, J.R., Jain, R.K. 2006. Bacterial community structure of a pesticide-contaminated site and assessment of charges induced in community structure during bioremediation. *FEMS Microbiol. Ecol.* 57(1): 116–127. https://doi.org/10.1111/j.1574-6941.2006.00103.x.

Pieper, D., Reineke, W. 2000. Engineering bacteria for bioremediation. *Curr. Opin. Biotechnol.* 11(3): 267–270. https://doi.org/10.1016/S0958-1669(00)00094-X.

Popowska, M., Krawczyk-Balska, A. 2013. Borad-host-range IncP-1 plasmids and their resistance potential. *Front. Microbiol.* 4: 44. http://dx.doi.org/10.3389/fmicb.2013.00044.

Rodrigues, J.L., Kachel, C.A., Aiello, M.R., Quensen, J.F., Maltseva, O.V., Tsoi, T.V., Tiedje, J.M. 2006. Degradation of Aroclor 1242 dechlorination products in sediments by *Burkholderia xenovorans* LB400 (ohb) and *Rhodococcus* sp. strain RHA1 (fcb). *Appl. Environ. Microbiol.* 72(4): 2476–2482. https://doi.org/10.1128/AEM.72.4.2476-2482.2006.

Rojas, L., Yanez, C., Gonzaalez M., Lobos, S., Smalla, K., Seeger, M. 2011. Characterization of the metabolically modified heavy metal resistant *Cupriavidus metallidurans* strain MSR33 generated for mercury bioremediation. *PLoS One.* 6(3): e17555. https://doi.org/10.1371/journal.pone.0017555.

Ronchel, M.C., Ramos, J.L. 2001. Dual system to reinforce biological containment of recombinant bacteria designed for rhizoremediation. *Appl. Environ. Microbiol.* 67: 2649–2656. doi: 10.1128/AEM.67.6.2649–2656.2001.

Russo, F., Ceci, A., Pinzari, F., Siciliano, A., Guida, M., Malusa, E., Tartanus, M., Miszczak, A., Maggi, O., Perisiani, A. 2019. Bioremediation of dichlorodiphenyltrichloroethane (DDT)-contaminated agricultural soils. Potential of two autochthonous saprtrophic fungal strains. *Appl. Environ. Microbiol.* 85(21): e01720–19. https://doi.org/10.1128/AEM.01720-19.

Saavedra, J.M., Acevedo, F., González, M., Seeger, M. 2010. Mineralization of PCBs by the genetically modified strain *Cupriavid usnecator* JMS34 and its application for bioremediation of PCBs in soil. *Appl. Microbiol. Biotechnol.* 87(4): 1543–1554. doi: 10.1007/s00253-010-2575-6.

Sakaki, T., Shinkyo, R., Takita, T., Ohta. M., Inouye, K. 2002. Biodegradation of polychlorinated dibenzo-p-dioxins by recombinant yeast expressing rat CYP1A subfamily. *Arch. Biochem. Biophys.* 401(1): 91–98. https://doi. org/10.1016/S0003–9861(02) 00036-X.

Salvi, M., Prabhakaran, R., Karuppiah, P., Paul, E., Albert, A., Selvam, S. 2017. Sequence based homology study of metallregulatory protein AsrR and cloning of *ars*R gene from *Enterobacter cloaae* in *E. coli* for arsenic bioremediation. *IJCRT.* 6(1): 2320–2882. http://doi.one/10.1729/IJCRT.17414.

Samin, G., Pavlova, M., Arif, M.I., Postema, C.P., Damborsky, J., Janssen, D.B. 2014. A *Pseudomonas putida* strain genetically engineered for 1,2,3-trichloropropane bioremediation. *Appl. Environ. Microbiol.* 80(17): 5467–5476. https://doi. org/10.1128/AEM. 01620–14.

Sato, S.I., Nam, J.W., Kasuga, K., Nojiri, H., Omori, T. 1997. Identification and characterization of genes encoding carbazole 1,9a-dioxygenase in *Pseudomonas* sp. strain CA10. *J. Bacteriol.* 179(15): 4850–4858. https://doi.org/10.1128/jb.179.15.4850-4858.1997.

Sayler, G.S. 1991. Contribution of molecular biology to bioremediation. *J. Hazard. Mater.* 28: 13–27.

Sayler, G.S., Layton, A.C. 1990. Environmental application of nucleic acid hybridization. *Annu. Rev. Microbiol.* 44: 625–648. https://doi.org/10.1146/annurev.mi.44.100190.003205.

Sayler, G.S., Ripp, S. 2000. Field applications of genetically engineered microorganisms for bioremediation processes. *Curr. Opin. Biotechnol.* 11(3): 286–289. https://doi.org/10.1016/S0958-1669(00)00097-5.

Schneegurt-Mark, A., Kulpa-Charler, F. Jr. 1998. The application of molecular techniques in environmental biotechnology for monitoring microbial systems. *Biotechnol. Appl. Biochem.* 27: 73–79. https://doi.org/10.1111/j.1470-8744.1998.tb01377.x.

Singh, M., Vishwakarma, S., Srivastava, A. 2013. Bioremediation of direct blue 14 and extracellular lignolytic enzyme production by white rot fungi: *Pleurotus* Spp. *Biomed. Res. Int.* 180156. https://doi.org/10.1155/2013/180156.

Singh, P., Singh, K.A., Borthakur, A. Ed. 2020. Bioremediation: A sustainable approach for management of environmental contaminants. *Abatement Env. Pollu. Trends Strategies.* 1–23. https://doi.org/10.1016/B978-0-12-818095-2.00001-1.

Strong, L.C., McTavish, H., Sadowsky, M.J., Wackett, L.P. 2000. Field-scale remediation of atrazine-contaminated soil using recombinant *Escherichia coli* expressing atrazine chlorohydrolase. *Environ. Microbiol.* 2(1): 91–98. https://doi.org/10.1046/j.1462-2920.2000.00079.x.

Szafranski, P., Mello, C.M., Sano, T., Smith, C.L., Kaplan, D.L., Cantor, C.R. 1997. A new approach for containment of microorganisms: Dual control of streptavidin expression by antisense RNA and the T7 transcription system. *Proc. Natl. Acad. Sci. USA.* 94: 1059–1063. https://doi.org/10.1073/pnas.94.4.1059.

Taghavi, S., Barac, T., Greenberg, B., Borremans, B., Vangronsveld, J., van der Lelie, D. 2005. Horizontal gene transfer to endogenous endophytic bacteria from poplar improves phytoremediation of toluene. *Appl. Environ. Microbiol.* 71(12): 8500–8505. https://doi.org.10.1128/AEM.71.12.8500-8505.2005.

Van der Bij, A.J., de Weger, L.A., Tucker, W.T., Lugtenberg, B. 1996. Plasmid stability in *Pseudomonas fluorescens* in the rhizosphere. *Appl. Environ. Microbiol.* 62(3): 1076–1080. https://doi.org/10.1128/aem.62.3.1076-1080.1996.

Van Dillewijn, P., Couselo, J.L., Corredoira, E., Delgado, E., Wittich, R.M., Ballester, A. 2008. Bioremediation of 2, 4, 6-trinitrotoluene by bacterial nitroreductase expressing transgenic aspen. *Environ. Sci. Technol.* 42: 7405–7410. https://doi.org/10.1021/es801231w.

Wang, M., Xu, Z., Dong, B., Zeng, Y., Chen, S., Zhang, Y., Huang, Y., Pei, X. 2022. An efficient manganese-oxidizing fungus *Cladosporium halotolerans* strain XM01: Mn (II) oxidization and Cd adsorption behaviour. *Chemosphere* 287(Part-1): 132026. https://doi.org/10.1016/j.chemosphere.2021.132026.

Wasilkowski, D., Swedziol, Z., Mrozik, A. 2012. The applicability of genetically modified microorganisms in bioremediation of contaminated environments. *Sci. Chemik.* 66(8): 817–826. https://doi.org/10.1080/07388550600842794.

Widada, J., Nojiri, H., Omori, T. 2002. Recent developments in molecular techniques for identification and monitoring of xenobiotic-degrading bacteria and their catabolic genes in bioremediation. *Appl. Microbiol. Biotechnol.* 60: 45–59. https://doi.org/10.1007/s00253-002-1072-y.

Wilson, M.S., Bakerman, C., Madsen, E.L. 1999. In situ, real-time catabolic gene expression: Extraction and characterization of naphthalene dioxygenase mRNA transcripts from groundwater. *Appl. Environ. Microbiol.* 65: 80–87. doi: 10.1128/AEM.65.1.80–87.1999.

Yang, C., Liu, N., Guo, X., Qiao, C. 2006. Cloning of mpd gene from a chlorpyrifos-degrading bacterium and use of this strain in bioremediation of contaminated soil. *FEMS Microbiol. Letter.* 265(1): 118–125. https://doi.org/10.1111/j.1574-6968.2006.00478.x.

Yuanfan, H., Jin, Z., Qing, H., Qing, W., Jiandong, J., Shunpeng, L. 2010. Characterization of Fenpropathrin-degrading strain and construction of a genetically engineered microorganism for simultaneous degradation of methyl parathion and Fenpropathrin. *J. Environ. Manag.* 91(11): 2295–2300. https://doi.org/10.1016/j.jenvman.2010.06.010.

Zang, R., Xu, X., Chen, W., Huang, Q. 2016. Genetically engineered *Pseudomonas putida* X3 strain and its potential ability to bioremediate soil microcosms contaminated with methyl parathion and cadmium. *Appl. Microbiol. Biotechnol.* 100: 1987–1997. doi: 10.1007/s00253-015-7099-7.

Zapana-Huarache, S., Romero-Sanchez, C., Duenas-Gonza, A., Torres-Huacpo, F., Rivera, A. 2020. Chromium (VI) bioremediation potential of filamentous fungi from Peruvian tannery industry effluents. *Environ. Microbiol.* 51: 271–278. https://doi.org/10.1007/s42770-019-00209-9.

Zhao, X.W., Zhou, M.H., Li, Q.B., Lu, Y.H., He, N., Sun, D.H., Deng, X. 2005. Simultaneous mercury bioaccumulation and cell propagation by genetically engineered *Escherichia coli*. *Process. Biochem.* 40(5): 1611–1616. https://doi.org/10.1016/j.procbio.2004.06.014.

Zuo, Z., Gong, T., Che, Y., Liu, R., Xu, P., Jiang, H., Qiao, C., Song, C., Yang, C. 2015. Engineering *Pseudomonas putida*KT2440 for simultaneous degradation of organophosphates and pyrethroids and its application in bioremediation of soil. *Biodegrad.* 26: 223–233. doi: 10.1007/s10532-015-9729-2.

9 Application of Genetically Modified Microorganisms for Effective Removal of Heavy Metals

Biplab Roy, Ajita Tiwari, Shamim Ahmed Khan,
Ajit Debnath and Pinku Chandra Nath

ABBREVIATIONS

HM   Heavy Metal
GMOs Genetically Modified Microorganisms
GS   Glutathione Synthetase
MT   Metallothionein
PCs   Phytochelatins
ROS   Reactive Oxygen Species

INTRODUCTION

Human activities like advancements in technology, hazardous agriculture practices, urbanization, and industrialization have increased pollution at a disquieting rate and degraded the global environment. Hazardous chemicals and potentially toxic heavy metals have caused widespread contamination of groundwater, surface water, air, and soil, posing a major threat to all living organisms on earth[1, 2]. Physicochemical and biological approaches to nontoxic byproducts cannot decompose hazardous toxic metals. Their durability in an ecosystem can be considerable and present challenges in being transformed to less injurious forms [3].Some heavy metals (HMs) such as copper (Cu), magnesium (Mg), calcium (Ca), nickel (Ni), sodium (Na), and manganese (Mn), as well as zinc (Zn) are essential nutrients required in low concentrations for various physiological and biochemical functions [4]. Other non-essential HMs like silver (Ag), Al, Cd, Pb, Hg, and gold (Au) possess no biological importance and are harmful to living organisms at even low concentrations [5]. The sources of heavy metals can be classified as illustrated in Figure 9.1.

When noxious metals are not digested by human organisms and accumulate in sensitive tissues, they pose a serious concern [6]. Heavy metals may enter the human body through drinking water and the food chain. In our body, toxicity of heavy metals can cause lung damage, cancer, skin rashes, neurological and cardiovascular diseases, sensory disturbances, nausea, and increased blood pressure [7]. In plants, heavy metals can cause low biomass accumulation, reduced enzyme activities, chlorosis, altered water balance and nutrient assimilation, inhibition of growth and photosynthesis, and senescence [8]. Besides, toxicity of metals also increases production of reactive oxygen species (ROS), which reduce antioxidant molecules, leading to cell death by disturbing the usual functioning of the organism [9]. The toxicity of Heavy metals can also be a major intimidation to microbes. It causes nucleic acid and protein denaturation, oxidative stress, inhibition of enzyme

DOI: 10.1201/9781003188568-9

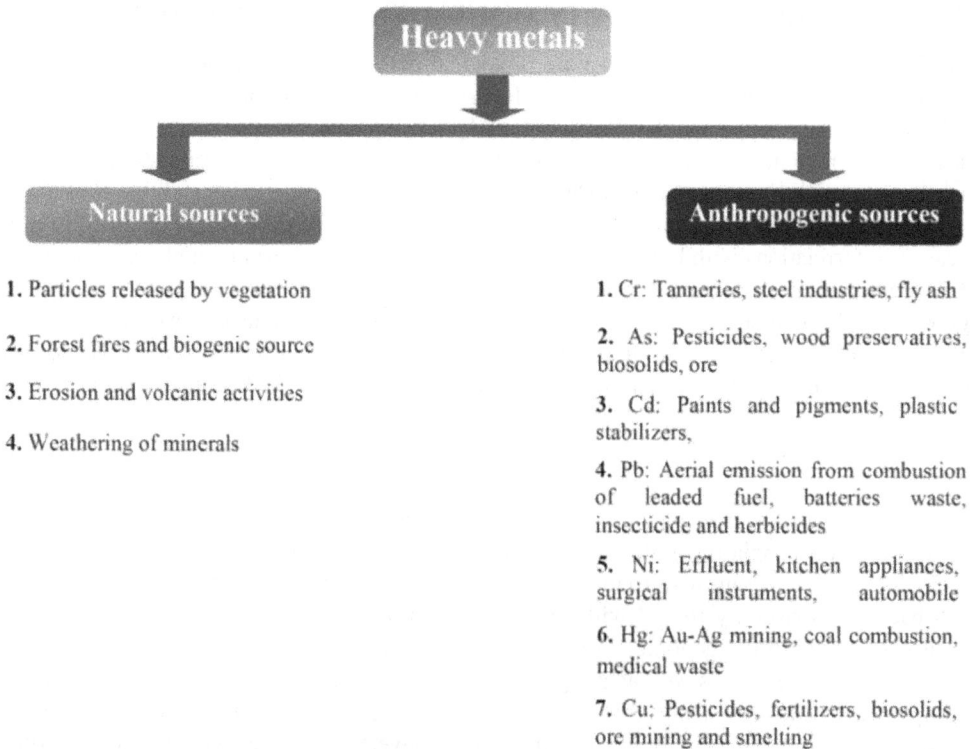

Heavy metals

Natural sources

1. Particles released by vegetation

2. Forest fires and biogenic source

3. Erosion and volcanic activities

4. Weathering of minerals

Anthropogenic sources

1. Cr: Tanneries, steel industries, fly ash

2. As: Pesticides, wood preservatives, biosolids, ore

3. Cd: Paints and pigments, plastic stabilizers,

4. Pb: Aerial emission from combustion of leaded fuel, batteries waste, insecticide and herbicides

5. Ni: Effluent, kitchen appliances, surgical instruments, automobile

6. Hg: Au-Ag mining, coal combustion, medical waste

7. Cu: Pesticides, fertilizers, biosolids, ore mining and smelting

FIGURE 9.1 Heavy metal origins in the environment.

functions, chromosomal aberrations, disruption of cellular activities and cell membranes, mutation, chromosomal aberrations, and more [10]. Therefore, considering the destructive impacts of heavy metal toxicity to living organisms, instant actions are required to address the detoxification of heavy metals from contaminated sites.

There are numerous scientific techniques applied to the recovery of heavy metals and their removal from contaminated sites. The most extensively used techniques involve ion exchange resin, adsorption, reverse osmosis, metal precipitation, electrochemical treatments, evaporation, membrane filtration, bioremediation, and flotation processes [11–17]. However, some of those technologies are hazardous, expensive, ineffective at heavy metal removal and produce secondary toxic sludge [18, 19]. Among some of these stated techniques, bioremediation is a good and appropriate method for the elimination of heavy metals [20–24]. It is a cost-effective and green technique, as this method utilizes microbes which can be available naturally in polluted web sites and can assist in the restoration of toxic heavy metals [1]. In contaminated sites, utilization of microbes including fungi, bacterial species and algae to detoxify non-essential heavy metals has appeared as a prominent generation. Microbes are omnipresent; they are tiny and multiply unexpectedly and decorate in significant numbers once they graft onto polluted sites [25]. When they are exposed continuously to pollutants, they become more tolerant and exhibit admirable capability levels to exchange pollutants as their energy source. Microorganisms can also adapt genetically to degrade the contaminants in an ecosystem. These ascribes can be exploited to build microorganisms a promising technology for an inexpensive and more eco-friendly biological process [26]. In this chapter, the most probable mechanisms of remediation processes are studied while the main focus lies on the technique of bioremediation and its field applicability, possibilities, and mechanism.

HEAVY METALS AND GENETICALLY MODIFIED MICROORGANISMS

Genetically modified organisms (GMOs) are biological entities for which the genetic substance has been altered utilizing recombinant DNA methodology. Genetically modified microorganisms are only useful if they not only thrive and develop in polluted environments, but also convey the desired genes for substantial remediation. As a result, the size of the inoculums, climatic circumstances, geographical distribution, growth rate, and the existence of competitive microorganisms all have an important impact on the features of the organisms. The geographical extent of GMOs implemented in polluted areas causes interrelations with local microflora and perhaps other environmental constituents, which affects bioremediation. Toxic heavy metals are naturally occurring components of the earth's surface. The geochemical processes and biological stability of these toxic substances have been significantly altered as a result of numerous anthropogenic activities that result in the generation of biologically active toxic substances including mercury, copper, lead, arsenic, zinc, and cadmium into ecosystems.

MERCURY

Mercury is amongst the most hazardous components on the globe. Various anthropogenic influences including electroplating industries, power plants, and chloro-alkali industries lead to the generation of mercury along with their compounds into the environment. Mercury bioremediation through bacteria is strongly linked with the representation of bacterial *mer* genes [27]. The mer operon is discovered in the genomes of mercury-resistant microbial species. This contains several properly functioning genes as well as a regulator, promoter, and operator. merA and merB are two common types of functional genes that generate organomercurial lyase and mercuric ion reductase. With the assistance of enzyme reductase, the lyase converts extremely poisonous organomercurial substances into almost environmentally benign combustible elemental mercury [28]. Mercuric reductase enzyme converts a biologically active Hg^{2+} into completely non-bio-available mercury (Hg). The genes that code for *mer*operons have been found to be extremely diverse. A second merB has already been discovered in certain bacteria, conferring wide spectrum rigidity to mercuric substances [29]. MerR, a regulatory gene found throughout many mer operons, is transposed independently as well as divergently from functional mer genes. In the appearance of stimulating concentrations of Hg^{2+}, MerR protein initiates operon transcription. *Deinococcus radiodurans*, one of the most investigated radiation-resistant organisms for remediation of blended radioactive waste products, was genetically altered to effectively address ionic mercury-contaminated sites by cloning and expressing the merA gene from *Escherichia coli* BL308 [30].

ARSENIC

When arsenic is oxidized, it becomes extremely toxic. Reducing As(V)-As(III) is not really a viable remediation strategy because As(III) is much additional hazardous than As(V). As a consequence, bioremediation of As is strongly linked to the transformation of these biologically active solubilized forms into highly volatile As compounds. Although a wide range of microbes have already been discovered as providing the ability to volatilize As, native microflora has also been shown to solubilize 2.2–4.5% of As in only one month under natural soil conditions [31]. Before As(III) may volatilize, it must undergo a series of methylation interactions, most notably with trimethylarsine, which is the final result of As(III) reduction [32]. Furthermore, when transgenic bacteria are contrasted to wild-category bacteria, there is substantial growth in the creation of volatilized As. The arsM gene isolated from *Rhodopseudomonas palustris* has been cloned and demonstrated in *E. coli* throughout the last century in order to methylate inorganic substances when it comes to highly flammable trimethylarsine [33]. Cloning the arsM gene from *Bacillus idriensis* and *Sphingomonas desiccabilis* results in a 10% rise in the discharge of methylated As gas in a water-soluble system in comparison to wild species.

LEAD

Lead is among the most toxic and harmful heavy metals on the planet. It is predominantly introduced into the atmosphere as a result of human activities, including discharge of Pb-containing batteries, smelting, mining, and combustion of fossil fuels containing Pb [34]. Lead (Pb) bioremediation is linked to the specific uptake and accretion of Pb in numerous microbes, with the prospect to transform it into inaccessible forms. It has been observed that extra chromosomal as well as chromosomal genetic material contain genes whose interpretation promotes Pb uptake and acquisition inside the microbial cell. *Pseudomonas aeruginosa*, *Salmonella choleraesuis*, and *Proteus penneri* having strains 4EA, 4A and GM10, containing genes that encode microbial metallothionein's (smtA and smtAB), which are mainly accountable for lead impedance on genomic DNA and plasmids [35].

CADMIUM

Cd is a highly poisonous metal found primarily in ores containing Pb, Zn, and Cu. Cadmium is comfortably absorbed by plant species and converted via the food supply chain, causing negative impacts on human health [36]. Phytochelatins (PCs) are inherently emerging peptides composed of a reiterating -Glu-Cys dipeptide component discharged by a certain Gly stains with a strong potential to attach toxic metals including Cd, As, Hg, and Pb, particularly Cd via thiolate complexes [37]. *Schizosaccharomyces pombe* was used in cloning and representation of PC synthase gene through *E. coli* which resulted in a sevenfold rise in Cd uptake and aggregation. Cloning and conveying plant PC genes in *E. coli* led to a significant rise in intracellular Cd composition when contrasted to the monitoring strain [38]. Sriprang et al. (2003) [39] discovered that *Mesorhizobium huakuii* modified with an *Arabidopsis thaliana* gene that codes for PCs, accumulated more Cd^{2+}. Deng et al. (2007) [40] revealed uptake of Cd^{2+} explicitly from multifactorial metal-polluted sites owing to the existence of a specialized cadmium transportation mechanism and Metallothionein (MT) protein in genetically engineered microorganisms.

HEAVY METALS AND GENETICALLY MODIFIED PLANTS

Various microbial genes have already been effectively inserted into plants including *Populus angustifolia*, *Brassica juncea*, *Nicotiana tabacum*, *Arabidopsis thaliana*, and *Silene cucubalus*, which have resulted, when compared to their wild counterparts, in increased toxic heavy metal deposition and modification. Phytochelatins are heavy metal-binding peptides that are produced by glutathione synthetase (GS), which is the primary enzyme associated with the removal of toxic metals by plants and their subsequent sequestration [41]. Thus, the *E. coli gshII* gene that codes GS was added to *Brassica juncea* so that the first rate-limiting variables for glutathione manufacturing could be overcome. The acceptance and acquisition of Cd can go up by as much as three times after the transmission and affirmation of genes that encode for gshI and gshII in *Brassica juncea* [42]. In addition to the increased concentration levels of *E. coli* enzymes, a rise in non-protein thiols and phytochelatins has even been observed, both of which result in increased biomass manufacture when contrasted to wild seedlings. It has already been observed that *Arabidopsis thaliana* modified with the microbial genes merA and merB showed a substantial rise in acceptance and transformation of methyl mercury into highly volatile as well as less hazardous elemental Hg [43]. Genetically modified *Arabidopsis thaliana* plants were transformed by cloning and expressing microbial genes that encode for gamma-glutamyl-cysteine synthetase (gamma-ECS) and arsenate reductase (arsC) that contributed to significantly higher As tolerance as well as a substantial rise in As accumulation compared to wild seedlings [44].

HEAVY METAL TOXICOLOGY IN MICROBES

Heavy metal toxicity is the functionality of an element to cause harmful consequences on microbes, and it relies upon the absorbed quantity and the bioavailability of heavy metal [45]. The toxicity of heavy metals entails numerous mechanisms, particularly destructing ion law, interacting like redox mediators throughout the technology of reactive oxygen species (ROS), interrupting poisonous enzymatic moves, and immediately affecting the production of protein and DNA [46]. The biochemical and physiological features of microbes can be transformed by the appearance of heavy metals. The non-essential HM such as chromium (Cr(III)) has the potential to alter the activity as well as structure of enzymes through interacting with their thiol and carboxyl groups [47]. Electrostatic interactions between Cr(III) complexes and weakly charged DNA phosphate groups may impact reproduction, transcription, and the generation of mutagenesis in cells [47]. Cadmium (Cd) and Chromium (Cr) are competent at inducing denaturation and oxidative destruction of microbes as well as deteriorating the bioremediation ability of microorganisms.

Heavy metals such as aluminium (Al) complexes may stabilize oxide radicals, which are liable for the destruction of DNA [48]. Cuprous oxide (Cu(I)) and Cupric oxide (Cu(II)) have the potential to catalyze the generation of ROS through Fenton and Haber-Weis processes, acting as permeable electron carriers. This has the potential to inflict significant damage to DNA structure, cytoplasmic compounds, lipids, as well as other proteins [49]. Heavy metals may additionally perform enzymatic actions through interactions with substrates, to be able to purpose configurational change in enzymes [50]. Aside from sticking to cell barriers and entering through ion channels, heavy metals can also cause imbalances in ion concentrations [47]. Lead (Pb) and cadmium (Cd) pose poisonous effects on microbes, obliterate the DNA structure, and injure cell membranes. Such toxicity is caused by metals being displaced from their inhabitant binding locations [51]. The development, metabolism, and morphology of microorganisms are all significantly influenced by altering the structure of nucleic acids, leading to functional disruption, oxidative phosphorylation, limiting enzymatic activity, and disturbing cell membranes, among other factors [52].

FACTORS INFLUENCING THE REMOVAL OF HEAVY METALS BY MICROBES

The inclination of heavy elements to be inhibitory or stimulatory to microbes is evaluated via the chemical forms of the metals, concentrations of total metal ions, and redox potential. Changes in heavy metal transit, transformation, microbial bioavailability, and valance states are all influenced by environmental conditions including pH, temperature, and humic acids. The adsorbent interface becomes more strongly charged at increasing hydrogen ion concentrations, which reduces the affinity between metallic ions and the adsorbent, increasing its toxicity. Heavy metals at acidic pH degrees have a tendency to form free ionic species with more protons to fill steel-binding sites.

Temperature plays a significant function in heavy metal adsorption. The adsorbate diffusion rate throughout the external boundary surface increases with temperature. The dissolution of heavy metals will increase because the temperature rises, which improves their bioavailability [53]. However, the activities of microbes rise with temperature at the appropriate variety, and it improves enzyme pastime and microbial metabolism, with the effect of hastening the bioremediation procedure. The conclusion of the decay technique relies upon on a variety of environmental factors as well as the substrate.

HEAVY METAL DETOXIFICATION MECHANISMS IN MICROORGANISMS

Microorganisms interact with metal ions by various mechanisms in the presence of heavy metals. As a means of resisting metal toxicity, microorganisms use a variety of methods including bioremediation and extrusion as well as enzyme-mediated reduction of exopolysaccharides and Metallothionein [54–56]. For metals in the environment, an ingenious mechanism has been developed by microbes for metal detoxification and resistance. The mechanism consists of precipitation, ion exchange, electrostatic interaction, oxidation, reduction, and surface complexation [57]. Key processes that microorganisms follow during metal resistance are metal oxidation, metal-organic complexion, metal-ligand degradation, extracellular metal sequestration, metal efflux pumps,

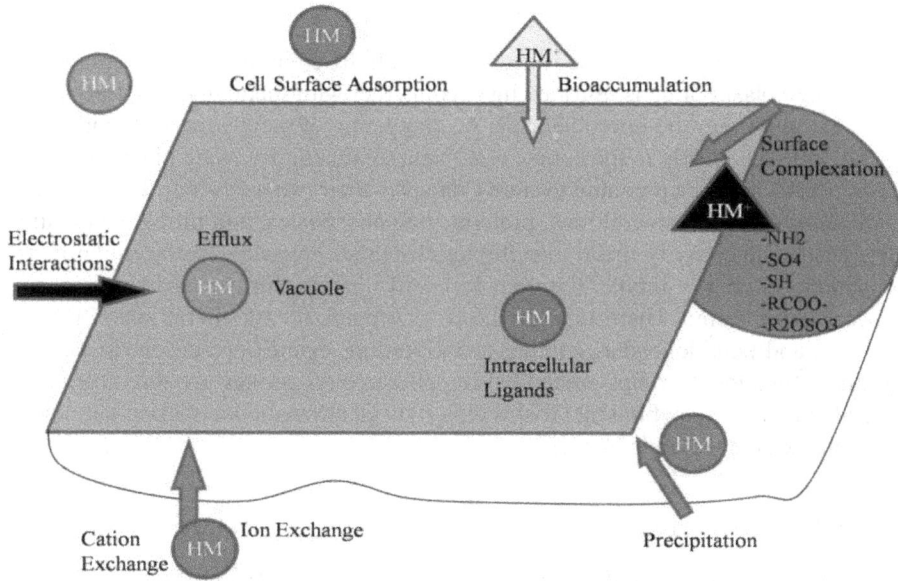

FIGURE 9.2 Mechanisms of heavy metal accumulation by microbes.

biosurfactants, methylation, and demethylation [57, 58]. Further, the decontamination of metal from microorganisms is done by volatilization, valence conversion or chemical precipitation. Processes of heavy element accumulation by microbes are illustrated in Figure 9.2.

BIOSORPTION EQUILIBRIUM AND KINETICS

Microbes as biosorbents assist in the binding of metal ions from aqueous solutions. To determine the stability of microbes as a biosorbent, it is quite important to evaluate the physical nature, sorption kinetics, sorption capacity, regeneration, and recovery of bound metal. For biosorbent selection, certain parameters should be followed: (i) the biosorbent should be cheap and reusable; (ii) it should be easily separable from the underlying solution; (iii) the rate at which binding occurs must be quick. The three major categories of biosorbents that are used for biosorption of metals from solutions are (a) living cultures, (b) exopolysaccharides, and (c) dead biomass. In comparison, living cells are less affected than dead cell in terms of metal absorbing capacity. Fungi of a filamentous nature have very high metal absorbing capacity, such as *Penicillium, Rhizopus, Mucor*, and *Aspergillus*.

Microorganisms have very complex structures and are capable of absorbing metal ions in many ways by the microbial cell. On the basis of cellular metabolism, biosorption mechanisms are divided into metabolism-established and non-metabolism based. An intercellular accumulation results from the motion of steel throughout the cell membrane depending on the cell's metabolism. It is typically associated with the lively protection system of the organism, which responds to the occurrence of heavy poisonous metals [59, 60]. However, in non-metabolism-based biosorption, the interplay of the functional businesses present in the microbial cellular surface with toxic metal by means of a physio-chemical procedure is the basis of metal uptake. It is based on bodily adsorption, chemical adsorption with ion alternate procedure, and is unbiased regarding cell metabolism [59, 60]. The fundamental additives inside the cell wall of microbial biomass are composed of proteins, lipids, and polysaccharides. They possess wealthy metallic binding capacities, as in the cases of hydroxyl, phosphoryl, carboxyl, ester, amine, sulphate, thioethers, and thiol corporations [61]. The non-metabolism established biosorption is exceptionally speedy and can be reversible [62].

Intracellular Sequestration

Intracellular sequestration refers to the process by which different chemicals combine with metal ions within the cytoplasm of cells. Surface ligands interact with metal, causing the slow transport of metal concentration into the microbial cell. Accumulation of metals intracellularly by bacterial cells has been exploited mostly in the course of effluent treatment. By using glutathione, *Rhizobium leguminosarum* cells have the potential to store Cd ions in their own cells [63]. The stiff cell walls of fungi contain chitin, lipids, mineral ions, proteins, polyphosphates, and nitrogen-containing polysaccharides, which can detoxify metal ions through valence conversion, intracellular and extracellular precipitation, or energy intake. External cell wall ligands eliminate inorganic heavy metals that have been administered where ligands act as scavengers for the metal ions. Proteins, lipids, polysaccharides, and peptidoglycans contain metal-binding ligands present in the cell walls [61]. Among so many functional groups embedded in cell barriers, amines are the most prominent in terms of metal uptake, as they bind metal ions either via electrostatic interaction (anionic metal) or by surface complexation (cationic metal).

Extracellular Sequestration

When an insoluble substance contains metal ion and they accumulate in the periplasm of cells, it is referred to as extracellular sequestration. In order for bacteria to sequentially deposit metallic ions in the periplasm, metal ions are expelled from the microbe's cytoplasm. By efflux system, a zinc ion crosses the cytoplasm and accumulates in the periplasm of Synechocyst is PCC-6803 strain. Metal precipitation occurs as a result of extracellular sequestration. *Geobacter* sp. and *Desulfuromonas* sp. (sulfur-reducing bacteria) are very effective in reducing heavy and harmful metals. An anaerobe, *G. metallireducens* reduces uranium (U) from toxic U(VI)-U(IV) and poisonous Mn(IV)-Mn(II) [64]. *G. metallireducens* and *G. sulfurreducens* are capable of converting the very fatal Cr(VI) into a less toxic Cr(III) [65]. Hydrogen sulfide (H_2S) liberated by sulfate-reducing bacteria initiate the precipitation of metal cations. Under anaerobic conditions, *Klebsiella planticola* strain produces H_2S, which causes precipitation of insoluble cadmium sulfide complex. The *Vibrio harveyi* strain generates soluble bivalent lead phosphate salt as a precipitate [66, 67].

Extracellular Metal-Resistant Barrier in Microbial Cells

Microbial cell walls, plasma membranes, and capsules are capable of inhibiting metal ions from entering cells. An ionizable group (hydroxyl, carboxyl, amino, and phosphate) present in the cell wall of bacteria absorbs metal ions [68]. Nonviable cells *Brevibacterium* sp. and *P. putida* biofilm cells show high levels of biosorption to Cu, Zn, and Pb elements than planktonic ones, while peripheral biofilms cells were killed [69]. The bacterial cells privileged in the biofilm were protected by extracellular polymers by accumulating metal ions on the outer surface of cell walls [70].

Methylation of Metals

The mechanism of methylation of metal is considered a phenomenon of resistance of metal, since only certain metal involves in such processes. Methylation increases the lipophilicity and hence the toxicity consequences of saturation in cell membranes. However, the process of volatilization results in the diffusion of metal away from the cell, and thus decreases the metal toxicity. Mercury, lead, selenium, arsenic, and tin can be effectively eliminated by the volatilization method. For example, poisonous mercury (Hg^{2+}) can be readily oxidized to volatile methyl mercury and dimethyl mercury and can be easily diffused away from the cell [71]. The metal methylation technique has been widely applied for the deduction of metal from water with various impurities. Bio methylation of arsenic

(As) to volatile arsines, selenium (Se) to gaseous dimethyl selenide, and lead (Pb) to dimethyl lead was also archived in literature [72].

MICROBIAL CELLS' REDUCTION OF HEAVY METAL IONS

The toxicity of destructive heavy elements can be decreased or even eliminated by manipulating their oxidation state [73]. For energy generation, bacteria employ metalloids and metals as electron acceptors or donors. Metals in reductive form can act as electron donors and electron acceptors in their oxidized form during anaerobic respiration. Reduction of metal ions via enzymatic processes generates its less toxic form such as $Cr(VI)$-$Cr(III)$ and $Hg(II)$-$Hg(I)$ [74].

MICROORGANISMS' BIOREMEDIATION CAPACITY FOR HEAVY METALS

Heavy metals are absorbed via microorganisms through bioaccumulation (lively approach) or adsorption (passive method). Several microorganisms including algae and fungi have been employed to clean up enflamed heavy steel surroundings, according to the EPA.

HEAVY METAL REMEDIATION BY BACTERIA

The biosorptive abilities of microbial biomass vary greatly between microbes (Table 9.1) [75–79]. Pretreatment and experimental conditions, on the other hand, determine each microbial cell's biosorption ability. Microbiological cells must be able to respond to changes in the bioreactor's and physicochemical properties in addition to enhancing biosorption [80]. Microorganisms such as bacteria are crucial biosorbents because of their widespread dispersion, long lifespan, capacity to proliferate in uncontrolled environments, and tolerance to environmental variables [81, 82].

De Jaysankar et al. (2008) [83] utilize mercury-resistant microorganisms together with *B. pumilus*, *A. faecalis*, *B. iodinium*, and *P. aeruginosa* to dispose of Pb and Cd. *P. Aeruginosa* and *A. Faecalis* reduced Cd concentrations by 70% and 75%, respectively, in this study, with reductions of 1000–17.5 mg/L through employing *P. Aeruginosa* and 19.4 mg/L by utilizing *A. Faecalis* after 72 hours. The removal of Pb by *B. iodinium* and *B. pumilus*is above 89% and 90%, correspondingly, with a decrease of 1000–1.9 mg/L after 96 hours. *Bacillus cereus*, an indigenous facultative anaerobic *Bacillus*, was utilized in a study for the detoxification of Cr(VI). *B. cereus* has a top-notch ability of 73% Cr(VI) deduction at 1000 g/mL chromate content. The microorganisms were able to minimize Cr(VI) under a wide range of temperatures (35–45°C) and pH (6–10), showing the most effectiveness at 37°C and a preliminary pH of 8.0 [84, 85].

Pseudomonas, Bacillus, Enterobacter, Flavobacterium, and *Micrococcus* sp. were used to test numerous heavy metals. Their significant surface-to-quantity ratios and dynamic chemosorption sites on the cell wall all contribute to their high biosorption potential. Bacteria in blended cultures

TABLE 9.1
Different Bacterial Species Utilized for Deduction of Heavy Metal

Bacterial strains	Metal ion	Deduction (%)	Highest adsorption rate (mg/g)	Ref.
Micrococcus sp.	Cr^{6+}, Ni^{2+}	55–92	NA	75
Enterobacter cloacae	Cu^{2+}, Co^{2+}	1.44–66.93	1.23–16.1	76
Citrobacter freudii	U^{6+}	NA	48.02	77
Pseudomonas aeruginosa	Cd^{2+}	99	NA	78
Bacillus laterosporus	Cr^{6+}, Cd^{2+}	84	61.0–158.6	79

are stronger and stay longer [86]. Because of this metabolic improvement, life consortia are more field-appropriate for metal biosorption. De Jaysankar et al. (2008) [83] found that with the aid of the use of a microbe grouping of *Arthrobacter* sp. and *Acinetobacter* sp. with a metallic ion attention of 16 mg/L, they may reduce chromium (Cr) by 78%. *Micrococcus luteus* was used to do away with a large amount of Pb from an artificial medium. In optimal situations, the removal capability became 1966 mg/g [87].

HEAVY METAL REMEDIATION BY FUNGI

Because of excessive steel absorption and repair powers, fungal species are mostly employed as biosorbents for the elimination of poisonous metals from the surroundings [88]. The majority of investigations found that both energetic and latent fungal cells have an essential position in the adherence of inorganic chemicals to inorganic substrates [89–91]. Srivastava and Thakur [92] moreover referred to the effectiveness of *Aspergillus* sp. in putting off chromium from tannery wastewater. The bioreactor's synthesized media has a higher success rate of removing chromium at a pH of 6.86 compared to the 65% success rate seen in tannery effluents. Due to natural contamination, an organism's ability to grow may be hindered [93].

Investigations are being conducted into the capability of *Coprinopsis atramentaria* to bioaccumulation 76% of Cd^{2+} at 1 mg/L concentration and 93.7% of Pb^{2+} at 800 mg/L concentration, respectively. As a consequence of this, it has been identified as a potent heavy metal ion accumulator for the purpose of mycoremediation. There are several dead fungi that can be utilized to transfer hazardous Cr(VI) to a much lesser carcinogenic or risk-free Cr(III), such as *R. oryzae*, *A. niger*, *P. chrysogenum*, and *S. cerevisiae*, in step with Park et al. (2005) [94]. Luna et al. (2016) [95] also located that bio-surfactants produced by *C. sphaerica* eliminate Fe, Zn, and Pb with elimination efficiencies of 95%, 90%, and 79%, respectively. Before detaching from the soil, those surfactants may generate compounds with metallic ions and engage with toxic substances immediately. *Candida* sp. gather sufficient amounts of Ni (58–72%) and Cu (53–68%); however, the method was developed as a result of preliminary measurements of metal ion concentration and pH (with a range of 3–5) [96]. A variety of fungal species are employed to eliminate heavy metals from the soil (Table 9.2) [97–101].

REMOVAL OF HEAVY METAL BY BIOFILM

Several researches on using biofilms to do away with heavy metals were published. Biofilm is a bioremediation tool that also features as a biological stabilizer. Biofilms have an excessive tolerance for poisonous inorganic factors, even at lethal concentrations. According to a study on *Rhodotorula*

TABLE 9.2
Several Fungal Species Utilized for Deduction of Heavy Metal

Fungi	Metal ions	Deduction (%)	Highest adsorption capacity (mg/g)	Ref.
A. niger	Cu^{2+}, Pb^{2+}	47.01–48.85	10.75–14.42	97
Candida albicans	Hg^{2+}	77.2–94.2	NA	98
Aspergillus sp.	Cr^{6+}, Ni^{2+}	90–92	NA	97
Cunninghamella elegans	Cd^{2+}	70–81	280	99
Mucor rouxii	Pb^{2+}, Cd^{2+}, Zn^{2+}	95	6.42–54.81	100
Pleurotus platypus	Cd^{2+}	76.55	10.07–13.27	101

mucilaginosa, planktonic cells were able to get rid of metals at a rate of 4.65 to 10.35%. Biofilm cells were able to get rid of metals at a rate of 91.69 to 95.37% [102]. Biosorbent or exopolymeric materials found in biofilms that have molecules with emulsifier or surfactant properties can be used to bioremediate through biosorbent or exopolymeric materials found in biofilms that have molecules with these properties.

HEAVY METAL REMEDIATION BY ALGAE

Algae, unlike other microbial biosorbents, are self-sustaining, requiring just a small amount of nutrients and producing a large proportion of biomass. Heavy metal removal with excessive sorption ability has additionally been achieved with these biosorbents [103]. It is possible to bioremediate heavy steel-polluted wastewater by adsorbing or integrating algal biomass into the cells. Heavy metal remediation is made possible via Phycoremediation, which makes use of several types of microalgae and cyanobacteria to eliminate or decompose toxicants. Carboxyl, amide, and phosphate moieties on the interface of algae serve as metal-binding sites because of their chemical properties [104].

Sheekh et al. (2019) [105] used dead *C. vulgaris* cells to eliminate Cd^{2+}, Cu^{2+}, and Pb^{2+} ions from aqueous solution underneath varied pH, biosorbents dosages, and different contact time conditions [106]. An analysis of *C. vulgaris* biomass showed that it could effectively remove Cd^{2+}, Cu^{2+}, and Pb^{2+} from the environment from blended solutions containing 50 mg dm; 3 for every steel ion at 95.5%, 97.7%, and 99.4%, respectively. To eliminate heavy metals from the environment, a variety of algae species have been employed (Table 9.3) [107–113].

CONCLUSION

Although conventional physicochemical and biological algorithms have been extensively used to remediate sites polluted with heavy metals, the utilization and implementation of GMOs has substantially increased their significance and participation in the separation of these contaminants. This is due in large part to a number of inherent problems associated with using biological processes including their ability to connect pollutant sites and compete for resources. Therefore, for establishing GMOs for environmental bioremediation, modifications in GMOs should also be taken into consideration regarding their continued existence, compatibility with native populations, and motility to contaminants.

TABLE 9.3
Different Algal Varieties Utilized for Deduction of Heavy Metals

Algal variety	Metal ion	Deduction (%)	Highest adsorption capacity (mg/g)	Ref.
Spirogyra sp.	Pb^{2+}	80	140	107
Padina sp.	Cd^{2+}	90	59.57	107
Sargassum tenerrimmum	Cd^{2+}	80	42.41	108
Chlorella vulgaris, *Anabaena spiroides*	Pb^{2+}, Cu^{2+}, Cd^{2+}	80	0.32–14.61	109
Bifurcaria bifurcata	Cd^{2+}	>90	64–95	110
Durvillaea potatorum	Pb^{2+}, Cu^{2+}	90	82.62–321.16	111
Ecklonia radiata	Pb^{2+}	60	282	112
Sargassum hystrix	Pb^{2+}	98	41.8–285	113

REFERENCES

[1] Ayangbenro, Ayansina, and Olubukola Babalola. 2017. "A New Strategy for Heavy Metal Polluted Environments: A Review of Microbial Biosorbents." *International Journal of Environmental Research and Public Health* 14 (1): 94.

[2] Soni, Ravindra, Biplab Dash, Prahalad Kumar, Udit Nandan Mishra, and Reeta Goel. 2019. "Microbes for Bioremediation of Heavy Metals." In *Microbial Interventions in Agriculture and Environment*, 129–141. Singapore: Springer.

[3] Jaishankar, Monisha, Tenzin Tseten, Naresh Anbalagan, Blessy B. Mathew, and Krishnamurthy N. Beeregowda. 2014. "Toxicity, Mechanism and Health Effects of Some Heavy Metals." *Interdisciplinary Toxicology* 7 (2): 60–72.

[4] Khalid, B. Y., and J. Tinsley. 1980. "Some Effects of Nickel Toxicity on Rye Grass." *Plant and Soil* 55 (1): 139–144.

[5] Tchounwou, Paul B., Clement G. Yedjou, Anita K. Patlolla, and Dwayne J. Sutton. 2012. "Heavy Metal Toxicity and the Environment." *Experientia Supplementum* 101: 133–164.

[6] Valko, Marian, Klaudia Jomova, Christopher J. Rhodes, Kamil Kuča, and Kamil Musílek. 2016. "Redox- and Non-Redox-Metal-Induced Formation of Free Radicals and Their Role in Human Disease." *Archives of Toxicology* 90 (1): 1–37.

[7] Balali-Mood, Mahdi, Kobra Naseri, Zoya Tahergorabi, Mohammad Reza Khazdair, and Mahmood Sadeghi. 2021. "Toxic Mechanisms of Five Heavy Metals: Mercury, Lead, Chromium, Cadmium, and Arsenic." *Frontiers in Pharmacology* 12 (April).

[8] Nagajyoti, P. C., K. D. Lee, and T. V. M. Sreekanth. 2010. "Heavy Metals, Occurrence and Toxicity for Plants: A Review." *Environmental Chemistry Letters* 8 (3): 199–216.

[9] Fashola, Muibat, Veronica Ngole-Jeme, and Olubukola Babalola. 2016. "Heavy Metal Pollution from Gold Mines: Environmental Effects and Bacterial Strategies for Resistance." *International Journal of Environmental Research and Public Health* 13 (11): 1047.

[10] Gordon, T. 2003. "Beryllium: Genotoxicity and Carcinogenicity." *Mutation Research/Fundamental and Molecular Mechanisms of Mutagenesis* 533 (1–2): 99–105.

[11] Annadurai, G., R.S. Juang, and D.J. Lee. 2003. "Adsorption of Heavy Metals from Water Using Banana and Orange Peels." *Water Science and Technology* 47 (1): 185–90.

[12] Chen, Quanyuan, Zhou Luo, Colin Hills, Gang Xue, and Mark Tyrer. 2009. "Precipitation of Heavy Metals from Wastewater Using Simulated Flue Gas: Sequent Additions of Fly Ash, Lime and Carbon Dioxide." *Water Research* 43 (10): 2605–2614.

[13] Tran, Thien-Khanh, Kuo-Feng Chiu, Chiu-Yue Lin, and Hoang-Jyh Leu. 2017. "Electrochemical Treatment of Wastewater: Selectivity of the Heavy Metals Removal Process." *International Journal of Hydrogen Energy* 42 (45): 27741–27748.

[14] Mnif, Amine, Imen Bejaoui, Meral Mouelhi, and Béchir Hamrouni. 2017. "Hexavalent Chromium Removal from Model Water and Car Shock Absorber Factory Effluent by Nanofiltration and Reverse Osmosis Membrane." *International Journal of Analytical Chemistry* 2017: 1–10.

[15] Blöcher, C., J. Dorda, V. Mavrov, H. Chmiel, N.K. Lazaridis, and K.A. Matis. 2003. "Hybrid Flotation—Membrane Filtration Process for the Removal of Heavy Metal Ions from Wastewater." *Water Research* 37 (16): 4018–4026.

[16] Jakob, A., S. Stucki, and P. Kuhn. 1995. "Evaporation of Heavy Metals During the Heat Treatment of Municipal Solid Waste Incinerator Fly Ash." *Environmental Science & Technology* 29 (9): 2429–2436.

[17] Aldrich, C., and D. Feng. 2000. "Removal of Heavy Metals from Wastewater Effluents by Biosorptive Flotation." *Minerals Engineering* 13 (10–11): 1129–1138.

[18] Aziz, Hamidi A., Mohd N. Adlan, and Kamar S. Ariffin. 2008. "Heavy Metals (Cd, Pb, Zn, Ni, Cu and Cr(III)) Removal from Water in Malaysia: Post Treatment by High Quality Limestone." *Bioresource Technology* 99 (6): 1578–1583.

[19] Hakizimana, Jean Nepo, Bouchaib Gourich, Mohammed Chafi, Youssef Stiriba, Christophe Vial, Patrick Drogui, and Jamal Naja. 2017. "Electrocoagulation Process in Water Treatment: A Review of Electrocoagulation Modeling Approaches." *Desalination* 404 (February): 1–21.

[20] Zeyad, Mohammad Tarique, Waquar Akhter Ansari, Mohd Aamir, Ram Krishna, Sushil Kumar Singh, Neelam Atri, and Abdul Malik. 2021. "Heavy Metals Toxicity to Food Crops and Application of Microorganisms in Bioremediation." In *Microbe Mediated Remediation of Environmental Contaminants*, 421–434. Sawston, Cambridge: Elsevier.

[21] Girma, Gosa. 2015. "Microbial Bioremediation of Some Heavy Metals in Soils: An Updated Review." *Egyptian Academic Journal of Biological Sciences, G. Microbiology* 7 (1): 29–45.

[22] Wang, Shipei, Ting Liu, Xiao Xiao, and Shenglian Luo. 2021. "Advances in Microbial Remediation for Heavy Metal Treatment: A Mini Review." *Journal of Leather Science and Engineering* 3 (1): 1.

[23] Kapoor, A., and T. Viraraghavan. 1995. "Fungal Biosorption—an Alternative Treatment Option for Heavy Metal Bearing Wastewaters: A Review." *Bioresource Technology* 53 (3): 195–206.

[24] Verma, Samakshi, and Arindam Kuila. 2019. "Bioremediation of Heavy Metals by Microbial Process." *Environmental Technology & Innovation* 14 (May): 100369.

[25] Barkay, Tamar, and Jeffra Schaefer. 2001. "Metal and Radionuclide Bioremediation: Issues, Considerations and Potentials." *Current Opinion in Microbiology* 4 (3): 318–323.

[26] Mandal, A., M. Madourie, R. Maharagh, and M. Voutchkov. 2015. "Erratum to 'Heavy Metals in Soils around the Cement Factory in Rockfort, Kingston, Jamaica' [International Journal of Geosciences 2 (2011) 48–54]." *International Journal of Geosciences* 6 (3): 246–246.

[27] Jackson, W. J., and A. O. Summers. 1982. "Biochemical Characterization of HgCl2-Inducible Polypeptides Encoded by the Mer Operon of Plasmid R100." *Journal of Bacteriology* 151 (2): 962–970.

[28] Dash, Hirak R., and Surajit Das. 2012. "Bioremediation of Mercury and the Importance of Bacterial Mer Genes." *International Biodeterioration & Biodegradation* 75 (November): 207–213.

[29] Barkay, Tamar, Susan M. Miller, and Anne O. Summers. 2003. "Bacterial Mercury Resistance from Atoms to Ecosystems." *FEMS Microbiology Reviews* 27 (2–3): 355–384.

[30] Brim, Hassan, Sara C. McFarlan, James K. Fredrickson, Kenneth W. Minton, Min Zhai, Lawrence P. Wackett, and Michael J. Daly. 2000. "Engineering Deinococcus Radiodurans for Metal Remediation in Radioactive Mixed Waste Environments." *Nature Biotechnology* 18 (1): 85–90.

[31] Liu, Shuang, Fan Zhang, Jian Chen, and Guoxin Sun. 2011. "Arsenic Removal from Contaminated Soil via Biovolatilization by Genetically Engineered Bacteria Under Laboratory Conditions." *Journal of Environmental Sciences* 23 (9): 1544–1550.

[32] Turpeinen, Riina, Mari Pantsar-Kallio, and Timo Kairesalo. 2002. "Role of Microbes in Controlling the Speciation of Arsenic and Production of Arsines in Contaminated Soils." *Science of The Total Environment* 285 (1–3): 133–145.

[33] Yuan, Chungang, Xiufen Lu, Jie Qin, Barry P. Rosen, and X. Chris Le. 2008. "Volatile Arsenic Species Released from Escherichia Coli Expressing the AsIII S-Adenosylmethionine Methyltransferase Gene." *Environmental Science & Technology* 42 (9): 3201–3206.

[34] Cho, Dae Haeng, and Eui Yong Kim. 2003. "Characterization of Pb2+ Biosorption from Aqueous Solution by Rhodotorula Glutinis." *Bioprocess and Biosystems Engineering* 25 (5): 271–277.

[35] Naik, Milind Mohan, and Santosh Kumar Dubey. 2013. "Lead Resistant Bacteria: Lead Resistance Mechanisms, Their Applications in Lead Bioremediation and Biomonitoring." *Ecotoxicology and Environmental Safety* 98 (December): 1–7.

[36] Kumar, Abhijit, Swaranjit Singh Cameotra, and Saurabh Gupta. 2012. "Screening and Characterization of Potential Cadmium Biosorbent Alcaligenes Strain from Industrial Effluent." *Journal of Basic Microbiology* 52 (2): 160–166.

[37] Mejáre, Malin, and Leif Bülow. 2001. "Metal-Binding Proteins and Peptides in Bioremediation and Phytoremediation of Heavy Metals." *Trends in Biotechnology* 19 (2): 67–73.

[38] Sauge-Merle, Sandrine, Stéphan Cuiné, Patrick Carrier, Catherine Lecomte-Pradines, Doan-Trung Luu, and Gilles Peltier. 2003. "Enhanced Toxic Metal Accumulation in Engineered Bacterial Cells Expressing Arabidopsis Thaliana Phytochelatin Synthase." *Applied and Environmental Microbiology* 69 (1): 490–494.

[39] Sriprang, Rutchadaporn, Makoto Hayashi, Hisayo Ono, Masahiro Takagi, Kazumasa Hirata, and Yoshikatsu Murooka. 2003. "Enhanced Accumulation of Cd 2+ by a Mesorhizobium Sp. Transformed with a Gene from Arabidopsis Thaliana Coding for Phytochelatin Synthase." *Applied and Environmental Microbiology* 69 (3): 1791–1796.

[40] Deng, X., X.E. Yi, and G. Liu. 2007. "Cadmium Removal from Aqueous Solution by Gene-Modified Escherichia Coli JM109." *Journal of Hazardous Materials* 139 (2): 340–44.

[41] Liang Zhu, Yong, Elizabeth A.H. Pilon-Smits, Lise Jouanin, and Norman Terry. 1999. "Overexpression of Glutathione Synthetase in Indian Mustard Enhances Cadmium Accumulation and Tolerance1." *Plant Physiology* 119 (1): 73–80.

[42] Ow, David W. 1996. "Heavy Metal Tolerance Genes: Prospective Tools for Bioremediation." *Resources, Conservation and Recycling* 18 (1–4): 135–149.

[43] Eapen, Susan, and S.F. D'Souza. 2005. "Prospects of Genetic Engineering of Plants for Phytoremediation of Toxic Metals." *Biotechnology Advances* 23 (2): 97–114.

[44] Pilon-Smits, Elizabeth, and Marinus Pilon. 2002. "Phytoremediation of Metals Using Transgenic Plants." *Critical Reviews in Plant Sciences* 21 (5): 439–456.

[45] Rasmussen, Lasse D., Søren J. Sørensen, Ralph R. Turner, and Tamar Barkay. 2000. "Application of a Mer-Lux Biosensor for Estimating Bioavailable Mercury in Soil." *Soil Biology and Biochemistry* 32 (5): 639–646.

[46] Hildebrandt, Ulrich, Marjana Regvar, and Hermann Bothe. 2007. "Arbuscular Mycorrhiza and Heavy Metal Tolerance." *Phytochemistry* 68 (1): 139–146.

[47] Igiri, Bernard E., Stanley I. R. Okoduwa, Grace O. Idoko, Ebere P. Akabuogu, Abraham O. Adeyi, and Ibe K. Ejiogu. 2018. "Toxicity and Bioremediation of Heavy Metals Contaminated Ecosystem from Tannery Wastewater: A Review." *Journal of Toxicology* 2018 (September): 1–16.

[48] Booth, Sean C., Aalim M. Weljie, and Raymond J. Turner. 2015. "Metabolomics Reveals Differences of Metal Toxicity in Cultures of Pseudomonas Pseudoalcaligenes KF707 Grown on Different Carbon Sources." *Frontiers in Microbiology* 6 (August).

[49] Argüello, José M., Daniel Raimunda, and Teresita Padilla-Benavides. 2013. "Mechanisms of Copper Homeostasis in Bacteria." *Frontiers in Cellular and Infection Microbiology* 3.

[50] Gauthier, Patrick T., Warren P. Norwood, Ellie E. Prepas, and Greg G. Pyle. 2014. "Metal–PAH Mixtures in the Aquatic Environment: A Review of Co-Toxic Mechanisms Leading to More-than-Additive Outcomes." *Aquatic Toxicology* 154 (September): 253–269.

[51] Olaniran, Ademola, Adhika Balgobind, and Balakrishna Pillay. 2013. "Bioavailability of Heavy Metals in Soil: Impact on Microbial Biodegradation of Organic Compounds and Possible Improvement Strategies." *International Journal of Molecular Sciences* 14 (5): 10197–10228.

[52] Bissen, Monique, and Fritz H. Frimmel. 2003. "Arsenic—a Review. Part I: Occurrence, Toxicity, Speciation, Mobility." *Acta Hydrochimica et Hydrobiologica* 31 (1): 9–18.

[53] Bissen, Monique, and Fritz H. Frimmel. 2003. "Arsenic—a Review. Part I: Occurrence, Toxicity, Speciation, Mobility." *Acta Hydrochimica et Hydrobiologica* 31 (1): 9–18.

[54] Gillespie, Iain M.M., and Jim C. Philp. 2013. "Bioremediation, an Environmental Remediation Technology for the Bioeconomy." *Trends in Biotechnology* 31 (6): 329–332.

[55] Dixit, Ruchita, Wasiullah Deepti Malaviya, Kuppusamy Pandiyan, Udai Singh, Asha Sahu, Renu Shukla et al. 2015. "Bioremediation of Heavy Metals from Soil and Aquatic Environment: An Overview of Principles and Criteria of Fundamental Processes." *Sustainability* 7 (2): 2189–2212.

[56] Thakare, Mayur, Hemen Sarma, Shraddha Datar, Arpita Roy, Prajakta Pawar, Kanupriya Gupta, Soumya Pandit, and Ram Prasad. 2021. "Understanding the Holistic Approach to Plant-Microbe Remediation Technologies for Removing Heavy Metals and Radionuclides from Soil." *Current Research in Biotechnology* 3: 84–98.

[57] Gholizadeh, Mortaza, and Xun Hu. 2021. "Removal of Heavy Metals from Soil with Biochar Composite: A Critical Review of the Mechanism." *Journal of Environmental Chemical Engineering* 9 (5): 105830.

[58] Yin, Kun, Qiaoning Wang, Min Lv, and Lingxin Chen. 2019. "Microorganism Remediation Strategies Towards Heavy Metals." *Chemical Engineering Journal* 360 (March): 1553–1563.

[59] Priyadarshanee, Monika, and Surajit Das. 2021. "Biosorption and Removal of Toxic Heavy Metals by Metal Tolerating Bacteria for Bioremediation of Metal Contamination: A Comprehensive Review." *Journal of Environmental Chemical Engineering* 9 (1): 104686.

[60] Ubando, Aristotle T., Aaron Don M. Africa, Marla C. Maniquiz-Redillas, Alvin B. Culaba, Wei-Hsin Chen, and Jo-Shu Chang. 2021. "Microalgal Biosorption of Heavy Metals: A Comprehensive Bibliometric Review." *Journal of Hazardous Materials* 402 (January): 123431.

[61] Lu, Ningqin, Tianjue Hu, Yunbo Zhai, Huaqing Qin, Jamila Aliyeva, and Hao Zhang. 2020. "Fungal Cell with Artificial Metal Container for Heavy Metals Biosorption: Equilibrium, Kinetics Study and Mechanisms Analysis." *Environmental Research* 182 (March): 109061.

[62] Kuyucak, N., and B. Volesky. 1988. "Biosorbents for Recovery of Metals from Industrial Solutions." *Biotechnology Letters* 10 (2): 137–142.

[63] Beni, Ali Aghababai, and Akbar Esmaeili. 2020. "Biosorption, an Efficient Method for Removing Heavy Metals from Industrial Effluents: A Review." *Environmental Technology & Innovation* 17 (February): 100503.

[64] Gavrilescu, M. 2004. "Removal of Heavy Metals from the Environment by Biosorption." *Engineering in Life Sciences* 4 (3): 219–232.

[65] Bruschi, Mireille, and Florence Goulhen. n.d. "New Bioremediation Technologies to Remove Heavy Metals and Radionuclides Using Fe(III)-, Sulfate- and Sulfur- Reducing Bacteria." In *Environmental Bioremediation Technologies*, 35–55. Berlin, Heidelberg: Springer.

[66] Sharma, Pramod K., David L. Balkwill, Anatoly Frenkel, and Murthy A. Vairavamurthy. 2000. "A New Klebsiella Planticola Strain (Cd-1) Grows Anaerobically at High Cadmium Concentrations and Precipitates Cadmium Sulfide." *Applied and Environmental Microbiology* 66 (7): 3083–3087.

[67] Mire, Chad E., Jeanette A. Tourjee, William F. O'Brien, Kandalam V. Ramanujachary, and Gregory B. Hecht. 2004. "Lead Precipitation by Vibrio Harveyi: Evidence for Novel Quorum-Sensing Interactions." *Applied and Environmental Microbiology* 70 (2): 855–864.

[68] Gupta, Pratima, and Batul Diwan. 2017. "Bacterial Exopolysaccharide Mediated Heavy Metal Removal: A Review on Biosynthesis, Mechanism and Remediation Strategies." *Biotechnology Reports* 13 (March): 58–71.

[69] Pardo, Rafael, Mar Herguedas, Enrique Barrado, and Marisol Vega. 2003. "Biosorption of Cadmium, Copper, Lead and Zinc by Inactive Biomass of Pseudomonas Putida." *Analytical and Bioanalytical Chemistry* 376 (1): 26–32.

[70] Gadd, Geoffrey M. 2008. "Accumulation and Transformation of Metals by Microorganisms." In *Biotechnology*, 225–264. Weinheim, Germany: Wiley-VCH Verlag GmbH.

[71] Park, Jung-Duck, and Wei Zheng. 2012. "Human Exposure and Health Effects of Inorganic and Elemental Mercury." *Journal of Preventive Medicine & Public Health* 45 (6): 344–352.

[72] Wood, J. M., A. Cheh, L. J. Dizikes, W. P. Ridley, S. Rakow, and J. R. Lakowicz. 1978. "Mechanisms for the Biomethylation of Metals and Metalloids." *Federation Proceedings* 37 (1): 16–21.

[73] Jaishankar, Monisha, Tenzin Tseten, Naresh Anbalagan, Blessy B. Mathew, and Krishnamurthy N. Beeregowda. 2014. "Toxicity, Mechanism and Health Effects of Some Heavy Metals." *Interdisciplinary Toxicology* 7 (2): 60–72.

[74] Lloyd, Jonathan R. 2003. "Microbial Reduction of Metals and Radionuclides." *FEMS Microbiology Reviews* 27 (2–3): 411–425.

[75] Kaliannan Thamaraiselvi. 2007. "Biosorption of Chromium and Nickel by Heavy Metal Resistant Fungal and Bacterial Isolates." *Journal of Hazardous Materials* 146 (1–2): 270–277.

[76] Iyer, Anita, Kalpana Mody, and Bhavanath Jha. 2005. "Biosorption of Heavy Metals by a Marine Bacterium." *Marine Pollution Bulletin* 50 (3): 340–343.

[77] Xie, Shuibo, Jing Yang, Chao Chen, Xiaojian Zhang, Qingliang Wang, and Chun Zhang. 2008. "Study on Biosorption Kinetics and Thermodynamics of Uranium by Citrobacter Freudii." *Journal of Environmental Radioactivity* 99 (1): 126–133.

[78] Wang, C. L., P. C. Michels, S. C. Dawson, S. Kitisakkul, J. A. Baross, J. D. Keasling, and D. S. Clark. 1997. "Cadmium Removal by a New Strain of Pseudomonas Aeruginosa in Aerobic Culture." *Applied and Environmental Microbiology* 63 (10): 4075–4078.

[79] Zouboulis, A.I., M.X. Loukidou, and K.A. Matis. 2004. "Biosorption of Toxic Metals from Aqueous Solutions by Bacteria Strains Isolated from Metal-Polluted Soils." *Process Biochemistry* 39 (8): 909–916.

[80] Fomina, Marina, and Geoffrey Michael Gadd. 2014. "Biosorption: Current Perspectives on Concept, Definition and Application." *Bioresource Technology* 160 (May): 3–14.

[81] Srivastava, Shalini, S. B. Agrawal, and M. K. Mondal. 2015. "A Review on Progress of Heavy Metal Removal Using Adsorbents of Microbial and Plant Origin." *Environmental Science and Pollution Research* 22 (20): 15386–15415.

[82] Wang, Jianlong, and Can Chen. 2009. "Biosorbents for Heavy Metals Removal and Their Future." *Biotechnology Advances* 27 (2): 195–226.

[83] De, Jaysankar, N. Ramaiah, and L. Vardanyan. 2008. "Detoxification of Toxic Heavy Metals by Marine Bacteria Highly Resistant to Mercury." *Marine Biotechnology* 10 (4): 471–477.

[84] Ilias, Mohammad, Iftekhar Md. Rafiqullah, Bejoy Chandra Debnath, Khanjada Shahnewaj Bin Mannan, and Md. Mozammel Hoq. 2011. "Isolation and Characterization of Chromium(VI)-Reducing Bacteria from Tannery Effluents." *Indian Journal of Microbiology* 51 (1): 76–81.

[85] Mosa, Kareem A., Ismail Saadoun, Kundan Kumar, Mohamed Helmy, and Om Parkash Dhankher. 2016. "Potential Biotechnological Strategies for the Cleanup of Heavy Metals and Metalloids." *Frontiers in Plant Science* 7 (March).

[86] Garbisu, C., J.H. Allica, O. Barrutia, I. Alkorta, and J.M. Becerril. 2002. "Phytoremediation: A Technology Using Green Plants to Remove Contaminants from Polluted Areas." *Reviews on Environmental Health* 17 (3).

[87] Akar, Tamer, and Sibel Tunali. 2006. "Biosorption Characteristics of Aspergillus Flavus Biomass for Removal of Pb(II) and Cu(II) Ions from an Aqueous Solution." *Bioresource Technology* 97 (15): 1780–1787.

[88] Martínez-Juárez, Víctor M., Juan F. Cárdenas-González, María Eugenia Torre-Bouscoulet, and Ismael Acosta-Rodríguez. 2012. "Biosorption of Mercury (II) from Aqueous Solutions onto Fungal Biomass." *Bioinorganic Chemistry and Applications* 2012: 1–5.

[89] Lima, Marcos de, Luciana Franco, Patrícia de Souza, Aline do Nascimento, Carlos da Silva, Rita Maia, Hercília Rolim, and Galba Takaki. 2013. "Cadmium Tolerance and Removal from Cunninghamella Elegans Related to the Polyphosphate Metabolism." *International Journal of Molecular Sciences* 14 (4): 7180–7192.

[90] Yan, Guangyu, and Thiruvenkatachari Viraraghavan. 2003. "Heavy-Metal Removal from Aqueous Solution by Fungus Mucor Rouxii." *Water Research* 37 (18): 4486–4496.

[91] Vimala, R., D. Charumathi, and N. Das. 2011. "Packed Bed Column Studies on Cd(II) Removal from Industrial Wastewater by Macrofungus Pleurotus Platypus." *Desalination* 275 (1–3): 291–296.

[92] Srivastava, Shaili, and Indu Shekhar Thakur. 2006. "Isolation and Process Parameter Optimization of Aspergillus Sp. for Removal of Chromium from Tannery Effluent." *Bioresource Technology* 97 (10): 1167–1173.

[93] Tiwari, S., S. N. Singh, and S. K. Garg. 2013. "Microbially Enhanced Phytoextraction of Heavy-Metal Fly-Ash Amended Soil." *Communications in Soil Science and Plant Analysis* 44 (21): 3161–3176.

[94] Park, Donghee, Yeoung-Sang Yun, Ji Hye Jo, and Jong Moon Park. 2005. "Mechanism of Hexavalent Chromium Removal by Dead Fungal Biomass of Aspergillus Niger." *Water Research* 39 (4): 533–540.

[95] Luna, Juliana Moura, Raquel Diniz Rufino, and Leonie Asfora Sarubbo. 2016. "Biosurfactant from Candida Sphaerica UCP0995 Exhibiting Heavy Metal Remediation Properties." *Process Safety and Environmental Protection* 102 (July): 558–566.

[96] Yong-Qian Fu. 2012. "Biosorption of Copper (II) from Aqueous Solution by Mycelial Pellets of Rhizopus Oryzae." *African Journal of Biotechnology* 11 (6).

[97] Dar, R.A., M. Parmar, E.A. Dar, R.K. Sani, and U.G. Phutela. 2021. "Biomethanation of Agricultural Residues: Potential, Limitations and Possible Solutions." *Renewable and Sustainable Energy Reviews* 135 (January): 110217.

[98] Karakagh, Rahim Mohammadzadeh, Mostafa Chorom, Hossein Motamedi, Yusef Kianpoor Kalkhajeh, and Shahin Oustan. 2012. "Biosorption of Cd and Ni by Inactivated Bacteria Isolated from Agricultural Soil Treated with Sewage Sludge." *Ecohydrology & Hydrobiology* 12 (3): 191–198.

[99] Vankar, Padma S., and Dhara Bajpai. 2008. "Phyto-Remediation of Chrome-VI of Tannery Effluent by Trichoderma Species." *Desalination* 222 (1–3): 255–262.

[100] Dönmez, G., and Z. Aksu. 2001. "Bioaccumulation of Copper(Ii) and Nickel(Ii) by the Non-Adapted and Adapted Growing CANDIDA SP." *Water Research* 35 (6): 1425–1434.

[101] Grujić, Sandra, Sava Vasić, Ivana Radojević, Ljiljana Čomić, and Aleksandar Ostojić. 2017. "Comparison of the Rhodotorula Mucilaginosa Biofilm and Planktonic Culture on Heavy Metal Susceptibility and Removal Potential." *Water, Air, & Soil Pollution* 228 (2): 73.

[102] Suseela, Lanka, and Pydipalli Muralidhar. 2018. "Reduction of Organic Load from Palm Oil Mill Effluent (POME) Using Selected Fungal Strains Isolated from POME Dump Sites." *African Journal of Biotechnology* 17 (36): 1138–1145.

[103] Gupta, V.K., and A. Rastogi. 2008. "Biosorption of Lead from Aqueous Solutions by Green Algae Spirogyra Species: Kinetics and Equilibrium Studies." *Journal of Hazardous Materials* 152 (1): 407–414.

[104] Kaewsarn, Pairat, and Qiming Yu. 2001. "Cadmium(II) Removal from Aqueous Solutions by Pre-Treated Biomass of Marine Alga Padina Sp." *Environmental Pollution* 112 (2): 209–213.

[105] El-Sheekh, Mostafa, Sabha El Sabagh, Ghada Abou El-Souod, and Amany Elbeltagy. 2019. "Biosorption of Cadmium from Aqueous Solution by Free and Immobilized Dry Biomass of Chlorella Vulgaris." *International Journal of Environmental Research* 13 (3): 511–521.

[106] Tien, C.-J. 2002. "Biosorption of Metal Ions by Freshwater Algae with Different Surface Characteristics." *Process Biochemistry* 38 (4): 605–613.

[107] Lodeiro, P., B. Cordero, J. Barriada, R. Herrero, and M. Sastredevicente. 2005. "Biosorption of Cadmium by Biomass of Brown Marine Macroalgae." *Bioresource Technology* 96 (16): 1796–1803.

[108] Matheickal, Jose T., and Qiming Yu. 1999. "Biosorption of Lead(II) and Copper(II) from Aqueous Solutions by Pre-Treated Biomass of Australian Marine Algae." *Bioresource Technology* 69 (3): 223–229.

[109] Matheickal, Jose T., and Qiming Yu. 1996. "Biosorption of Lead from Aqueous Solutions by Marine Algae." *Water Science and Technology* 34 (9): 1–7.

[110] Jalali, R., H. Ghafourian, Y. Asef, S. Davarpanah, and S. Sepehr. 2002. "Removal and Recovery of Lead Using Nonliving Biomass of Marine Algae." *Journal of Hazardous Materials* 92 (3): 253–262.

[111] Chabukdhara, Mayuri, Sanjay Kumar Gupta, and Manashjit Gogoi. 2017. "Phycoremediation of Heavy Metals Coupled with Generation of Bioenergy." In *Algal Biofuels*, 163–188. Cham: Springer International Publishing.

[112] He, Jinsong, and J. Paul Chen. 2014. "A Comprehensive Review on Biosorption of Heavy Metals by Algal Biomass: Materials, Performances, Chemistry, and Modeling Simulation Tools." *Bioresource Technology* 160 (May): 67–78.

[113] Napiórkowska-Krzebietke, Agnieszka, Agnieszka Napiórkowska-Krzebietke, Abd-Ellatif M. Hussian, Ahmed M. Abd El-Monem, Mohamed E. Goher, Amaal M. Abdel-Satar, and Mohamed H. Ali. 2016. "Biosorption of Some Toxic Metals from Aqueous Solution Using Non-Living Algal Cells of Chlorella Vulgaris." *Journal of Elementology* (3) (May).

10 Application of Genetically Modified Microorganisms for the Reduction of the Toxicity of Hazardous Compounds

Bhagwan Toksha, Saurabh Tayde, Ajinkya Satdive, Shyam Tonde and Aniruddha Chatterjee

INTRODUCTION

The ubiquitous environmental pollution is a major global concern over the past century causing adverse effects on human health. Accidental spillage in seawater or the intermittent discharge of contaminants in the form of toxic chemicals from various industries contribute greatly towards pollution (Ivshina Irena et al. 2015; Islamoglu et al. 2020; Wolok et al. 2020). Ameliorating the health of the environment and ecosystem has become a mandate for every country today. Generally, removal of such pollutants takes place through different routes such as capping the contaminated site or landfilling, chemical decomposition, and high-temperature incineration. Landfilling or capping is the traditional method that functions by forming a barrier between the surface and the contaminated site, consequently safeguarding the environment and the living system from its harmful effects. The high-temperature incineration and chemical decomposition techniques are more effective than the traditional ones but pose some disadvantages, such as high cost and higher exposure of contaminants to civilians and workers present in the vicinity of the site. Therefore to placate the cleansing of contaminated sites, bioremediation is an advanced technology that involves microorganisms for effectively removing contaminants (Jaiswal and Shukla 2020; Janssen and Stucki 2020). A wide variety of microbes have the capacity to dismember organic substances into their primitive condition through enzymatic or metabolic processes. But various chemical elements present in the toxic pollutants are resilient against the biodegradation capability of the microorganisms and hence some microorganisms exhibit a slower degradation cycle (Kumar et al. 2018). To overcome this, a genetic engineering approach is sought to design novel microbial strains with unique characteristics and a wide array of bioremediation potential as compared to indigenous microbes. For example, remediation of heavy metals like mercury is not possible using natural bacteria, but genetically modified microorganisms (GMOs) can overcome this issue. Genetically modified microorganisms have been employed for the remediation of various heavy metals such as As, Ni, Fe, Cd, and Hg (Chellaiah 2018; Diep et al. 2018; Gupta and Singh 2017). Organophosphates have been used as pesticides in the agricultural sector and their use has become a serious environmental concern. Chlorinated organic compounds such as trichloroethylene and lindane can be easily metabolized using GMOs (Kobayashi and Rittmann 1982; Bhatt et al. 2007; Azad et al. 2014; Janssen and Stucki 2020).

The focus of this chapter is on the bioremediation process of different types of microorganisms utilized in reducing the toxicity of hazardous compounds. We will try to describe different bioremediation processes (bioaugmentation, bioattenuation, biostimulation), factors affecting GMOs, and their mechanisms. Microorganisms can be genetically modified for various purposes and the effective

DOI: 10.1201/9781003188568-10

remediation of toxic compounds is one of them. So, we will also be focusing on drawing on GMOs in the removal of heavy metals, phytoremediation, and environmental challenges using GMOs.

BIODEGRADATION AND BIOREMEDIATION

The rate of consumption of natural resources by human beings has increased to a much higher extent than nature can replenish or regenerate. With technological advancement, the natural resources per capita consumption is also growing (Sharma, A. et al. 2013). Human activities lead to the exploitation of natural resources in various ways. So, there arises a need to conserve and preserve the ecosystem along with the process of replenishment that will help in preventing the issues such as wastage, global warming, soil imbalance, and more (Okpokwasili, G.C. 2007).

The natural decomposition of hazardous substances is known as biodegradation, whereas employing biological agents with a view to converting toxic substances into innocuous substances is called bioremediation (Harshavardhan K and Jha B. 2013). Different environmental parameters like moisture content, temperature, and pH are generally manipulated in the bioremediation process to achieve optimum growth of microorganisms and higher degradation rates (Bento FM et al. 2005). The decomposition results in breaking down the organic and inorganic nutrients which are originally connected with the fundamental unit of any substance into simple reusable forms (Andreolli M et al. 2015).

The biological balance in life is maintained by microorganisms playing catalytic roles in the nutritional chains of ecosystems. Bioremediation employs fungi, bacteria, algae, and yeast for removing contaminated materials (Fritsche W and Hofrichter M. 2000). Microbes can grow below 0 °C and at extremely high temperatures in the presence of hazardous compounds. Adaptability and the biological system are the characteristics of microbes making them suitable for the remediation process. For any microbial activity to take place the presence of a carbon compound is required. Microorganisms that help in the bioremediation process comprise soil, water, and nonfermenting bacteria such as Achromobacter, Arthrobacter, Alcaligenes, and so on.

BIOAUGMENTATION

The remediation of petroleum hydrocarbons from polluted sites can be accomplished through the incorporation of microorganisms confiscated from the polluted habitat. Such microorganisms are carefully selected and genetically modified to support the remediation process (Mrozik A and Piotrowska-Seget Z. 2010). Compared to the physicochemical approaches, bioaugmentation is a cheap and environmentally friendly technique. By employing GMOs, one can expedite the process of removing undesired compounds to bioaugment toxic waste sites (Madhurankhi Goswami et al. 2018).

Bioaugmentation is generally carried out in oil-contaminated environments as a substitute for bioremediation (Mrozik A et al. 2003). It is an addition of pollutant-degrading microorganisms that have the ability to biodegrade molecules in a polluted environment. Bioaugmentation is generally carried out in two ways: (i) by employing indigenous microorganisms; and (ii) by employing non-indigenous microorganisms (Silva E et al. 2004). Indigenous bioaugmentation is a process of isolating previously acclimated indigenous microorganisms from the site and reinoculating them in the soil. In some cases where the actively polluting degrading microbes are absent from the sites, then exogenous microbes are used as an alternate solution (Menn FM et al. 2008). Compared to other techniques used for water remediation such as using activated carbon, filtration, and chemical oxidation, the biological treatment is cost-effective and versatile, with the capacity to trigger complete mineralisation. However, due to the high concentrations of pollutants these systems often contain, these methods have often failed. Bioaugmentation promotes the development of more powerful microbial systems that can fit in the large irregular fluctuations to meet yielding in a more coherent manner.

In the bioremediation of pollutants, a microorganism employs various processes such as oxidative degradation, volatilization, and immobilization followed by chemical progressions initiated in the adulterant (Seeda et al. 2017). The putrefaction of adulterants via catabolic stir of microbes such as hydrocarbons, polychlorinated biphenyls (PCBs), polyaromatic hydrocarbons (PAHs), heavy metals, and hydrocarbons can be achieved through biotransformation and bioremediation. Hydrocarbons are carbon and hydrogen-containing organic compounds present in linear, branch or cyclic forms that may be aromatic or aliphatic (Kafilzadeh, F. et al. 2011). Polycyclic aromatic hydrocarbons (PAHs) are hydrophobic organic pollutants (HOCs) generated from industrial production (Mrozik et al. 2003).

PAHs are found to absorb in soil containing organic compound sediment, as well as accumulate in the bodies of fish and other aquatic organisms. They may get transferred into the human body after the consumption of such seafood (Mrozik A et al. 2003). Polychlorinated biphenyls (PCBs) are organic compounds. PCBs possess excellent thermal, chemical, and insulating properties and hence are widely used in paints, plastics, and rubber products as well as various industrial applications. The degradation of PCBs is a prime concern as they are carcinogenic (Seeger M et al. 2010). Synthetic pesticides are organic and resist degradation, and are known as persistent organic pollutants (Vargas JM et al. 1975). Dyes are colouring agents; organic and aromatic (azo dyes) are used ubiquitously in paper and clothing printing, colour photography, pharmaceuticals, cosmetics, and many other industries (Raffi F et al. 1997). As these organic synthetic compounds are difficult to degrade naturally, they are separated by different separation methods like filtration, adsorption, electrochemical separation, and coagulation (Verma P et al. 2003). These separation methods can be more effective by the utilization of microorganisms which are beneficial in the decolourization of organic dyes. Heavy metals are non-biodegradable hazardous materials; hence they must convert to a stable form after separation. Biotransformation is used to perform bioremediation of heavy metals through biosorption, biomineralization, bioleaching, and enzyme-catalysed transformation (Lloyd JR and Lovley DR 2001).

BIOATTENUATION

Attenuation employs natural processes to prevent contamination from spreading through chemical spills and reduces the number of pollutants at contaminated sites. Bioattenuation as an *in-situ* treatment method is also called intrinsic remediation. This can be understood as the natural attenuation of environmental contaminants as they are left in place (PinakiSar et al. 2004). Natural attenuation also helps in removing or controlling the contamination source and is often employed as one part of site cleanup.

Bioattenuation includes processes such as aerobic/anaerobic biodegradation, plant and animal consumption and digestion, physical volatilization through various means, and chemical transformations such as ion exchange, and complex formation (Singh JS et al. 2011). Our ecosystem upon coming into contact with toxic chemicals can operate via four different routes to clean up: (i) the leaked chemicals are consumed and digested by the tiny bugs or microbes present in the soil and groundwater. Upon complete digestion, these chemicals are converted into harmless gases and water. (ii) Chemicals tend to stick or sorbs to the soil holding them in place. Consequently, this keeps them from leaving the site and prevents groundwater pollution. (iii) Even if the chemicals or pollutants move with the soil or groundwater, they get further mixed with the clean water, subsequently reducing or diluting the pollutants. (iv) Some chemicals or pollutants are volatile, which means they get converted from a liquid state to a gaseous state within the soil. When these gases escape into the air at the ground surface, sunlight may destroy them (Chen SL and Wilson DB 1997). When bioattenuation lacks speed or full completion, the bioremediation will be carried out either by bioaugmentation or biostimulation.

BIOSTIMULATION

This remediation technique is cost-effective, highly efficient, and eco-friendly. In this technique, sites badly affected by toxins stimulate the prevalent bacterial species to degrade the hazardous contaminants by incorporating rate-limiting nutrients like oxygen, phosphorus, nitrogen, and electron donors (Madhurankhi Goswami et al. 2018).

BIOREMEDIATION OF POLLUTANTS USING MICROORGANISMS

In this day in age, due to globalization and vast development in every sector, environmental pollution is created from various sources as shown in Figure 10.1. These environmental pollutants are threatening biodiversity, causing various health-related problems, and disturbing ecological balance. These toxic compounds hamper human health along with biotic and abiotic components (Raghunandan et al. 2018). The removal of these pollutants is not easy and cannot be completely solved by conventional techniques, as these are costly, inefficient, and time-consuming. By conventional methods, the separation of pollutants is possible, but their full degradation/destruction is not achieved. Hence researchers focus on the usage of genetically engineered microorganisms for bioremediation. Bioremediation is an eco-friendly and low-cost technique used for the degradation of pollutants at the site of contamination. DNA and RNA recombinant technologies are used to develop genetically engineered microbes. These microbes are used to destroy heavy metals and toxic substances from pollutants (Md. Abul Kalam Azad et al. 2014). Biodegradation is known as the biological catalytic reduction of chemical compounds by the action of microorganisms (Alexander M. 1994). The biodegradation process refers to the breakdown of organic compounds into their fundamental building

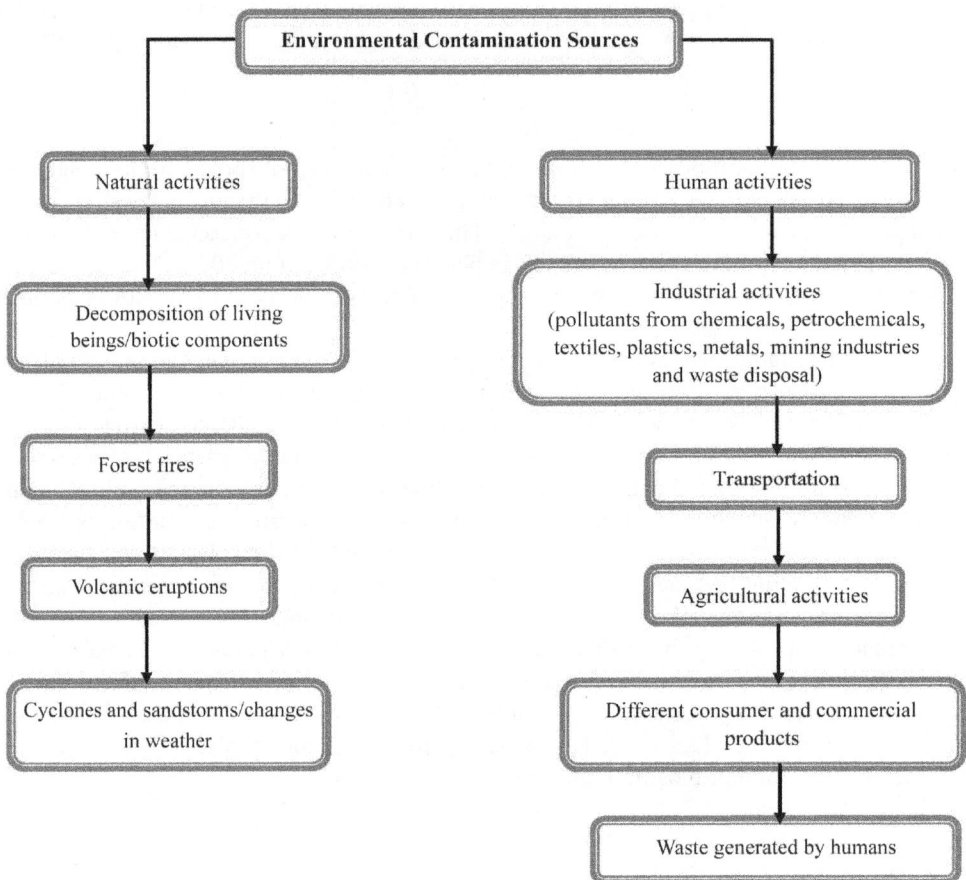

FIGURE 10.1 Sources of environmental contamination.

blocks by living microbial organisms (Marinescu M et al. 2009). The process of completion of biodegradation is known as "mineralisation". Researchers are always in search of tools that are able to degrade pollutants in an environmentally friendly manner. Recently, researchers are focusing on the advanced multi-omics approach rather than single omics in microbial-mediated bioremediation, where the data sets from different omic groups are combined during analysis. Bioremediation is one of the prominent low-cost processes by which microorganisms remove toxic compounds from the environment (Demnerová K et al. 2005).

The following figures depict the common sources of environmental contamination.

Bioremediation follows different natural processes such as attenuation, bioaugmentation, and biostimulation (Olaniran AO et al. 2006). Natural attenuation is the simple method to check soil properties that the pollutant transformation is taking place actively or not (Kaplan CW and Kitts CL 2004). In the bioaugmentation process, pollutant-degrading microorganisms are added to enhance the degradation rate of pollutants in which naturally active communities of microorganisms are absent, even if present in low numbers (El Fantroussi S and Agathos SN 2005). The bioremediation carried out by microorganisms has been shown in Figure 10.2.

DEGRADATION BY BACTERIA

It has been found that various bacteria are helpful in the degradation of environmental pollutants. Many bacteria that only feed on hydrocarbons are known as hydrocarbon-degrading bacteria (Yakimov MM et al. 2007). Hydrocarbons undergo aerobic and anaerobic degradation under the influence of various hydrocarbon-degrading bacteria (Kafilzadeh F et al. 2011). PCBs can be biodegraded in an anaerobic as well as aerobic way. Anaerobic as well as aerobic bacteria are capable of biotransforming PCBs. There are reports available about various genera of bacteria being involved in PCB degradation (Petrić I et al. 2007). Pesticides are degraded by chlorpyrifos degrading bacterium *Providencia stuartii* separated from agricultural soil (Surekha Rani M et al. 2008) and *Bacillus*, *Staphylococcus*, and *Stenotrophomonas* obtained from cultivated and uncultivated soil can degrade dichlorodiphenyltrichloroethane (DDT) (Kanade SN et al. 2012; Mónica P et al. 2016). Some researchers have found that individual bacterial strains (*Shewanella decolourations*) show high efficacy in carrying away azo dyes (Hong Y et al. 2007). Under a given environmental condition, the use of pure culture is beneficial to predict execution and thoroughly understand the

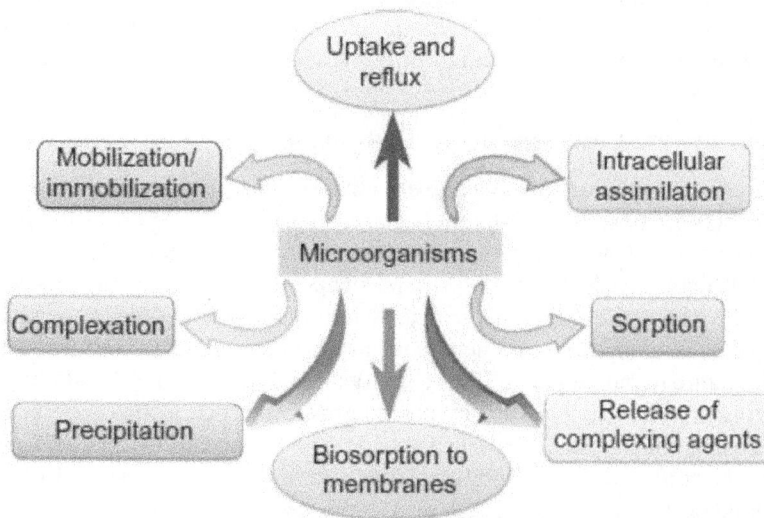

FIGURE 10.2 Bioremediation using microorganisms (source: Surajit Das and Hirak R. Dash 2014).

responsive degradation pathways followed for dyes, leading to the formation of nontoxic end products. The metal reduction can be obtained through a dissimilatory metal reduction (Fernández PM et al. 2012), where the metal as a terminal electron acceptor is utilized by the bacteria for anaerobic respiration, which is the terminal electron acceptor reduced to water (H_2O) in aerobic respiration. Additionally, the bacteria may undergo reduction mechanisms that are not associated with respiration but indeed are imparting metal resistance. For instance, under the aerobic or anaerobic condition, reduction of Cr(VI) to Cr(III) takes place (Zhu W. et al. 2008), obtaining element Se by reduction of Se (VI) (Yee N et al. 2007), reduction of U(VI) takes place to U(IV) (Gao W and Francis AJ 2008); similarly, Hg(II) reduces to Hg(0) (Brim H et al. 2000). In the bioremediation of heavy metals, microbial methylation plays a vital role, as methylated compounds are highly volatile. For example, biomethylation of Mercury, Hg(II) can be done by several bacterial species converting to gaseous methyl mercury (De Jaysankar et al. 2008). Along with these two routes, iron bacteria such as *Acidithiobacillus ferrooxidans* are able to thrive in acidic environments (Takeuchi F and Sugio T. 2006), and bacteria able to oxidize sulphur are able to lixiviate high concentrations of heavy metals from contaminated soils.

PLANT GROWTH-PROMOTING BACTERIAL DEGRADATION

The culturable bacterial cells in the soil have an efficiency of about 1%. The soil consists of about 10^8 to 10^9 per gram volume of bacteria possessing only 10^5 to 10^6 culturable bacterial cells (L. Schoenborn et al. 2004). To combat the heavy metal and toxic contaminants, it is necessary to use plants combined with some plant growth-promoting microorganisms; such a technique is called rhizoremediation. Various microscopic lives such as algae, fungi, bacteria, protozoa, and actinomycetes are part of the soil. The utilization of plant growth-promoting bacteria is a vital phytoremediation strategy to treat soil containing heavy metals. The bacteria promoting plant growth (PGPB), including rhizospheric, endophytic, and other bacteria which perform phytoremediation are present naturally in soil. Rhizospheric bacteria grow rapidly and form colonies near plant roots whereas endophytic bacteria, which are non-pathogenic, present naturally in plants and promote plant growth by degrading toxic heavy metals in contaminated soil (Divya B and Deepak Kumar M. 2011; Zhaoyu Kong and Bernard R. Glick 2017). PGPB improves plant growth by increasing nitrogen and phosphorus uptake, producing plant-growth-promoting components such as auxin, ACC deaminase, cytokinin, gibberellin, and antibiotics in the rhizosphere to cope with the root pathogens. The concentration of the bacteria varies in the soil and is more present near the roots (Oluwaseyi Samuel Olanrewaju et al. 2017). *Pseudomonas spp.* bacteria have the ability to degrade hydrocarbon and help in plant growth-promoting activities (Hontzeas N et al. 2004).

PLANT GROWTH-PROMOTING RHIZOBACTERIAL DEGRADATION

Rhizobacteria are proven to be an efficient and effective environment-friendly strategic tool for bioremediation at contamination sites. Different bacterial colonies obtained near the plant roots have the capability to improve plant growth and are subsequently referred to as plant growth-promoting bacteria (PGPB). Rhizobacteria form nodules on particular plant roots whereas endophytes are present within the interior plant tissues and contribute an important role in phytoremediation (Santoyo G et al. 2016; Leigh, M.B. et al. 2006). Some plants have the natural facility to form compounds that have a similar structure to that of phenol; this activity helps to degrade PAH and ultimately helps to improve plant growth. *Rhodococcus* spp. are able to grow rhizospheric atmosphere at the contaminated site and are thus found capable to degrade PCB. The rhizobacteria *Azospirillum lipoferum* is a nitrogen-fixing bacteria with the capability to degrade organophosphorus and Marathon, which are commonly used pesticides (Kanade, S.N. et al. 2012). Numerous investigations have proved that the rhizobacteria responsible for plant growth have a substantial propensity for bioremediation of adulterated soil with high concentrations of heavy metals (Glick, B.R. 2010).

DEGRADATION USING ALGAE AND PROTOZOA

Algae and protozoa are a crucial class of microbes beneficial in biodegradation in aquatic as well as terrestrial ecosystems. Algae have the propensity to degrade aliphatic and aromatic hydrocarbons. Cerniglia and Gibson reported that different algae strains (cyanobacteria, green/red/brown alga) are used to degrade naphthalene (Cerniglia, C.E. and Gibson, D.T. 1977). It has been also reported by many researchers that protozoa cannot degrade hydrocarbon nor crude oil. They have been found to have a negative effect on the biodegradation of hydrocarbon (Stapleton Jr, R.D. and Singh, V.P. (Eds.) 2002). Algae cannot degrade pesticides but can degrade various other contaminants; however, algae can bioaccumulate the pesticides and biotransform some contaminants.

It has been reported by some research work that algae belonging to genera of Green and Chlorella are able to uptake and degrade PAHs. Algae such as *Chlorella, Anabaena, Westiellopsis, Stigeoclonium,* and *Synechoccus,* along with marine algae, can adsorb and remove heavy metals, but on-site constraints restrict their use (Dwivedi S. 2012). Brown algae show the capacity of biosorption of heavy metals. Protozoa are capable of enhancing the biodegradation of contaminants like naphthalene and PAH through different mechanisms such as nutrient mineralization, bacterial activation, selective grazing, direct degradation, and sym-metabolism (Chen X et al. 2007).

FACTORS AFFECTING GENETICALLY MODIFIED MICROORGANISMS

The rate of genetic modification of crops has always been much higher than the production of transgenic animals. The bovines (Cattle, Buffalo, Bison, and Hybrid) and pharmaceutical interlinks could be greatly influenced by increasing the productivity rate of transgenic animals. Economic benefits for farmers, processors, and consumers could be achieved through the advancement in animal biotechnology. Apart from the direct economic effects, other pros and cons must also be factored in during the overall evaluation (Moss, B. and Buller, M. L. 1985). In the case of fishery, industry growth and perfected feed conversion ratios contribute to cost reduction and lower market prices, and have driven an interest in genetically modifying aquatic animals, rather than terrestrial animals. This also explains the significance of the economic impact of introducing GM fish. Simultaneously, human health-related concerns and associated environmental factors should also be factored in before introducing GM fish.

For the agriculture and pharmaceutical industries, biopharming is a new territory presenting novel challenges for government regulators. To be a feasible economic investment, the high cost in the production of transgenic animals such as goats, sheep, cattle, and pigs must be either brought down or the resulting profit must be elevated. Due to this reason, the production of high-value pharmaceutical substances corresponding to a market value of billions of dollars is currently the most promising and principal application for the transgenesis of animals (Migheli, Q. 2001). The development phase spans a longer duration and demands higher financial commitment. This has hampered attempts at commercial exploitation and, currently, only two drugs produced with these approaches are market ready. With the rapid development of the technologies involved and blistering scientific deliberations on GM, government bodies are beginning to create legal and supervisory systems for the marketing of GM animals (Frewer et al. 2013; Ahuja 2018). Criminal usage as well as negative consequences from DNA modifications through GM to a human chromosome via food consumption are also imposing serious threats (Walters 2006).

BIOREMEDIATION USING GENETICALLY MODIFIED MICROORGANISMS

The advancements of technologies in recombinant deoxyribonucleic acid (DNA) technology have paved the way for the conceivable metabolic changes in organisms. The role of genetically modified microorganisms has been critical in waste bioremediation. The process of bioremediation functions as an environmental key to rejuvenating contaminated water, soil, and subsurface sediments, by elimination and/or degradation of the target pollutant under the stimulated growth of microorganisms. This stimulated growth of microorganisms is achieved through genetic engineering, which is advantageous to increase molecular ambit and chemical preference. The genetic engineering approach for bioremediation is one of the current focuses of research. The notable achievements in this direction lead to the assured availability of desired products at much lower production costs along with safer processing.

GENETICALLY MODIFIED ORGANISMS BIOREMEDIATION MECHANISMS

Bioremediation requires substantial populations of microorganisms effective at the removal of pollutants. Naturally occurring microorganisms work inherently for the removal of pollutants, but face challenges such as time scale and availability in quantity. The alternative to this situation is bioaugmentation that can be performed via approaches such as adding microorganisms that are enabled with catabolic genes on their own or have been genetically modified organisms (GMOs). The genetic modifications may also involve the carryover of plasmids containing vital genetic material among the various bacterial populations. The plasmids transfer would occur through gene transfer from one microorganism to another either by conjugation or horizontal gene transfer. Microbiology research has been applied in order to produce novel functionalities in microorganisms for the bioremediation processes. Catabolic pathways are one such approach. A study involving the mineralization of Polychlorobiphenyls in soil by the genetically modified strain *Cupriavidus necator* was reported by Saavedra et al. (Saavedra et al. 2010). The enhancement in the catalytic activity of horseradish peroxidase has been demonstrated as an advancement in the genetic stability of catabolic gaieties by Aimin Huang (Huang et al. 2018).

There are numerous mechanisms currently being explored in the production and implementation approaches of genetically engineered microorganisms (GEMs) that could exhibit better bioremediation efficiency. These approaches include genetically modified microorganisms realistically designed for their enzymes that degrade persistent organic pollutants (POPs) such as PAHs, PCBs, and pesticides. The role of ligninolytic enzymes and mechanisms involved in the degradation of lignocellulosic waste in the environment is an enzyme-based bioremediation approach (Kumar and Chandra 2020). The enzyme reactive Red-120 Acinetobacter junii FA10 being used for elemental and dye remediation simultaneously has been reported on by Faiza Anwar (Anwar et al. 2014). The directed evolution scaffolds are based on the dynamic structures of genomes and genomic substructures and functionality to evolve in all directions exhibiting stimulated response. *Shewanella* microorganisms are known to have uneven metal reduction model bacteria with multifaceted extracellular electron exchange mechanisms. The use of species in microbial electrochemical systems for bioremediation is reported by Long Zou (Zou, L. et al. 2018). The other approach involved is the recombinant DNA technologies such as the development of "suicidal-GEMs" (S-GEMs) to achieve laborsaving and complete bioremediation of contaminated sites. The incorporation of suicide systems provide for the death of microorganisms after the bioremediation process is completed. The "suicidal genetically engineered microorganisms" minimize the anticipated hazards and achieve efficacious and unassailable bioremediation of contaminated sites (Adetunji and Anthony 2021).

The controlled activation of suicide genes in the host-microbe activated in the absence of the pollutant at the contaminant site is the logical flow of this approach (Paul et al. 2005). The gene encoding streptavidin acting as a suicide gene was incorporated along with the substrate. The streptavidin gene gets activated to produce streptavidin at the end of the substrate. The binding of streptavidin with D-biotin causes inhibition of thiamine production, which leads to cell death. The strategy of designing such an entity susceptible to programmed cell death after detoxification of any given contaminated site was reported by Pandey et al (Pandey et al. 2005).

GENETICALLY MODIFIED MICROORGANISMS FOR REMOVAL OF HEAVY METALS

The elements with high atomic density and/or atomic numbers are known as heavy elements. In general, the elements categorised from the physiological point of view are (i) essential, nontoxic,

(ii) essential but menacing at high concentrations, and (iii) highly toxic (Valls and de Lorenzo 2002; Tekere 2020). Pollution due to exposure and accumulation of radioactive/non-radioactive heavy metals such as Sr, Cs, U, Co, Cu, and Cr has become one of the most serious environmental concerns of modern times (Briffa et al. 2020). The Lead element is a cumulative toxicant which is the so-called "silent cause" of one of the longest epidemics attacking multiple body organs, being particularly harmful to young children (Markowitz 2016). The accumulation of lead in the vital organs such as the brain, liver, kidney, bones and teeth, make it dangerous to human health. Arsenic is an environmental pollutant at the root cause of a global epidemic of arsenic poisoning, leading to thousands of people suffering skin-related ailments and other medical conditions (Pearce et al. 2003; Thomas Charrier et al. 2010). Anthropogenic processes such as the mining of rocks, smelting, power stations, metal ion-based pesticides and fertilizers, irresponsible disposal of wastes, the colored pigment industry, batteries, erosion of rocks involved in construction, combustion byproducts, traffic, and more, all act as seedbeds of metal pollutants in the environment (Aksu 2015).

The main challenge with heavy metal contaminants in ecosystems is that they are more stable and non-biodegradable compared to other organic pollutants (Wei et al. 2015). It is difficult to nullify these metals and their compounds from the environment, as they do not degrade chemically or biologically. The physio/chemical removal methods of these pollutants involve complex and expensive methodologies. The possibility of using genetically modified microorganisms to aid in the remediation of metal-polluted environments would be a boon to address the critical problem of the removal of metals. Genetically modified organisms initiating biosorption and/or bioaccumulation for the removal of heavy metals is one of the optimal solutions. The survival instinct of these microorganisms leads them to develop various strategies in heavy metal-polluted habitats. The detoxifying mechanisms can be adopted *ex situ* or *in situ*, and biosorption, bioaccumulation, biotransformation, and biomineralization can result in effective bioremediation of heavy metals.

The cell walls of microorganisms rich in polysaccharides, lipids, and proteins, have carboxylate, hydroxyl, amino, and phosphate functional groups that bind to heavy metal contaminants (Nanda et al. 2019). The microorganisms uptake heavy metal ions by using proteins and thus isolate these metal ions inside cells to be consumed in enzyme catalysis, signalling, and stabilizing charges on biomolecules. The GEMs with a recombinant expression of storing tendencies permit higher consumption and isolation of heavy metal ions. The biosorption process is feasible compared to the bioaccumulation process at large volumes, as microbes need the addition of nutrients for their active uptake of heavy metals, creating the needs of "BOD/COD"; in other words, biological oxygen demand or chemical oxygen demand in the waste. The fungi genera have been reported as promising microbial agents for the bioremediation of heavy metals in the soil and aqueous sources (Ahmad et al. 2006; Zafar et al. 2007; Srivastava et al. 2011). The genetically modified organisms performing heavy metal isolation have concentrated on ameliorating uptake from the periplasm into the cytoplasm of gram-negative bacteria using recombinantly expressed inner-membrane importers through various possible means (Reddy and Saier Jr 2016). The channels are mono-componential α-helical proteins that open the door for the passive diffusion of heavy metals as per their concentration gradient across the inner membrane. They don't become energy-independent because of the force mechanisms that move their substrates (Reddy and Saier Jr 2016). These approaches have been found beneficial for the isolation of As^{3+} and Hg. Secondary carrier proteins are single components classified as uniporters, symporters, and antiporters (Forrest and Rudnick 2009). The approaches involving these secondary carriers have been found beneficial for the isolation of Ni, Co, and As^{4+}. Primary active transporters with multi-component protein complexes contain a transmembrane component for the translocation pathway that uses phosphoanhydride bond hydrolysis to drive the translocation of substrates. The schematic system governing the Cd^{2+} resistance in gram-positive bacteria in response to the accumulation of Cd^{2+} in the cell walls is depicted in Figure 10.3. The entry of cadmium ions in the bacterial cell is permitted through the metal ion transporter. The plasmid-contained system generates the resistance that actively drums out bacterial resistance to Cadmium (Nanda et al. 2019).

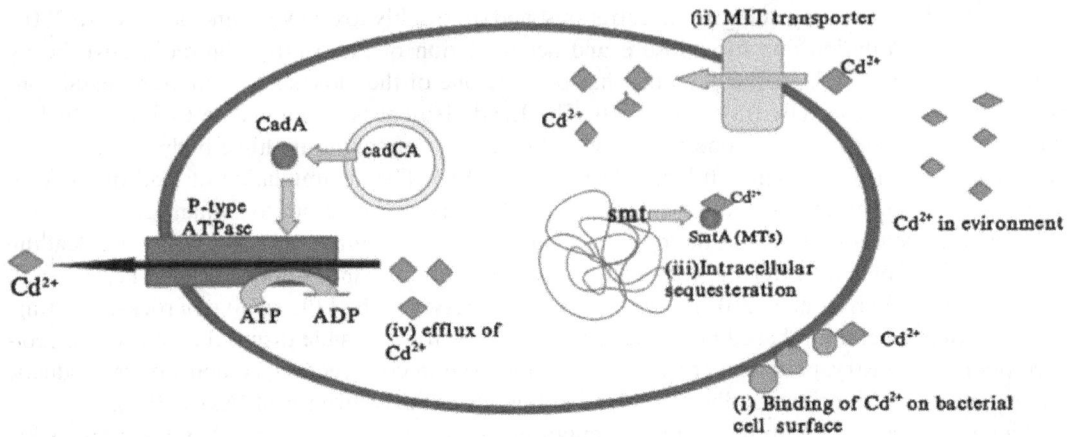

FIGURE 10.3 A schematic depiction of the plasmid mode bacterial resistance to Cadmium depicting four sections as (i) cadmium ion bound on the bacterial cell surface, (ii) metal ion transporter, (iii) intracellular isolation by bacterial mtA protein, (iv) efflux of cadmium ion by CadA transporter (source: Nanda et al. 2019).

GMOs in Phytoremediation

The bioremediation of heavy metals is also performed by mother nature through a process termed phytoremediation. This word phytoremediation is a portmanteau of the Greek word "phyto" and the Latin word "remedium" to represent the meanings "plant" and "balance or remediation". The approach is based on the feature of plants capable of typical and matchless consumption skills. An indicative schematic of different phytoremediation approaches involving contaminants push out and the physiological reactions carried out in foliage during phytoremediation is presented in Figure 10.4. The plants consume the heavy metals from the soil through plant root systems, and consequently, translocation, bioaccumulation, and decomposition of contaminants occur throughout the foliage (Galal and Shehata 2015). The natural process of soil feeding to plants thus may be an effective bioremediation approach applied to the environmental setting to eliminate a variety of contaminants, such as organic compounds and metals (Pilon-Smits and Freeman 2006; Ahmadpour et al. 2012). This green alternative has great potential to address heavy metal pollution, with 400 plant species identified as prospective phytoremediators (Lone et al. 2008; Ali et al. 2013).

The functionalities required by severe heavy metal pollution can be attained through genetically modified plants since phytoremediation capabilities can be achieved (Agarwal et al. 2020; Ozyigit et al. 2021), revealing promising results in both abiotic stress and the environmental presence of metals. The healing of polluted environments can be achieved with less economic stress and low collateral impacts (Ibanez et al. 2016). This approach of using plants for the cleansing of heavy metals is suitable in different ecosystems and has the added benefit of increasing foliage growth (Pilon-Smits and Freeman 2006). The shortcomings of this approach lie with the time scale involved in the growth of the plant and how to nullify the toxicity accumulated in the plant left over after phytoremediation.

The phytoremediation approaches involved in heavy metal bioremediation are classified into five categories: (i) phytofiltration, in which heavy metal pollutants are absorbed or adsorbed from a contaminated site, restricting their further movement (Moreno et al. 2008); (ii) phytostabilization: stabilizing the precipitation of target heavy metal within the plant root proximity by complex formation (Guo et al. 2014); (iii) phytoextraction: isolation of heavy metal through higher transfer rates from root to shoot, making it possible to reclaim metals from shoots (Lampis et al. 2015); (iv) phytovolatilization: the removal of target heavy metals from the soil through gas exchange by the mediating plants with conversion and expulsion in the atmosphere (Martínez et al. 2006);

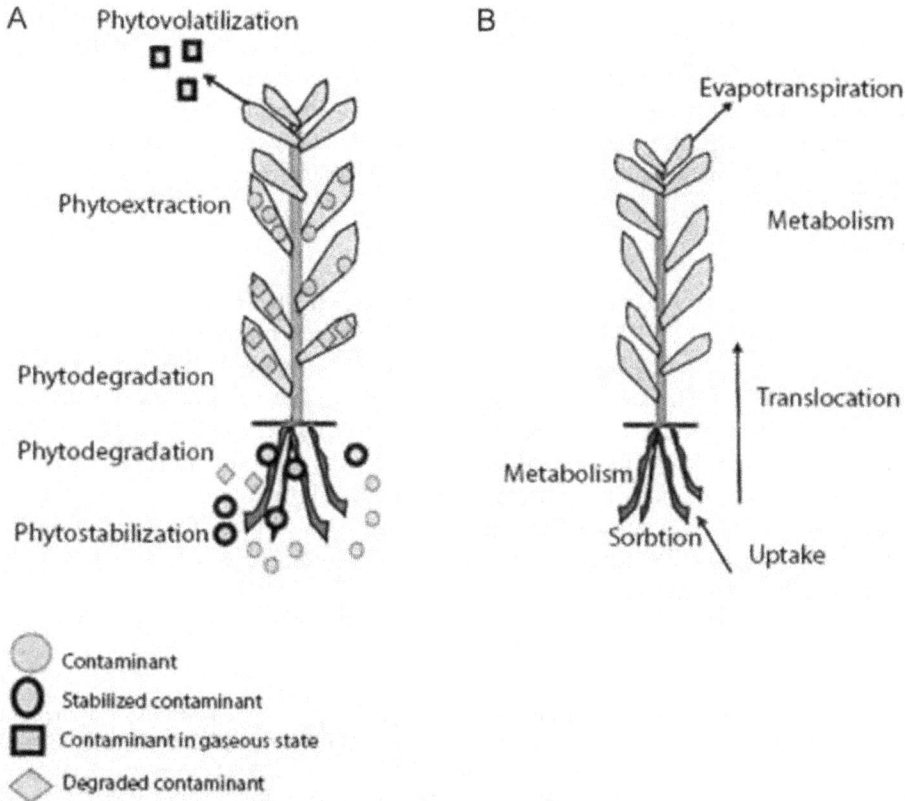

FIGURE 10.4 Various strategies to achieve phytoremediation (Figure A) and physiological processes in a plant during phytoremediation (Figure B) (source: Gomes et al. 2016).

(v) phytotransformation: the capture and chemical modification of target heavy metal pollutants from contaminated sites as a result of plant metabolism resulting in toxic elements inactivation, degradation or immobilization mediated through plant enzymes metabolism and transformation into less toxic and more stable forms (Rout et al. 2019). The plants in heavy metal bioremediation must pass the criteria of having a high biomass production along with a superior capacity for pollutant tolerance, accumulation, and degradation. The hyperaccumulation tendencies make the plants promising candidates for the removal of heavy metals. The phytoremediation efficiency of genetically modified plants is reportedly higher as compared to conventional plants (Koźmińska et al. 2018). The genes responsible for inculcating hyperaccumulation tendencies to higher shoot-biomass-producing plants have modification potential in phytoremediation for viable remediation solutions (Grennan 2009; Sarma et al. 2021). Plants such as *Lepidium sativum* or Alyssum species, Thlaspi species and *Brassica juncea*, *Viola calaminaria*, *Astragalus racemosus*, and *Prosopis laevigata* are reported to be useful in bioremediation heavy metals (Diaconu et al. 2020; Raj et al. 2020; Muro-González et al. 2020).

The research in the direction of genetically modified initiated phytoremediation of heavy metals needs to address issues such as longer-scale effects on plants, soil, and overall environmental conditions. The other concern is related to slowing rates of detoxification and being seasonally effective. The total amount of effort expended in terms of both time and economic resources is not offset by the result; there is no substantial reduction in pollutant concentrations (Bestawy et al. 2013). There are chances of genetically modified plants containing genes that develop resistance towards certain antibiotics, affecting human immunity towards certain diseases and continuous resistance towards antibiotics

(Raman 2017). Allergic reactions and genetically modified plants causing cancer are other possible threats (Bennett et al. 2004; Breyer et al. 2014). The concerns regarding the possibilities of the formation of harmful intermediates during remediation need to be eliminated. The mechanism of involving GMOs for bioremediation involves refashioning and monitoring for better enzyme affinity and specificity, spawn, and efficacies. Besides their stability, there are risks involved in GMOs in the environment, as the transfer of genetic matter is considered to be a negative interference (Pant et al. 2021).

SUMMARY

The recent status of GMOs used as bioremediation agents of the toxicity of hazardous compounds is reviewed in the present study. The chapter shows much promise for the bioremediation of hazardous compounds through genetically modified plants and microorganisms. The techniques involving microbial gene transfer, algae, and protozoa-based techniques have evolved promisingly in recent years. Future research on gene transfer-based bioremediation must extend to heavy metals and other contaminants. The research aiming at systematically analyzing the purported downsides with profound evidence regarding human health issues associated with the remediation practices is found to be a knowledge gap and a prominent fact of the present book chapter. The need of the hour is to balance ecosystems while still being effective in bioremediation. This would set the stage for the development of a balanced ecosystem.

REFERENCES

Adetunji, Charles Oluwaseun, and Osikemekha Anthony Anani. 2021. "Recent Advances in the Application of Genetically Engineered Microorganisms for Microbial Rejuvenation of Contaminated Environment." *Microbial Rejuvenation of Polluted Environment*: 303–324.

Agarwal, Priyanka, Balendu Giri, and Radha Rani. 2020. "Unravelling the Role of Rhizospheric Plant-Microbe Synergy in Phytoremediation: A Genomic Perspective." *Current Genomics* 21 (5): 334–342.

Ahmad, Iqbal, Mohd Ikram Ansari, and Farrukh Aqil. 2006. "Biosorption of Ni, Cr and Cd by Metal Tolerant Aspergillus Niger Penicillium Sp. Using Single and Multi-Metal Solution." *Indian Journal of Experimental Biology* 44 (1): 73–76.

Ahmadpour, P., F. Ahmadpour, M. Mahmud, A. Arifin, S. Mohsen, and T. Farhad. 2012. "Phytoremediation of Heavy Metals: A Green Technology." *African Journal of Biotechnology* 11 (76): 14036–14043.

Ahuja, V. 2018. "Regulation of Emerging Gene Technologies in India." *BMC Proceedings* 12 (8): 14. https://doi.org/10.1186/s12919-018-0106-0.

Aksu, Abdullah. 2015. "Sources of Metal Pollution in the Urban Atmosphere (A Case Study: Tuzla, Istanbul)". *Journal of Environmental Health Science and Engineering* 13 (1): 79.

Alexander, M. 1994. *Biodegradation and Bioremediation*. San Diego, CA: Academic Press.

Ali, Hazrat, Ezzat Khan, and Muhammad Anwar Sajad. 2013. "Phytoremediation of Heavy Metals—Concepts and Applications." *Chemosphere* 91 (7): 869–881.

Andreolli, Marco, Silvia Lampis, Pierlorenzo Brignoli, and Giovanni Vallini. 2015. "Bioaugmentation and Biostimulation as Strategies for the Bioremediation of a Burned Woodland Soil Contaminated by Toxic Hydrocarbons: A Comparative Study." *Journal of Environmental Management* 153: 121–131.

Anwar, Faiza, Sabir Hussain, Shahla Ramzan, Farhan Hafeez, Muhammad Arshad, Muhammad Imran, Zahid Maqbool, and Naila Abbas. 2014. "Characterization of Reactive Red-120 Decolorizing Bacterial Strain Acinetobacter Junii FA10 Capable of Simultaneous Removal of Azo Dyes and Hexavalent Chromium." *Water, Air, & Soil Pollution* 225 (8): 2017.

Azad, Md. Abul Kalam, Latifah Amin, and Nik Marzuki Sidik. 2014. "Genetically Engineered Organisms for Bioremediation of Pollutants in Contaminated Sites." *Chinese Science Bulletin* 59 (8): 703–714.

Bennett, P. M., C. T. Livesey, D. Nathwani, D. S. Reeves, J. R. Saunders, and R. Wise. 2004. "An Assessment of the Risks Associated with the Use of Antibiotic Resistance Genes in Genetically Modified Plants: Report of the Working Party of the British Society for Antimicrobial Chemotherapy." *Journal of Antimicrobial Chemotherapy* 53: 418–431. https://doi.org/10.1093/jac/dkh087.

Bento, F. M., F. A. Camargo, B. C. Okeke, and W. T. Frankenberger. 2005. "Comparative Bioremediation of Soils Contaminated with Diesel Oil by Natural Attenuation, Biostimulation and Bioaugmentation." *Bioresource Technology* 96 (9): 1049–1055.

Bestawy, Ebtesam El., Shacker Helmy, Hany Hussien, Mohamed Fahmy, and Ranya Amer. 2013. "Bioremediation of Heavy Metal-Contaminated Effluent Using Optimized Activated Sludge Bacteria." *Applied Water Science* 3 (1): 181–192.

Bhatia, Divya, and Malik Kumar. 2011. "Plant -Microbe Interaction with Enhanced Bioremediation." *Research Journal of Biotechnology* 6: 72–79.

Bhatt, Praveena, M. Suresh Kumar, Sandeep Mudliar, and Tapan Chakrabarti. 2007. "Biodegradation of Chlorinated Compounds—A Review." *Critical Reviews in Environmental Science and Technology* 37 (2): 165–198.

Breyer, Didier, Lilya Kopertekh, and Dirk Reheul. 2014. "Alternatives to Antibiotic Resistance Marker Genes for in Vitro Selection of Genetically Modified Plants—Scientific Developments, Current Use, Operational Access and Biosafety Considerations." *Critical Reviews in Plant Sciences* 33 (4): 286–330.

Briffa, Jessica, Emmanuel Sinagra, and Renald Blundell. 2020. "Heavy Metal Pollution in the Environment and Their Toxicological Effects on Humans." *Heliyon* 6 (9): e04691.

Brim, Hassan, Sara C. McFarlan, James K. Fredrickson, Kenneth W. Minton, Min Zhai, Lawrence P. Wackett, and Michael J. Daly. 2000. "Engineering Deinococcus Radiodurans for Metal Remediation in Radioactive Mixed Waste Environments." *Nature Biotechnology* 18 (1): 85–90.

Cerniglia, C. E., and D. T. Gibson. 1977. "Metabolism of Naphthalene by Cunninghamella Elegans." *Applied and Environmental Microbiology* 34 (4): 363–370.

Charrier, Thomas, Marie José Durand, Mahmoud Affi, Sulivan Jouanneau, Hélène Gezekel, and Gérald Thouand. 2010. "Bacterial Bioluminescent Biosensor Characterisation for On-Line Monitoring of Heavy Metals Pollutions in Waste Water Treatment Plant Effluents." *Biosensors*. IntechOpen.

Chellaiah, Edward Raja. 2018. "Cadmium (Heavy Metals) Bioremediation by Pseudomonas Aeruginosa: A Minireview." *Applied Water Science* 8 (6): 154.

Chen, Shaolin, and David B. Wilson. 1997. "Genetic Engineering of Bacteria and Their Potential for Hg2+ Bioremediation." *Biodegradation* 8 (2): 97–103.

Chen, Xiaoyun, Manqiang Liu, Feng Hu, Xiaofang Mao, and Huixin Li. 2007. "Contributions of Soil Micro-Fauna (Protozoa and Nematodes) to Rhizosphere Ecological Functions." *Acta Ecologica Sinica* 27 (8): 3132–3143.

Das, Surajit, and Hirak R. Dash. 2014. "1—Microbial Bioremediation: APotential Tool for Restoration of Contaminated Areas." In *Microbial Biodegradation and Bioremediation* edited by Surajit Das, Elsevier, 1–21.

De, Jaysankar, N. Ramaiah, and L. Vardanyan. 2008. "Detoxification of Toxic Heavy Metals by Marine Bacteria Highly Resistant to Mercury." *Marine Biotechnology* 10 (4): 471–477.

Demnerová, Katerina, Martina Mackova, Veronika Spevákova, Katarina Beranova, Lucie Kochánková, Petra Lovecká, Edita Ryslavá, and Tomas Macek. 2005. "Two Approaches to Biological Decontamination of Groundwater and Soil Polluted by Aromatics-Characterization of Microbial Populations." *International Microbiology: The Official Journal of the Spanish Society for Microbiology* 8 (3): 205–211.

Diaconu, Mariana, Lucian Vasile Pavel, Raluca-Maria Hlihor, Mihaela Rosca, Daniela Ionela Fertu, Markus Lenz, Philippe Xavier Corvini, and Maria Gavrilescu. 2020. "Characterization of Heavy Metal Toxicity in Some Plants and Microorganisms—A Preliminary Approach for Environmental Bioremediation." *New Biotechnology* 56: 130–139.

Diep, Patrick, Radhakrishnan Mahadevan, and Alexander F. Yakunin. 2018. "Heavy Metal Removal by Bioaccumulation Using Genetically Engineered Microorganisms." *Frontiers in Bioengineering and Biotechnology* 6: 157.

Dwivedi, Seema. 2012. "Bioremediation of Heavy Metal by Algae: Current and Future Perspective." *Journal of Advanced Laboratory Research in Biology* 3 (3): 195–199.

El Fantroussi, Saïd, and Spiros N. Agathos. 2005. "Is Bioaugmentation a Feasible Strategy for Pollutant Removal and Site Remediation?" *Current Opinion in Microbiology, Ecology and Industrial Microbiology/Edited by Sergio Sánchez and Betty Olson Techniques*, edited by Peter J. Peters and Joel Swanson 8 (3): 268–275.

Fernández, Pablo M., María M. Martorell, Julia I. Fariña, and Lucia I. C. Figueroa. 2012. "Removal Efficiency of Cr6+ by Indigenous Pichia Sp. Isolated from Textile Factory Effluent." *The Scientific World Journal* 2012: 6.

Forrest, Lucy R., and Gary Rudnick. 2009. "The Rocking Bundle: A Mechanism for Ion-Coupled Solute Flux by Symmetrical Transporters." *Physiology* 24 (6): 377–386.

Frewer, L. J., G. A. Kleter, M. Brennan, D. Coles, A. R. H. Fischer, L. M. Houdebine, C. Mora, K. Millar, and B. Salter. 2013. "Genetically Modified Animals from Life-Science, Socio-Economic and Ethical Perspectives: Examining Issues in an EU Policy Context." *New Biotechnology* 30 (5): 447–460. https://doi.org/10.1016/j.nbt.2013.03.010.

Fritsche, Wolfgang, and Martin Hofrichter. 2000. "Aerobic Degradation by Microorganisms." In *Biotechnology* edited by H. J. Rehm and G. Reed, Weinheim, Germany: John Wiley& Sons, Ltd., 144–167.

Galal, Tarek, and Hanaa Shehata. 2015. "Bioaccumulation and Translocation of Heavy Metals by Plantago Major L. Grown in Contaminated Soils under the Effect of Traffic Pollution." *Ecological Indicators* 48: 244–251.

Gao, Weimin, and Arokiasamy Francis. 2008. "Reduction of Uranium(VI) to Uranium(IV) by Clostridia." *Applied and Environmental Microbiology* 74 (14): 4580–4584.

Glick, Bernard R. 2010. "Using Soil Bacteria to Facilitate Phytoremediation." *Biotechnology Advances* 28 (3): 367–374.

Gomes, Maria Angélica da Conceição, Rachel Ann Hauser-Davis, Adriane Nunes de Souza, and Angela Pierre Vitória. 2016. "Metal Phytoremediation: General Strategies, Genetically Modified Plants and Applications in Metal Nanoparticle Contamination." *Ecotoxicology and Environmental Safety* 134: 133–147.

Grennan, Aleel K. 2009. "Identification of Genes Involved in Metal Transport in Plants." *Plant Physiology* 149 (4): 1623–1624.

Guo, Pan, Ting Wang, Yanli Liu, Yan Xia, Guiping Wang, Zhenguo Shen, and Yahua Chen. 2014. "Phytostabilization Potential of Evening Primrose (Oenothera Glazioviana) for Copper-Contaminated Sites." *Environmental Science and Pollution Research* 21 (1): 631–640.

Gupta, Saurabh, and Daljeet Singh. 2017. "Role of Genetically Modified Microorganisms in Heavy Metal Bioremediation." In *Advances in Environmental Biotechnology*, edited by Raman Kumar, Anil Kumar Sharma, and Sarabjeet Singh Ahluwalia. Singapore: Springer, 197–214.

Harshvardhan, Kumar, and Bhavanath Jha. 2013. "Biodegradation of Low-Density Polyethylene by Marine Bacteria from Pelagic Waters, Arabian Sea, India." *Marine Pollution Bulletin* 77 (1): 100–106.

Hong, Yiguo, Meiying Xu, Jun Guo, Zhicheng Xu, Xingjuan Chen, and Guoping Sun. 2007. "Respiration and Growth of Shewanella Decolorationis S12 with an Azo Compound as the Sole Electron Acceptor." *Applied and Environmental Microbiology* 73 (1): 64–72.

Hontzeas, Nikos, Jérôme Zoidakis, Bernard R. Glick, and Mahdi M. Abu-Omar. 2004. "Expression and Characterization of 1-Aminocyclopropane-1-Carboxylate Deaminase from the Rhizobacterium Pseudomonas Putida UW4: A Key Enzyme in Bacterial Plant Growth Promotion." *Biochimica et Biophysica Acta (BBA)—Proteins and Proteomics* 1703 (1): 11–19.

Huang, Aimin, Bangzhi Wei, Junyong Mo, Yajing Wang, and Lin Ma. 2018. "Conformation and Activity Alteration of Horseradish Peroxidase Induced by the Interaction with Gene Carrier Polyethyleneimines." *Spectrochimica Acta Part A: Molecular and Biomolecular Spectroscopy* 188: 90–98.

Ibañez, Sabrina, Melina Talano, Ornella Ontañon, Jachym Suman, María I. Medina, Tomas Macek, and Elizabeth Agostini. 2016. "Transgenic Plants and Hairy Roots: Exploiting the Potential of Plant Species to Remediate Contaminants." *New Biotechnology, Plant Biotechnology: Green for Good III, Olomouc* 33 (5, Part B): 625–635.

Islamoglu, Timur, Zhijie Chen, Megan C. Wasson, Cassandra T. Buru, Kent O. Kirlikovali, Unjila Afrin, Mohammad Rasel Mian, and Omar K. Farha. 2020. "Metal–Organic Frameworks Against Toxic Chemicals." *Chemical Reviews* 120 (16): 8130–8160.

Ivshina, Irena B., Maria S. Kuyukina, Anastasiya V. Krivoruchko, Andrey A. Elkin, Sergey O. Makarov, Colin J. Cunningham, Tatyana A. Peshkur, Ronald M. Atlas, and James C. Philp. 2015. "Oil Spill Problems and Sustainable Response Strategies Through New Technologies." *Environmental Science: Processes & Impacts* 17 (7): 1201–1219.

Jaiswal, Shweta, and Pratyoosh Shukla. 2020. "Alternative Strategies for Microbial Remediation of Pollutants via Synthetic Biology." *Frontiers in Microbiology* 11.

Janssen, Dick, and Gerhard Stucki. 2020. "Perspectives of Genetically Engineered Microbes for Groundwater Bioremediation." *Environmental Science: Processes & Impacts* 22 (3): 487–499.

Kafilzadeh, Farshid, Parvaneh Sahragard, Hooshang Jamali, and Yaghoob Tahery. 2011. "Isolation and Identification of Hydrocarbons Degrading Bacteria in Soil Around Shiraz Refinery." *African Journal of Microbiology Research* 5 (19): 3084–3089.

Kanade, S. N., A. B. Ade, and V. C. Khilare. 2012. "Malathion Degradation by Azospirillum Lipoferum Beijerinck." *Science Research Reporter* 2 (1): 94–103.

Kaplan, Christopher, and Christopher Kitts. 2004. "Bacterial Succession in a Petroleum Land Treatment Unit." *Applied and Environmental Microbiology* 70 (3): 1777–1786.

Kobayashi, Hester, and Bruce E. Rittmann. 1982. "Microbial Removal of Hazardous Organic Compounds." *Environmental Science and Technology* 16 (3): 170A–183A.

Kong, Zhaoyu, and Bernard R. Glick. 2017. "Chapter Two—The Role of Plant Growth-Promoting Bacteria in Metal Phytoremediation." *Advances in Microbial Physiology*, edited by Robert K. Poole, 71: 97–132. Academic Press.

Koźmińska, Aleksandra, Alina Wiszniewska, Ewa Hanus-Fajerska, and Ewa Muszyńska. 2018. "Recent Strategies of Increasing Metal Tolerance and Phytoremediation Potential Using Genetic Transformation of Plants." *Plant Biotechnology Reports* 12 (1): 1–14.

Kumar, Adarsh, and Ram Chandra. 2020. "Ligninolytic Enzymes and Its Mechanisms for Degradation of Lignocellulosic Waste in Environment." *Heliyon* 6 (2).

Kumar, Narasimhan Manoj, Chandrasekaran Muthukumaran, Govindasamy Sharmila, and Baskar Gurunathan. 2018. "Genetically Modified Organisms and Its Impact on the Enhancement of Bioremediation." In *Bioremediation: Applications for Environmental Protection and Management. Energy, Environment, and Sustainability*, edited by Sunita J. Varjani, Avinash Kumar Agarwal, Edgard Gnansounou, and Baskar Gurunathan. Singapore: Springer, 53–76.

Lampis, Silvia, Chiara Santi, Adriana Ciurli, Marco Andreolli, and Giovanni Vallini. 2015. "Promotion of Arsenic Phytoextraction Efficiency in the Fern Pteris Vittata by the Inoculation of As-Resistant Bacteria: A Soil Bioremediation Perspective." *Frontiers in Plant Science* 6.

Leigh, Mary Beth, Petra Prouzová, Martina Macková, Tomáš Macek, David P. Nagle, and John S. Fletcher. 2006. "Polychlorinated Biphenyl (PCB)-Degrading Bacteria Associated with Trees in a PCB-Contaminated Site." *Applied and Environmental Microbiology* 72 (4): 2331–2342.

Lloyd, Jonathan R., and Derek R. Lovley. 2001. "Microbial Detoxification of Metals and Radionuclides." *Current Opinion in Biotechnology* 12 (3): 248–253.

Lone, Mohammad Iqbal, Zhen-li He, Peter J. Stoffella, and Xiao-e Yang. 2008. "Phytoremediation of Heavy Metal Polluted Soils and Water: Progresses and Perspectives." *Journal of Zhejiang University Science B* 9 (3): 210–220.

Madhurankhi, Goswami, Poulomi Chakraborty, Koushik Mukherjee, Garbita Mitra, Purnita Bhattacharyya, Samrat Dey, Prosun Tribedi. 2018. "Bioaugmentation and Biostimulation: A Potential Strategy for Environmental Remediation." 6 (5): 223–231.

Marinescu, M., M. Dumitru, and A. Lacatusu. 2009. "Biodegradation of Petroleum Hydrocarbons in an Artificial Polluted Soil." *Research Journal of Agricultural Science* 41 (2).

Markowitz, Gerald. 2016. "The Childhood Lead Poisoning Epidemic in Historical Perspective." *Endeavour, Living in a Toxic World, 1800–2000* 40 (2): 93–101.

Martínez, Mar, Pilar Bernal, Concepción Almela, Dinoraz Vélez, Pilar García-Agustín, Ramón Serrano, and Juan Navarro-Aviñó. 2006. "An Engineered Plant That Accumulates Higher Levels of Heavy Metals Than Thlaspi Caerulescens, with Yields of 100 Times More Biomass in Mine Soils." *Chemosphere* 64 (3): 478–485.

Menn, Fu-Min, James P. Easter, and Gary S. Sayler. 2008. "Genetically Engineered Microorganisms and Bioremediation." In *Biotechnology*, edited by H-J. Rehm and G. Reed. Weinheim, Germany: Wiley, 441–463.

Migheli, Q. 2001. "Genetically Modified Biocontrol Agents: Environmental Impact and Risk Analysis." *Journal of Plant Pathology* 83: 47–56.

Mónica, Pérez, Rueda O. Darwin, Bangeppagari Manjunatha, Johana J. Zúñiga, Ríos Diego, Rueda B. Bryan, Sikandar I. Mulla, and Naga R. Maddela. 2016. "Evaluation of Various Pesticides-Degrading Pure Bacterial Cultures Isolated from Pesticide-Contaminated Soils in Ecuador." *African Journal of Biotechnology* 15 (40): 2224–2233.

Moreno, Fabio N., Christopher W. N. Anderson, Robert B. Stewart, and Brett H. Robinson. 2008. "Phytofiltration of Mercury-Contaminated Water: Volatilisation and Plant-Accumulation Aspects." *Environmental and Experimental Botany* 62 (1): 78–85.

Moss, B. 1985. "Genetically Engineered Poxviruses for Recombinant Gene Expression, Vaccination, and Safety." *Proceedings of the National Academy of Sciences of the United States of America* 93 (21): 11341–11348.

Mrozik, A., and Z. Piotrowska-Seget. 2010. "Bioaugmentation as a Strategy for Cleaning Up of Soils Contaminated with Aromatic Compounds." *Microbiological Research* 165 (5): 363–375.

Mrozik, A., Z. Piotrowska-Seget, and S. Labuzek. 2003. "Bacterial Degradation and Bioremediation of Polycyclic Aromatic Hydrocarbons." *Polish Journal of Environmental Studies* 12 (1): 15–25.

Muro-González, Dalia A., Patricia Mussali-Galante, Leticia Valencia-Cuevas, Karen Flores-Trujillo, and Efraín Tovar-Sánchez. 2020. "Morphological, Physiological, and Genotoxic Effects of Heavy Metal Bioaccumulation in Prosopis Laevigata Reveal Its Potential for Phytoremediation." *Environmental Science and Pollution Research* 27 (32): 40187–40204.

Nanda, M., V. Kumar, and D. K. Sharma. 2019. "Multimetal Tolerance Mechanisms in Bacteria: The Resistance Strategies Acquired by Bacteria That Can Be Exploited to 'Clean-up' Heavy Metal Contaminants from Water." *Aquatic Toxicology* 212: 1–10.

Okpokwasili, G. C. 2007. "Biotechnology and Clean Environment." In Proceedings of the 20th Annual Conference of the Biotechnology Society of Nigeria (BSN).

Olaniran, Ademola O., Dorsamy Pillay, and Balakrishna Pillay. 2006. "Biostimulation and Bioaugmentation Enhances Aerobic Biodegradation of Dichloroethenes." *Chemosphere* 63 (4): 600–608.

Olanrewaju, Oluwaseyi Samuel, Bernard R. Glick, and Olubukola Oluranti Babalola. 2017. "Mechanisms of Action of Plant Growth Promoting Bacteria." *World Journal of Microbiology and Biotechnology* 33 (11): 197.

Ozyigit, Ibrahim Ilker, Hasan Can, and Ilhan Dogan. 2021. "Phytoremediation Using Genetically Engineered Plants to Remove Metals: A Review." *Environmental Chemistry Letters* 19 (1): 669–698.

Pandey, Gunjan, Debarati Paul, and Rakesh K. Jain. 2005. "Conceptualizing 'Suicidal Genetically Engineered Microorganisms' for Bioremediation Applications." *Biochemical and Biophysical Research Communications* 327 (3): 637–639.

Pant, Gaurav, Deviram Garlapati, Urvashi Agrawal, R. Gyana Prasuna, Thangavel Mathimani, and Arivalagan Pugazhendhi. 2021. "Biological Approaches Practised Using Genetically Engineered Microbes for a Sustainable Environment: A Review." *Journal of Hazardous Materials* 405: 124631.

Paul, Debarati, Gunjan Pandey, and Rakesh K. Jain. 2005. "Suicidal Genetically Engineered Microorganisms for Bioremediation: Need and Perspectives." *BioEssays* 27 (5): 563–573.

Pearce, C. I., J. R. Lloyd, and J. T. Guthrie. 2003. "The Removal of Colour from Textile Wastewater Using Whole Bacterial Cells: A Review." *Dyes and Pigments* 58 (3): 179–196.

Petrić, Ines, Dubravka Hršak, Sanja Fingler, Ernest Vončina, Helena Ćetković, Ana Begonja Kolar, and Nikolina Udiković Kolić. 2007. "Enrichment and Characterization of PCB-Degrading Bacteria as Potential Seed Cultures for Bioremediation of Contaminated Soil." *Food Technology and Biotechnology* 45 (1): 11–20.

Pilon-Smits, Elizabeth A. H., and John L. Freeman. 2006. "Environmental Cleanup Using Plants: Biotechnological Advances and Ecological Considerations." *Frontiers in Ecology and the Environment* 4 (4): 203–210.

Pinaki, Sar, Sufia K. Kazy, S. F. D'Souza. 2004. "Radionuclide Remediation Using a Bacterial Biosorbent." *International Biodeterioration & Biodegradation* 54 (2–3): 193–202.

Raffi, F., J. D. Hall, and C. E. Cerniglia. 1997. "Mutagenicity of Azo Dyes Used in Foods, Drugs and Cosmetics Before and After Reduction by Clostridium Species from the Human Intestinal Tract." *Food and Chemical Toxicology* 35 (9): 897–901.

Raghunandan, K., A. Kumar, S. Kumar, K. Permaul, and S. Singh. 2018. "Production of Gellan Gum, an Exopolysaccharide, from Biodiesel-Derived Waste Glycerol by Sphingomonas spp." *3Biotech* 8: 71.

Raj, Deep, Adarsh Kumar, and Subodh Kumar Maiti. 2020. "Mercury Remediation Potential of Brassica Juncea (L.) Czern. for Clean-up of Flyash Contaminated Sites." *Chemosphere* 248: 125857.

Raman, R. 2017. "The Impact of Genetically Modified (GM) Crops in Modern Agriculture: A Review." *GM Crops and Food* 8: 195–208.

Reddy, Bhaskara L., and Milton H. Saier Jr. 2016. "Properties and Phylogeny of 76 Families of Bacterial and Eukaryotic Organellar Outer Membrane Pore-Forming Proteins." *PLoS One* 11 (4).

Rout, Gyana Ranjan, Dhaneswar Swain, and Bandita Deo. 2019. "Chapter 12—Restoration of Metalliferous Mine Waste Through Genetically Modified Crops." In *Transgenic Plant Technology for Remediation of Toxic Metals and Metalloids*, edited by Majeti Narasimha Vara Prasad. Academic Press, 257–278.

Saavedra, J., F. Acevedo, M. Vergara, and M. Seeger. 2010. "Mineralization of PCBs by the Genetically Modified Strain Cupriavidus Necator JMS34 and Its Application for Bioremediation of PCBs in Soil." *Applied Microbiology and Biotechnology* 87: 1543–1554.

Santoyo, Gustavo, Gabriel Moreno-Hagelsieb, Ma. del Carmen Orozco-Mosqueda, and Bernard R. Glick. 2016. "Plant Growth-Promoting Bacterial Endophytes." *Microbiological Research* 183: 92–99.

Sarma, Hemen, N. F. Islam, Ram Prasad, M. N. V. Prasad, Lena Q. Ma, and Jörg Rinklebe. 2021. "Enhancing Phytoremediation of Hazardous Metal(Loid)s Using Genome Engineering CRISPR–Cas9 Technology." *Journal of Hazardous Materials* 414: 125493.

Schoenborn, Liesbeth, Penelope S. Yates, Bronwyn E. Grinton, Philip Hugenholtz, and Peter H. Janssen. 2004. "Liquid Serial Dilution Is Inferior to Solid Media for Isolation of Cultures Representative of the Phylum-Level Diversity of Soil Bacteria." *Applied and Environmental Microbiology* 70 (7): 4363–4366.

Seeda, Abou, and Abou El-Nour El-Zanaty. 2017. "Microorganism as a Tool of Bioremediation Technology for Cleaning Waste and Industrial Water." *Bioscience Research* 14 (3): 633–644.

Seeger, Michael, Marcela Hernández, Valentina Méndez, Bernardita Ponce, Macarena Córdova, and Myriam González. 2010. "Bacterial Degradation and Bioremediation of Chlorinated Herbicides and Biphenyls." *Journal of Soil Science and Plant Nutrition* 10 (3): 320–332.

Sharma, Akash, Mohit Mishra, Simanta Sheet, and Mukesh Thite. 2013. "Role of Microbes as Cleaning Degrading Industrial Wastes for Environmental Sustainability-A Reveiw." *Recent Research in Science and Technology* 5: 21–25.

Silva, Elisabete, Arsénio M. Fialho, Isabel Sá-Correia, Richard G. Burns, and Liz J. Shaw. 2004. "Combined Bioaugmentation and Biostimulation to Cleanup Soil Contaminated with High Concentrations of Atrazine." *Environmental Science & Technology* 38 (2): 632–637.

Singh, Jay Shankar, P. C. Abhilash, H. B. Singh, Rana P. Singh, and D. P. Singh. 2011. "Genetically Engineered Bacteria: An Emerging Tool for Environmental Remediation and Future Research Perspectives." *Gene* 480 (1): 1–9.

Srivastava, Pankaj Kumar, Aradhana Vaish, Sanjay Dwivedi, Debasis Chakrabarty, Nandita Singh, and Rudra Deo Tripathi. 2011. "Biological Removal of Arsenic Pollution by Soil Fungi." *Science of The Total Environment* 409 (12): 2430–2442.

Stapleton, Jr, R. D., and V. P. Singh. 2002. "Biotransformations: Bioremediation Technology for Health and Environmental Protection". In *Progress in Industrial Microbiology*. edited by Singh Ved Pal and Stapleton Raymond D. Jr. Elsevier, 36, 1–614.

Surekha Rani, M., K. Vijaya Lakshmi, P. Suvarnalatha Devi, R. Jaya Madhuri, S. Aruna, K. Jyothi, and G. Narasimha. 2008. "Isolation and Characterization of a Chlorpyrifos-Degrading Bacterium from Agricultural Soil and Its Growth Response." *African Journal of Microbiology Research* 2 (2): 26–31.

Takeuchi, F., and T. Sugio. 2006. "Volatilization and Recovery of Mercury from Mercury-Polluted Soils and Wastewaters Using Mercury-Resistant Acidithiobacillus Ferrooxidans Strains SUG 2–2 and MON-1." *Environmental Sciences: An International Journal of Environmental Physiology and Toxicology* 13 (6): 305–316.

Tekere, M. 2020. "Biological Strategies for Heavy Metal Remediation." In *Methods for Bioremediation of Water and Wastewater Pollution, Environmental Chemistry for a Sustainable World*, edited by M. I. Ahamed Inamuddin, E. Lichtfouse, A. M. Asiri. Cham: Springer International Publishing, 393–413.

Valls, Marc, and Víctor de Lorenzo. 2002. "Exploiting the Genetic and Biochemical Capacities of Bacteria for the Remediation of Heavy Metal Pollution." *FEMS Microbiology Reviews* 26 (4): 327–338.

Vargas, J. M. 1975. "Pesticide Degradation." *Journal of Arboriculture* 1 (12): 232–233.

Verma, Pradeep, and Datta Madamwar. 2003. "Decolourization of Synthetic Dyes by a Newly Isolated Strain of Serratia Marcescens." *World Journal of Microbiology and Biotechnology* 19 (6): 615–618.

Walters, R. 2006. "Criminology and Genetically Modi ed Food." In *Green Criminology*, edited by Nigel South, London: Routledge.

Wei, Wei, Sok Kim, Myung-Hee Song, John Kwame Bediako, and Yeoung-Sang Yun. 2015. "Carboxymethyl Cellulose Fiber as a Fast Binding and Biodegradable Adsorbent of Heavy Metals." *Journal of the Taiwan Institute of Chemical Engineers* 57: 104–110.

Wolok, Eduart, Jamal Barafi, Navneet Joshi, Rossella Girimonte, and Sudip Chakraborty. 2020. "Study of Bio-Materials for Removal of the Oil Spill." *Arabian Journal of Geosciences* 13 (23): 1244.

Yakimov, Michail M., Kenneth N. Timmis, and Peter N. Golyshin. 2007. "Obligate Oil-Degrading Marine Bacteria." *Current Opinion in Biotechnology, Energy Biotechnology/Environmental Biotechnology* 18 (3): 257–266.

Yee, N., J. Ma, A. Dalia, T. Boonfueng, and D. Y. Kobayashi. 2007. "Se(VI) Reduction and the Precipitation of Se(0) by the Facultative Bacterium Enterobacter Cloacae SLD1a-1 Are Regulated by FNR." *Applied and Environmental Microbiology* 73 (6): 1914–1920.

Zafar, Shaheen, Farrukh Aqil, and Iqbal Ahmad. 2007. "Metal Tolerance and Biosorption Potential of Filamentous Fungi Isolated from Metal Contaminated Agricultural Soil." *Bioresource Technology* 98 (13): 2557–2561.

Zhu, Wenjie, Liyuan Chai, Zemin Ma, Yunyan Wang, Haijuan Xiao, and Kun Zhao. 2008. "Anaerobic Reduction of Hexavalent Chromium by Bacterial Cells of Achromobacter Sp. Strain Ch1." *Microbiological Research* 163 (6): 616–623.

Zou, Long, Yun-hong Huang, Zhong-er Long, and Yan Qiao. 2018. "On-Going Applications of Shewanella Species in Microbial Electrochemical System for Bioenergy, Bioremediation and Biosensing." *World Journal of Microbiology and Biotechnology* 35 (1): 9.

11 Bioremediation of Heavy Metals Using Microorganisms

M.S. Nagmote, A.R. Rai, R. Sharma,
M.F. Desimone, R.G. Chaudhary and N.B. Singh

INTRODUCTION

The basic requirements for the survival of all living beings on the earth are water and soil (Zaynab et al., 2021). However, these are being polluted continuously due to the industrial revolution and anthropogenic activities. Of the common pollutants, heavy metals (HMs) even in trace amounts are very toxic and harmful to human health and ecosystems, since they are non-biodegradable (Ferrey et al., 2018; Brodin et al., 2017). Elements with specific gravity higher than 5 and atomic mass higher than 65 are considered HMs. Some of the HMs include Hg, As, Ag, Cd, Zn, Fe, Pt, Cr and Cu. (Yin et al., 2019). When these contaminate water, soil, and environments, they have serious negative impacts on animals, humans, and plants (Wang L. et al., 2020; Gu et al., 2018). Some common toxic effects of HMs on plants, fish and humans are given in Figure 11.1 (Sharma et al., 2021).

A number of technologies have been developed for removing HMs from polluted areas (Yin et al., 2019). These methods have both advantages and disadvantages. However, in recent years lot of emphasis is being given to bioremediation technologies based on microorganisms due to their exceptional advantages, such as being cheap, ecologically sound and efficient (Potbhare et al., 2020; Chouke et al., 2022a). Microorganisms can resist environmental stress much more compared to plants and animals, through rapid mutation and evolution. When HMs are removed by viable microbial cells and modified microbial biomass, the processes are advantageous due to high adsorption capacity, low cost, large availability and simplicity of use. Among the microorganisms, fungi, bacteria, and algae are the most widely used. In addition to microorganisms, technology using genetically engineered microorganisms (GEMs) has been considered faster and better for the removal of HMs (Verma and Kuila, 2019).

Different forms of GEMs such as algae, bacteria, and fungi have been obtained by recombinant DNA and RNA methods, which are much more effective and safe for HMs remediation (Sharma et al., 2021). In this chapter, the sources and toxic effects of HMs and the role of GEMs in their removal have been discussed in detail.

HEAVY METALS AND POLLUTION

HMs are naturally present in the hydrosphere, atmosphere, lithosphere, and biosphere (Krishna et al., 2016). However, due to (1) unique chemical affinities to various vital cellular components, (2) capability of easy distribution in different biological systems and (3) lack of efficient physiological mechanisms for their excretion or removal from the body, HMs tend to gradually accumulate in living bodies and transmit as well as to animals of higher orders via the links of the food chain.

The primary source of HMs is natural or ecological activities since the formation of the earth. The trace elements of Earth's crust and chemicals from deeper layers of the earth are released in the atmosphere by active volcanoes and to the surrounding land and water bodies. The principal process which alters the HMs biogeochemistry in boreal, forest ecosystems is the weathering of rock and soil. Metal smelting and other extractive processes use intense heat, leaching in very

DOI: 10.1201/9781003188568-11

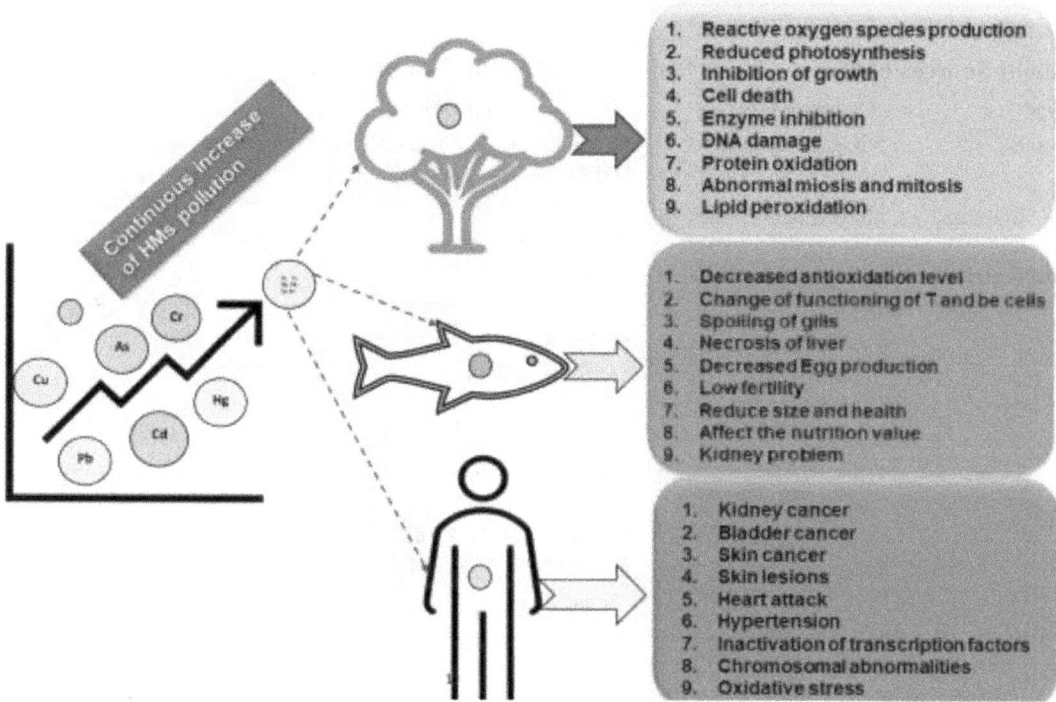

FIGURE 11.1   Toxic effects of HMs on human, fish and plants (Sharma et al., 2021).

strong acid or base or metallo-electrolytic processes. These processes generate huge amounts of numerous hazardous waste materials. In the mining and metallurgical extractive procedures, particularly in gold mining, over 99% of the extracted ores are admitted as waste to the environment (Adler and Rascher, 2007). Coal mining also contributes to contamination. At least 22 different types of HMs are believed to be present in coal (Ren et al., 1999). All these toxic elements are finally deposited in the organisms of ecosystems. Combustion of coal and petroleum-based fuel additionally mobilize HMs. About 50 various elements are reported in coal (Ruch et al., 1974; Gluskoter and Harold, 1975; Rao et al., 1973), 35 in crude oil (Bakirova et al., 1980; Shah et al., 1970), and about 20 in gasoline (Jungers et al., 1975). The chemical composition and structural identity of particulate emissions depend on fuel type and combustion processes; besides the toxic impact on health, some HMs can catalyze the atmospheric transformation of primary pollutants of air, and cause economic impacts due to corrosive and abrasive activity on metal surfaces. Water is one of the most important components of the ecosystem, which is being continuously polluted by the discharging of industrial and domestic waste. This component provides a crucial pathway for the transmission of HMs to living bodies of the ecosystem. Major HMs from various sources are listed in Table 11.1.

TOXICITY OF HMs TO LIFE

The contamination of HMs into air, soil and water results in serious impacts on the growth, productivity and quality of forest plants and cultivated crops. Plants growing in HMs-contaminated soil are constantly exposed to harmful elements and tend to bioaccumulate in higher amounts. The HMs exert toxic effects on plants through the following mechanisms.

TABLE 11.1

Major Sources of Heavy Metals from Various Natural and Anthropogenic Activities

Type	Sources/Activities	Major HMs	References
Natural	Volcanic eruptions	Ti, Mn, Fe, Cu, Zn, Cr, Mn, Ni, Pb, Cd, Hg	(Vigneri et al., 2017) (Ragnarsdottir et al., 1994)
	Weathering of rocks/soil	Cd, Cu, Ni, Pb, Zn, Ca, Mg, Zr As, Pb, Ni, Mo, Cd, Cu, Cr	(Starr et al., 2003) (Xu et al., 2022)
Anthropogenic	Smelting of metal/ minerals	As, Cd, Cr, Pb, Mn, Hg	(Sanders et al., 2014)
	Coal/Gold mining	As, Cu, Zn, Cd, Ni, Pb, Co, Hg Cr, Ni, Cu, Zn, Cd, Pb	(Fashola et al., 2016) (Li et al., 2018)
	Industrial waste	Cd, Cr, Cu, As, Pb, Ni, Zn, Hg As, Cd, Cr, Cu, Fe, Mn, Ni, Pb,	(Azimi et al., 2017) (Singh et al., 2013)
	Domestic waste	Cd, Zn, Cu, As, Pb, Cr, Fe, Ni, Mn	(Drozdova et al., 2019)
	Auto exhausts	Cr, Cu, Cd, Ni, Pb, Zn Pb, Ca, Fe, Cu, Zn, Sb, Ba	(Ndiokwere, 1984) (Kweon et al., 2002)
	Pesticides	Pb, Sb, Mn, Ni, V, As, Cd, Cu, Ga, Se and Zn, Ti	(Gimeno-García et al., 1996)

(i)   Generation of ROS or methylglyoxal formation by auto-oxidation or via Fenton reactions and Haber-Weiss reactions or by compromising antioxidant defense mechanism or disturbing the electron transport chain. These cytotoxic chemical species lead to biomacromolecular damage, membrane instability, nucleic acids damage, and more (Rascio et al., 2011; Romero-Puertas et al., 2002; Braconi et al., 2011);

(ii)   direct interaction with regulatory proteins because of having affinities for thiol-, carboxyl- and histidyl-groups, which ultimately paralyze the structural, transport and catalytic authorities of cells (Hossain et al., 2012);

(iii)   when essential metals are displaced from their binding sites that cause functional loss of proteins (Sharma et al., 2009; Schutzendubel et al., 2002); and

(iv)   competitive absorption of non-essential toxic HMs at root surfaces, which compete with the essential active metals. e.g., Cd and As compete with Zn and P respectively for absorption (Sharma et al., 2009; DalCorso et al., 2013).

The aquatic biota receives a major proportion of metal toxicants from the various activities on land and in the atmosphere, and are probably the ultimate depository for remobilized HMs. In general, aquatic species are more susceptible to metals than land-dwelling animals. The primary producers of aquatic ecosystems are equally affected by toxic exposure to metals. Primary producers' viz. algae and phytoplankton form the lowest trophic level, the base of the aquatic food web. Energy is generally derived from photosynthesis and used by primary consumers such as small fish, zooplanktons and crustaceans. The primary consumers are eaten by small sharks, large fish, corals and baleen whales. The top ocean predators include billfish, dolphins, large sharks, toothed whales and large seals. The phytoplanktons are the principal marine organic food producers. Any toxic alteration in their growth and production can produce serious consequences in an entire marine ecosystem. The influence on phytoplankton by bioactive toxicants differs because of their different relative toxicity. Hg and Cu are the most toxic elements, affecting ecosystems to different extents depending on the phytoplankton species-specific susceptibility of HMs (Thomas et al., 1980).

The HMs ultimately limit the production of plankton and their oceanic growth (González-Dávila and Melchor, 1995). Algae serve as one of the foremost producers in freshwater and marine ecosystems. Because of being continuously exposed in water, algae are generally adversely affected by HMs. The effect of HMs exposure is manifested by suppression of cell division, stunted growth, restraining of enzyme function and reduction of photosynthesis (Chen et al., 2009; Ismail et al., 2002; Baumann et al., 2009). This toxicity mainly results from stable bonding with sulfhydryl protein groups or altering the structure of proteins or relocating other essential heavy elements. The metals are known to modulate enzymes of antioxidant machinery viz. superoxide dismutase, ascorbate peroxidase and glutathione peroxidase (Arunakumara and Zhang, 2008). Membrane instability in cyanobacteria is also reported (Rangsayatorn et al., 2002).

Fish are one of the essential components of aquatic ecosystem. Their gills are easily contaminated by increased concentrations of HMs. Abnormally high concentrations of HMs cause inflammation, oedema and sloughing of the epithelial cell layer in gill epithelium. This causes reduced diffusion of respiratory gases to blood, and leads to the most frequent reason for their death due to hypoxia and/or acidosis (Mallatt, 1985; McDonald and Wood, 1993). For example, Cu and Ag inhibit the Na^+/K^+ ATPase activity in the membranes of gills (Morgan et al., 1995; Stagg and Shuttleworth, 1982). These result in an acute reduction of Na^+ and Cl- concentrations in the blood, which eventually promotes lethal cardio-vascular collapse (Laurén et al., 1986; Wood et al., 1996). Zn and Cd show powerful inhibitory action on Ca^{++}-ATPase, which ultimately leads to fetal hypocalcemia in fish (Verbost et al., 1988;Spry and Wood, 1989; Verbost et al., 1987). Another important effect of HMs in fish is suppressed reproductive capacity of long-term exposure of Cu (Mount and Donald, 1968) and Zn (Birge et al., 1981), and reduced viability of fish eggs even at low concentrations of Cd (Birge et al., 1981). In addition, patterns of accumulation of HMs in tissues or organs of fish vary widely. The liver, kidney, muscle, gills and heart are the usual locations of HMs bioaccumulation in fish (Garai et al., 2021). These HMs repositories are further transmitted partly or completely as trophic cascades that affect the organisms across different trophic levels, including birds and humans. The incidence of metal toxicity has been well documented in recent decades and the occurrence of deadly health impacts is alarmingly high. Metal intoxication, especially by lead and zinc, is frequently found in aviary birds (Puschner and Robert, 2009). Excessive dietary exposure and deregulated homeostatic mechanisms can be the prime reason for metal toxicity in birds, including ducks (Plumlee and Konnie, 2004; Wedekind and Baker, 1990), gray-headed chachalaca (Droual et al., 1991), lovebirds (Reece et al., 1986), Nicobar pigeon (Van der Zee et al., 1985) and Amazon parrots (Smith, 1995). The physiological and biochemical alterations due to chronic exposure to Hg, Pb, Cd and Al are well reviewed. The toxic HMs predominantly affect the reproductive capabilities of birds, which involve their suppressed egg production, reduced hatchability and enhanced hatching mortality indices (Scheuhammer, 1987).

In addition to impacting the important consumer of food chains of terrestrial ecosystems, HMs exposure in human beings is also associated with atmospheric sources, aquatic and marine ecosystems. Additional intake of metals, limited biological assimilation and limited physiological elimination have increased the incidences of metal-induced pathologies in humans. Mercury, lead, arsenic, cadmium, chromium and their oxides are foremost among toxic HMs. These metals give rise to diverse sets of physiological as well as anatomical anomalies, ranging from frequent allergies, neurological, hepatocellular immunological and genetic damage to a variety of life-threatening cancers. These metals and their oxides are very harmful to the life cycle of all organisms (Chouke et al., 2022b; Singh et al., 2022).

BIOREMEDIATION OF HMs EMPLOYING MICROORGANISMS

Actually, different methodologies and processes are being employed to decontaminate aqueous systems and soils contaminated with heavy metals (Kumar et al., 2021). Traditional methodologies involve the immobilization of metal ions in a variety of substrates through different mechanisms like adsorption in activated carbon or ionic exchange. Alternatively, the chemical reduction of metal ions and the subsequent precipitation was widely evaluated. However, bioremediation of heavy metals has emerged as a valuable tool to replace conventional remediation processes. Moreover, this technology can be adapted for the development of bioreactors and consequently further employed to decontaminate effluents from various industries. Indeed, microbial remediation of toxic HMs is eco-friendly and cost-effective. Microorganisms have been used for the efficient remediation of HMs from various ecosystems. The mechanisms involved include but are not limited to bioaccumulation, biosorption, intracellular or extracellular sequestration, methylation or metal ions reduction (Igiri et al., 2018).

Alvarez et al. (2011) reported Cr(VI) reduction from aqueous and solid samples employing free and silica matrix encapsulated *Burkholderia* sp bacteria. Interestingly, Cr(VI) (100 mg ml^{-1}) were completely reduced during 4 days of culture in liquid media. Moreover, a higher amount of Cr(VI) (200 mg ml^{-1}) were completely reduced during 7 days' culture in sterilized soil. Moreover, the Cr(III) obtained is less toxic and was subsequently sequestrated by adsorption in the silica matrix. Interestingly, improved rates of reduction were obtained by immobilized bacteria, which were effectively protected from the toxicological effects of elevated amounts of Cr(VI) (Alvarez et al., 2011).

Similarly, transforming Hg^{2+} to a reduced form like Hg(0) is highlighted as a valuable alternative to reduce its toxicity, especially because Hg^{2+} could be converted to methylmercury which is the highest harmful form (Wang N. et al., 2020). For this purpose, bacteria that possess the biochemical machinery (i.e., genes and gene products) involved in mercury reduction and tolerance are employed (Mathema et al., 2011). Recently, a mercury-resistant strain was efficiently employed for the remediation of mercury. The results revealed the activity of the mercuric reductase enzyme and the detection of volatilized mercury. Indeed, within 6 hours of culture, 44% and 23% mercury were volatilized and accumulated in live bacterial pellets, respectively (Mahbub et al., 2017).

Microorganisms can also reduce the toxicological effects of Pb(II) through biomineralization, adsorption, complexation or chelation, precipitation, active transport and intracellular uptake (Pan et al., 2017). Among these, complexation and biosorption of Pb(II) are the most common bioremediation mechanisms. Chen et al. (2015) employed *Bacillus thuringiensis* for lead (II) bio-sorption. The authors found that amide, phosphate and carboxyl functional groups of *Bacillus thuringiensis* were involved in lead (II) biosorption. The maximum lead (II) biosorption capacity achieved was 164.77 mg g^{-1} (dry weight) (Chen et al., 2015). Alternatively, a higher capacity of lead (II) biosorption (333.3 mg g^{-1}) was reported employing *Bacillus gibsonii* as a biosorbent (Zhang et al., 2013). It is reported that complexation and ion exchange with the chemical moieties present on the *Bacillus gibsonii* surface were responsible for biosorption.

Pb^{2+} and Cd^{2+} biosorption by *Lactobacillus acidophilus* are reported (Afraz et al., 2020) and the maximum biosorption capacity was 87.68% for Pb^{2+} and 69% for Cd^{2+}. In a different approach, the biosorption of Cd^{2+} by *Bacillus licheniformis* sp. dead bacteria was reported (Baran and Duz, 2021). The bacteria were obtained from soil in the Tigris River area and were employed in batch experiments using aqueous solutions. Under the optimum operational conditions, the highest adsorption capability was 24.51 mg/g. Alternatively, *Pseudoalteromonas* sp. bacteria performed better as demonstrated by the maximum biosorption capacity of 153.85 mg g^{-1}(Zhou et al., 2014). Interestingly, growing cells showed better performance than grown cells at low Cd^{2+} concentrations.

Recently, microbial remediation of arsenic from the environment has undergone great development as a promising biotechnology mainly due to its cost-effectiveness and public acceptance (Hayat et al., 2017). The mechanism of arsenic remediation is based on the transformation of arsenic aqueous species into methylated arsine species (i.e., trimethylarsine, dimethylarsine), which are volatile and can be released into the atmosphere with low toxicity (Maguffin et al., 2015). In a different approach, arsenic-oxidizing microorganisms were employed for the modification of the highly toxic As (III) to As (V), which possesses lower toxicity. This process was followed by the addition of FeCl$_3$, which stabilized As (V) (Karn et al., 2017). The integration of biological and chemical processes was efficient for the removal and stabilization of As from contaminated environments.

Thallium (Tl) is highly toxic but studied less than other metals (Liu et al., 2019). Microbial fuel cells were employed to oxidize Tl(I) and generate Tl(III). This process has a removal efficiency of

67%, followed by an elevated electrical power density (457.8 mW/m^2) during a 72 h period. In addition, the Tl(III) oxidation product was less mobile and could precipitate naturally in a wide range of pHs (Wang et al., 2018). Other work suggests that a great percentage of Tl(I) (80.5%) could be removed from groundwater in a period of 4 h by employing an aerated electrochemical reactor with microbial fuel cells (Tian et al., 2017).

BIOSORPTION BY MICROBIAL BIOMASS

Biosorption is a technique for eliminating easily non-biodegradable pollutants from water. Bacteria, algae, fungi and agricultural and industrial wastes are among the biomaterials known to bind these contaminants (Opeolu et al., 2010). The features of cell wall constituents including peptidoglycan, as well as functional groups including amine, carboxyl and phosphonate, are investigated based on their biosorption potentials. The binding processes that influence the passive intake of contaminants are investigated.

Bioaccumulation is characterized as a behavior that involves live organisms, whereas biosorption relies on dead matter. Bioaccumulation refers to the accumulation of toxicants by living cells. Toxins have the ability to enter cells, concentrate intracellular cross cell membranes and move through cells (Malik, 2004). This is the result of a number of metabolism-self-regulating actions that occur mostly in the cell wall, with different pollutant uptake pathways depending on the type of biomass. Biosorption has some advantages compared to bioaccumulation (Table 11.2). When the toxicant concentration is very high or the process is running for a long period, the toxicant will reach a saturation point (Vijayaraghavan and Yun, 2008). Fungi, algae, bacteria, agricultural wastes, industrial wastes and other forms of polysaccharide materials are used as biosorbents to remove metals/dyes.

In general, biomaterials of all types have shown good biosorption properties for metal ions. The bacterial category contains strains effective at metal biosorption, including *Bacillus* (Wen et al., 2018; Todorova et al., 2019), *Pseudomonas* (Tuzen et al., 2008; Xu et al., 2020) and *Streptomyces* (Kirova et al., 2021; Sedlakova-Kadukova et al., 2019). Important fungal biosorbents include *Aspergillus* (Liao et al., 2019; Paria et al., 2022; Liao et al., 2019; Aftab et al., 2020), *Rhizopus* (Yahya, 2021; Njoku et al., 2020) and *Penicillium* (Sundararaju et al., 2020; Coelho et al., 2020). These bacteria are produced as waste because they are widely used in the pharmaceutical and food sectors. Another biosorbent that has become important in current years is seaweed. Seaweeds, often called marine algae, are a type of biological resource found all over the world. Green, red, and brown seaweeds are among the algal divisions, and brown seaweeds have proven to be effective biosorbents (Bibak et al., 2020). This is because of alginate, found in gel form, in their cell walls. Furthermore, because of their macroscopic form, the sorption process is simple (Zhao et al., 2018). Many toxicological studies have been published, but they have all focused on the accumulation of metals in living cells (Nkwunonwo et al., 2020). In addition, it has also been found that deactivated microbial biomass can passively bind metal ions using a number of physicochemical methodologies. As a result of this, biosorption research became more active, with a range of biosorbents of various origins for the elimination of metals/dyes. Biosorption is influenced not only by the kind or chemical content of the biomass, as well as by external physicochemical parameters and solution chemistry. Several researchers have proposed the mechanisms for biosorption, including complexation, coordination, ion exchange, electrostatic interaction, adsorption, chelation and microprecipitation (Veglio and Beolchini, 1997; Volesky and Bohumil, 2003). While the same bacterium was used for a similar metal in different settings, there were some variances in the findings. This is owing to diverse experimental settings, as well as the biomass being processed or immobilized to increase biosorbent properties. A weak acidic condition resulted in maximal biosorption for most metal ions. Metal hydroxide and metal-ligand complexes reduce the number of metal ions that are sorbed at high pH (Malandrino et al., 2006). After a few minutes of contact time, the metal-bacterial biomass consolidates. The rapid kinetics found in bacterial biomasses is a good attribute to have

TABLE 11.2

Important Features and Comparison of Biosorption and Bioaccumulation

Description	Biosorption	Bioaccumulation
Expenditure of operation	Agricultural, industrial, and other types of waste biomass are the most common biosorbents used. The majority of the cost is made up of carrying to site and other processing fees.	Expenditure is high. Because the process includes living Prokaryotic cells, cell preservation is costly.
pH	Absorption capacity of dead biomass is highly influenced by the pH of the solution. The technique can be used in a broad range of pH conditions.	High pH levels have a significant impact on live cells.
Temperature	Temperature has no effect on the process because the biomass is dormant. In fact, multiple studies have found that when the temperature rises, absorption increases.	Temperature has a significant impact on the process.
Storage	Easy to use and store, as it is dead biomass.	Media and trace element sources are required for maintenance of the culture.
Degree of uptake	Extremely high. According to reports, some biomasses can hold nearly as much toxicant as their dry weight.	High toxicant concentrations are toxic to live cells, and absorption is usually modest.
Rate of uptake	Usually quick. The majority of biosorption mechanisms are quick.	Typically slower than biosorption because intracellular buildup takes time.
Toxicant recovery	Toxicant recovery is feasible with the right eluent selection. Alkaline or acidic solutions have proven to be effective in recovering toxicants in many cases.	Even if it were feasible, the biomass could not be used in the next cycle.

in a wastewater treatment system. There are a number of parameters responsible for biosorption as discussed subsequently.

STRUCTURE OF BACTERIAL CELLS FAVORING BIOSORPTION

Bacteria are single-celled organisms that live in soil, water, and as symbionts in other organisms. Cocci (like *Streptococcus*), rods (like *Bacillus*), spirals (like *Rhodospirillum*), and filamentous (like *Bacillus*) bacteria are among different types of bacteria. Eubacteria have simple cell walls (Salton, 1964) and provide structural stability but differ from other organisms in that they contain peptidoglycan (Baddiley, 1972). Peptidoglycan, which also controls cell shape, determines the stiffness of the bacterial cell wall (Kolenbrander and Ensign, 1968). Bacterial cell walls differ from one another. The cell wall composition is, in fact, an important feature in the research and classification of bacterial species. As a result, there are two types of bacteria: gram-negative and gram-positive bacteria. There is a thicker peptidoglycan layer in gram-positive bacteria (Beveridge, 1981; Dijkstra and Keck, 1996) linked by amino acid bridge. Poly-alcohols are embedded in the gram-positive cell wall. Lipoteichoic acids are responsible for attaching peptidoglycan to the cytoplasmic membrane because they are chemically bonded to lipids within the membrane. The cell is surrounded by a grid-like network of cross-linked peptidoglycan molecules. Teichoic acids provide gram-positive

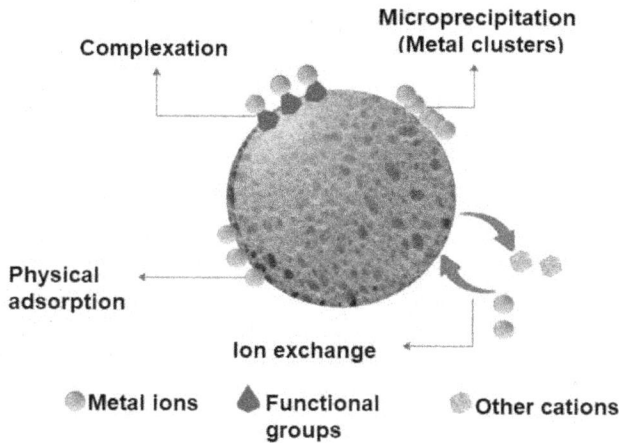

FIGURE 11.2 Biosorption mechanism (SáCosta et al., 2021).

cell wall (Sonnenfeld et al., 1985). Peptidoglycan, in general, comprises 90% of the gram-positive cell wall. Gram-negative bacteria have a thinner cell wall (10–20% peptidoglycan) (Kolenbrander and Ensign, 1968). An extra outer membrane made up of phospholipids and lipopolysaccharides is found on the cell wall. Due to the highly charged structure of lipopolysaccharides, the gram-negative cell wall has an overall negative charge. The anionic functional groups of gram-positive and gram-negative bacteria were found responsible for the cell wall's anionic character and metal-binding ability (Stoddart, 1979). Metals can also be bound by extracellular polysaccharides (Bazaka et al., 2011). On the other hand, bacterial species and growth circumstances determine their availability, and they are easily eliminated.

BACTERIAL BIOSORPTION METHOD

Several authors have discussed different mechanisms of biosorption (Abdi et al., 2015; Doyle et al., 1980; Golab et al., 1995; Vijayaraghavan and Yun, 2007; Hitch and Mesmer, 1976).

The overall mechanisms of biosorption can be represented by Figure 11.2 (SáCosta et al., 2021).

IDENTIFICATION AND CHARACTERIZATION OF BACTERIAL SURFACE

Characterization can be done by using potentiometric titrations (Pagnanelli et al., 2010), FTIR spectroscopy (Xu et al., 2020), X-ray diffraction (Sivashankar et al., 2021), scanning electron microscopy (Dhal et al., 2018), TEM (Kiran et al., 2017) and EDX microanalysis (Muñoz et al., 2021). In current years, there has been a lot of interest in boosting biomass sorption capability. Several biomasses that are considered industrial wastes as a result of particular procedures have limited biosorption capabilities. Because sorption occurs mostly on the surface of biomass, increasing binding sites would be a good way to boost biosorption ability.

CHEMICALLY PERSONALIZED BIOSORBENTS

Pre-treatment, binding site augmentation, modification and polymerization are examples of chemical alterations. The accomplishment of a chemical pre-treatment is highly dependent on the biomass's biological machinery. Acidic pre-treatment has been successful in many circumstances (Zafar et al., 2013; Sar et al., 1999; Jeon and Höll, 2003; Yu et al., 2007).

BIOREMEDIATION USING GENETICALLY MODIFIED MICROORGANISMS

The methods used to create recombinant strains for protein overproduction differ significantly from those used to develop a few microbial cells capable of degrading certain substances and surviving for many generations. Microorganisms have innate genetic, metabolic and physiological properties that make them excellent pollutant remediation proxies in the environment (Singh et al., 2017). The balance between organism growth and energy consumption shall be considered when devising any approach (Keasling and Bang, 1998). Different groups have discovered a multitude of expression systems containing numerous promoters and their regulators in order to build an efficient microorganism. Carrier and Keasling devised a method for improving mRNA stability by inserting DNA cassettes into the 5' untranslated region of a chosen gene. This inserted DNA cassette created a hairpin at the 5' end of the mRNA, boosting its G of formation threefold and hence mRNA stability (Carrier and Keasling, 1997). Different strategies for developing genetically modified microorganisms for bioremediation are shown in Figure 11.3.

According to environmental biotechnology, microorganisms such as bacteria, yeast, and filamentous fungi may eliminate heavy metals from aqueous solutions. The use of microbial metabolic potential to remove toxins from contaminated locations is a safe and cost-effective strategy. Bioremediation has been effective using GE microorganisms, recombinant DNA and RNA technology. To improve bioremediation processes, microorganism genes have been modified to build up new metabolic pathways.

BIOREMEDIATION OF HEAVY METALS

Heavy metal contamination, such as mercury, cannot be removed by normal bacteria, but metals such as Cd, Hg, Ni, Cu, As and Fe can be removed by genetically engineered bacteria (Ogden et al., 1989). *Pseudomonas putida* 0690 has been genetically engineered to produce MBP-EC20, a metal-binding peptide with a high affinity for Cd. The bacteria *P. putida* 06909 was Cd-resistant due to the presence of an efflux pump (Lee et al., 2001). Because of its specialized nature in converting toxicants and mitigating health concerns, the use of genetically modified bacteria for heavy metal bioremediation has gained popularity. Metal detoxification by bacteria can be a cost-effective, highly efficient, and

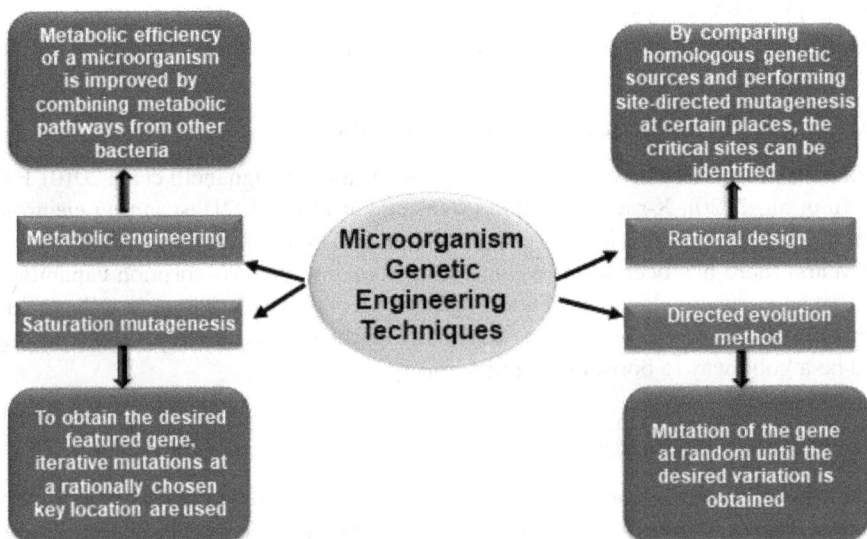

FIGURE 11.3　Various methods for creating genetically modified bacteria cells for bioremediation.

eco-friendly remedial method. Due to advances in molecular biology, metal-microbe interaction could be understood in a better way for metal bioremediation in the environment (Dixit et al., 2015).

Inoculating Cd^{2+} polluted soil with GE *Ralstonia* drastically decreased the deleterious effects of heavy metals on tobacco plant growth. The transfer of a mouse gene producing metallothionein, as found in *Ralstonia eutropha* (a soil dweller), improved the capacity to sequester heavy metals (for example, cadmium) (Valls et al., 2000). This could be the remedial breakthrough needed to minimize the effects of toxic heavy metals on the environment.

KINETICS AND MECHANISM OF BIOSORPTION

Biosorption, a comparatively rapid method of detoxifying HMs contaminants, involves binding heavy metals with metal-binding proteins located on the cell wall of algae, fungi and bacterial biomass. It has been found that to achieve their various important survival aspects, microorganisms exhibit HMs sorption via biosorption or bioaccumulation. For example, essential metals are up-taken for the proper functioning of microbial cell metabolism, whereas heavy metals like mercury, cadmium and lead are not utilized for cellular metabolism. Occurrence of carbonyl, imidazole, thiol, phosphate, sulphate, amino, amide, thioether and hydroxyl functional groups are creating negatively charged surfaces in biosorbents and thus they are presumed to enhance the heavy metal adsorption capacities and can be further enhanced by surface reactive modification (Gavrilescu, 2004; Yin et al., 2019).

The complex structures of microbial cells offer several metal uptake (biosorption) sites; according to these multiple possible mechanisms, they canthus can be categorized along different criteria as shown in Figure 11.4.

However, on the basis of cellular metabolic dependency, biosorbing mechanisms can be categorized grossly into metabolic-dependent mechanism and non-metabolic-dependent mechanism.

METABOLIC-DEPENDENT MECHANISM

Because of the energy-generating metabolic pathways in a viable cell, it exhibits active uptake of metals. In biosorption, metal transport across the cell membrane during the cellular metabolic pathway leads to

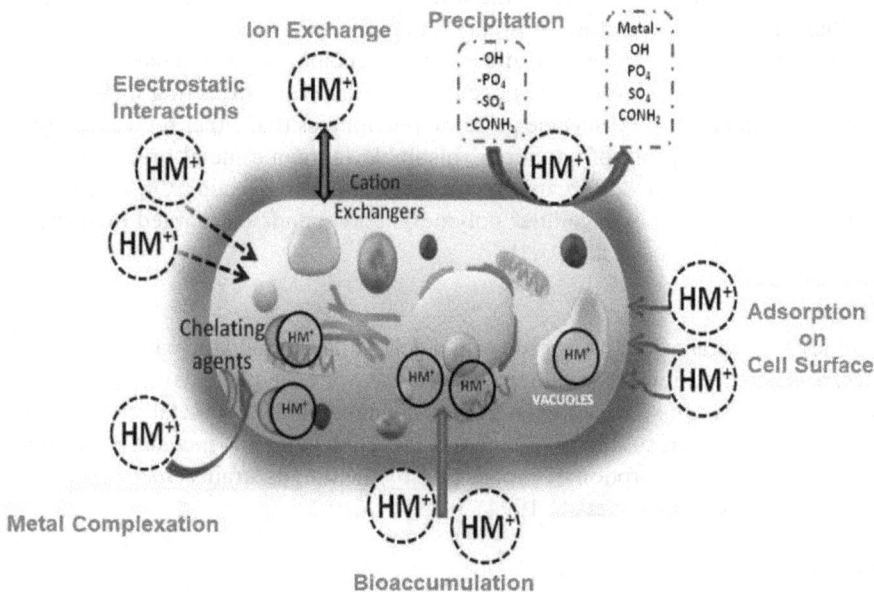

FIGURE 11.4 Various mechanisms of biosorption by microbial cells.

intracellular accumulation inside the viable cell only. A metabolic-dependent mechanism may involve metal transport across the cell membrane or else the precipitation of metal capture. This metal transportation across the cell membrane is often coupled with the microorganism's active defense mechanism.

NON-METABOLIC-DEPENDENT MECHANISM

It involves physical and chemical bonding amongst metal ions and active sites associated with cell walls. However, it is comparatively rapid and may be reversed. Non-metabolic-dependent mechanisms may involve physical adsorption, ion exchange and chemo-sorption, as these processes work independently from cell metabolic pathways.

Biosorption by dead biomass is the passive uptake process involving chemical interactions between functional groups on cellular biomass and adsorbate HMs. There could be a possibility of passive uptake in a metabolically active cell, but it might be restricted by the protection mechanism of the cell against HMs. Thus, dead biomass exhibits more significant sorption capabilities towards toxic HMs pollutants. In the case of biosorption by living cells, the mechanism becomes more complex and integrated due to several simultaneous cellular metabolic reactions occurring. Moreover, cell surface sorption interactions by living cells may be interfered with by the extracellular environment (Lu et al., 2016). Biosorption may occur in the following ways, as described next.

TRANSPORTING ACROSS CELL MEMBRANES

Transportation of HMs cations across cell membranes is a typical mechanistic model in microbial cells. It may first involve independent heavy metal ions binding on cell wall sites, which is then followed by transporting those bound metal ions across the cell membrane as part of the cell metabolism (Perpetuo et al., 2011). Heavy metal capturing may be carried out by the microbial cell as the part of essential metabolism of the cell or else it may occur because of false-selection of heavy metals instead of essential metals like Fe, K, and Na in metabolism due to the similarity in charges and ionic sizes (Redha, 2020).

Precipitation

Precipitation can be either reliant on or non-dependent on cellular metabolism. Removing metals from solutions is frequently related to the microorganisms' active defense mechanism. In the existence of harmful metal-generating substances, they react for favoring the precipitation process. It may be possible that instead of cellular metabolism, precipitation is caused by normal chemical reactions. The biosorption processes might occur at the same time. Reacting metal ions with functional groups of the microbial cell surface generate precipitates that either retain intactness or enter the microbial cell. In a majority of situations, insoluble inorganic metal precipitates are formed. When microbial cells are utilized, organic metal precipitates may occur. Organic precipitates are formed by the majority of the extracellular polymeric compounds discharged by bacteria (García-Mendieta et al., 2012; Perpetuo et al., 2011).

Ion Exchange

Polysaccharides are a vital active part of microbial cell walls and metallic cations are replaced with the polysaccharide counter-ions. Alginate from sea algae, for example, exists as potassium, sodium, calcium and magnesium salts. These ions may interchange with heavy metal ions including cobalt, copper, cadmium and zinc, resulting in effective biosorption. The ion-exchanging method was discovered to be involved in the sorption of copper ions by the fungus strains *Ganoderma lucidium* and *Aspergillus niger* (Muraleedharan et al., 1994).

Complexation

The extracellular complex formation or complexation process is achieved by electrostatic forces of attraction between a metal cations chelating agents and polymers expelled by a living or non-viable microorganism. Polysaccharides of cell walls, biosurfactants, nucleic acids and proteins can all induce it. These chelating agents include electron pairs that exhibit electrostatic forces, not involving electron transfer. Elimination of metals may be due to complexation on cell surfaces. According to many researchers, microalga *Chlorella vulgaris* and *Zoogloea ramigera bacteriumin* capture copper via both physical adsorption involving coordination interactions between metals and $-NH_2$, -COOH groups. In *Pseudomonas syringae*, metals accumulation was discovered to be solely due to the complexation mechanism. Organic acids produced by microorganisms can chelate harmful metals, producing organic compounds. Metals bound to polysaccharides are biosorbed or complexed by carboxyl groups (Perpetuo et al., 2011).

Physical Adsorption

The adsorption mechanism is a method where—with the aid of electrostatic forces, covalent bonding, vander Waals forces, redox interaction or biomineralization—metal ions connect with polyelectrolytes on microbial cell walls to achieve electroneutrality (Redha, 2020). It is a reversible and advantageous mechanism, especially for large-scale wastewater treatment at lower contamination concentrations (Nishitani et al., 2010). Adsorption is associated with affinity amongst cations and -ve potential charge on the cell wall and it is a pH-dependent mechanism (Bashir et al., 2019). On increased pH negative charge on cell wall binding sites increases, thus maximizes metal ions attraction, resulting in efficient surface adsorption. Consequently, the solubility of heavy metal cations weakens and results in precipitation (Kuroda and Ueda, 2011). Gupta and colleagues have reported the efficient involvement of extracellular polymeric substances (EPS) expelled by bacterial cells in the biosorption of HMs (Gupta et al., 2004).

Kinetic and Adsorption Models of Biosorption

For practical utilization of any sorption process, its design, operational control and biosorption kinetics play vital roles. However, biosorption kinetics provides insight into sorption mechanisms, metal uptake and reaction pathways. More than 25 models are reported to date and applied to describing the quantitative kinetic behavior of biosorption. But, amongst them, pseudo-first-order models and second-order models are frequently chosen to illustrate biosorption kinetic data.

For the detailed study of biosorption kinetics, the quality of a biosorbent should be keenly understood. To access the quality of a biosorbent, two parameters must be taken under consideration viz. (a) the extent of heavy metal ion fascinated by the biosorbent, (b) the amount of cations actually captured by the biosorbent upon immobilization (Abdi and Kazemi, 2015; Ahalya et al., 2003; Upadhyay et al., 2017).

Different adsorption isotherm models such as Freundlich and Langmuir have also been tested.

FACTORS INFLUENCING BIOSORPTION PROCESSES

Biosorption capabilities are influenced mainly by three parameters: properties of the heavy metal ions, environmental conditions and the nature of the biosorbents. However, growth conditions, physiological conditions, cell age and types of biomass are also considerable parameters in HM-capturing mechanisms (Chen and Wang, 2008). The maximum biosorption capacity can be achieved at optimal pH and medium temperature, these are the most influential factors. A pH range of 4–8 is moderately acceptable for heavy metal removal by almost all biomass types. Factors affecting biosorption processes are discussed subsequently and given in Figure 11.5.

TEMPERATURE

Temperature is an essential parameter for metal sorption. The biosorption capability of the biosorbent changes when the temperature rises or falls. Rising temperature improves biosorption capabilities,

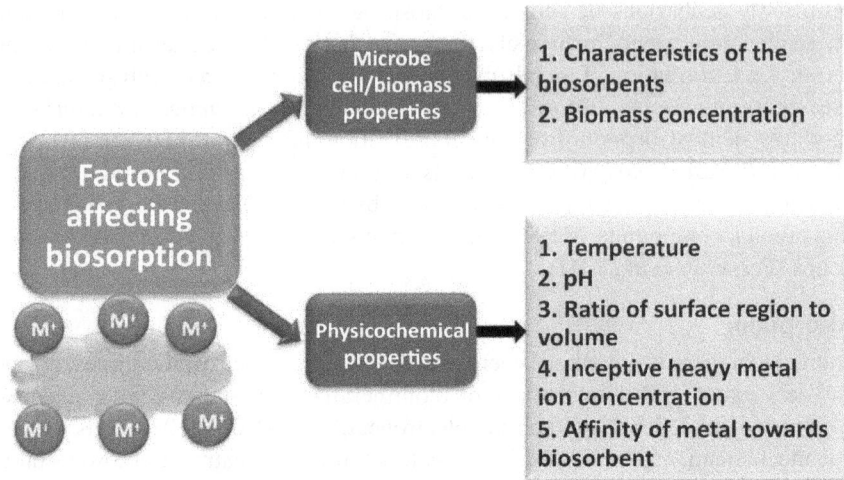

FIGURE 11.5   Factors affecting biosorption processes.

but it is also correlated to less structural damage to the biosorbent. Thus, for efficient biosorption, the optimal temperature must be chosen for maximum metal ion binding. *Saccharomyces cerevisiae* was used to obtain a maximum removal of 86% for Cd^{+2} ions at 40°C. The biosorption speed of $Cr^{6+}$by *Streptococcus equisimilis* increased dramatically on raising the temperature from 25°C to 40°C (Ramya, 2018).

pH

As pH influences complexing rates of inorganic as well as organic ligands, hydrolysis and precipitation, it constitutes a vital parameter. In addition, it influences the solubility of metal ions and the binding sites of biomass (Michalak et al., 2013). A pH range of 2.5 to 6 is optimal for metal uptake. However, protonation occurring at an acidic pH can limit the biosorption capacity in lower pH, and a pH above 6 compromises the biosorption ability (Abbas et al., 2014; Filote et al., 2021). pH also contributes to activating the functional groups on cell walls for metal binding. The –COOH group is involved in the pH range 2 to 5, whereas combined carboxyl and phosphate groups are activated in the range 5 to 9. Moreover, pH influences the selection of the biosorption mechanism by the adsorbent. It has been found that lower pHs promote the ion exchange method for capturing lead ions, and at higher pHs, it is achieved by electrostatic interactions among lead ions and –COOH and –OH groups on biomass (Bashir et al., 2019).

CHARACTERISTICS OF THE BIOSORBENTS

Metal uptake capacity has been documented in a variety of forms, including immobilization in microbial cells, biofilms and freely suspended microbial cells. Physical and chemical treatments can change it. Autoclaving, drying, boiling and sonication are examples of physical therapies. Chemical treatment, as the name implies, improves biosorption capacity by using chemicals such as acid or alkali. Wang and Chen have reported that deacetylation of a fungal strain alters the chitin structure, ensuing in the production of chitosan-glycan complexes with strong metal affinity (Wang and Chen, 2006). Abbas et al. also discussed the impact of age and growth medium components on biosorption, as these factors may influence cell size as well as cell wall composition (Abbas et al., 2014).

SURFACE AREA TO VOLUME RATIO

This feature is crucial for removing heavy metals from a medium efficiently. In the case of biofilms, the surface area attribute is essential. Despite the fact that internal metal adsorption is an energy-intensive process, bacteria prefer cell wall adsorption (Shamim, 2018).

BIOMASS CONCENTRATION

The quantities of metals adsorbed and biomass present are interrelated with each other. The electrostatic contact between cells is a vital factor for metal uptake. Low cell densities adsorb more metal ions than high cell densities at a given equilibrium. The amount of metal consumed is determined by the binding sites. Metal ions have less access to binding sites when there is an abundant biomass or heavy metal ion.

HMs CONCENTRATION

In heavy metal biosorption, preliminary cation concentration acts as a powerful thrust, overcoming metal mass transfer conflicts between the liquid and solid phases. With increasing metal concentrations, the metal uptake will rise. At low preliminary metal strength, the maximum metal removal can be achieved. As a result, at a given biomass concentration, metal uptake increases as the starting concentration increases.

AFFINITY OF METAL TOWARDS BIOSORBENT

The surface charges and permeability of the biomass are affected by physical or chemical pretreatment, which renders metal binding sites available for binding. Treatment of biomass using alkalis, acids, detergents and heat can enhance the quantity of metal capture.

Heavy metal capturing capabilities of biomass are enhanced by acidic, alkaline, soap or thermal treatment.

DESORPTION METHODS AND SORBENTS REGENERATION

Upon metal uptake, biosorbents become completely exhausted, and their revival becomes an essential footstep which associates desorption of metals from the biosorbents. Upon complete biosorption, removal and safe discard of the biosorbed metal ions from exhausted biosorbents is desorption. Here, the desorbing agent plays a vital role in the desorption of sorbed metal species followed by the regeneration of sorbents for reuse (Smily and Pasumalai, 2017).

A greater desorption percentage indicates an elevated regeneration rate of exhausted biosorbents to maintain the sustainability of the process. A variety of desorbing agents are reported to be implemented for regeneration, including organic acids, complexing agents, mineral acids and alkalis (NaOH, KOH and $NaCO_3$). In view of desorption percentage and speed, acidic recovery agents exhibit superior desorption than alkaline agents (Lata et al., 2015). While selecting recovery agents, the retention of biosorbents' physical characteristics with low damage to sorbents is a prior factor to consider (Dey U. et al., 2016; Shamim, 2018). Using 0.1 N NaOH, an alkaline desorbent, Cr^{+6} was almost entirely desorbed from *Mucor hiemalis* biomass. In addition, the recovered biomass kept its sorption and desorption activities for up to five sorption-desorption cycles. The Langmuir isotherm model met the experimental results well, and IR spectra revealed that the $-NH_2$ groups are associated in biosorption (Tewari et al., 2005).

Exhausted fungal biomass of *Aspergillus* species was recovered by desorbing nickel, calcium and iron by varied HCl concentrations. On rising HCl strength, the desorption rate was found to be greater, thus resulting in complete removal of iron and calcium, and approximately 78% recovery of nickel was achieved by 5M HCl (Sekhar et al., 1998). Arica and colleagues reported Hg^{+2} desorption from biosorbent-immobilized fungal biomass of *Trametes versicolor* and *Pleurotus sajur-caju* by batch method. These biosorbents loaded with Hg^{+2} by acidic desorbing agent HCl (10mmol/L) and 97% adsorbed Hg^{+2} ions were removed from exhausted biomass (Arıca et al., 2003).

APPLICATION OF BIOSORPTION FOR WASTE TREATMENT

Due to high solubility in aqueous environments, HMs are easily adsorbed and thus transferred through ecosystems. These combined or elemental forms of HMs are discharged notably from domestic and industrial effluents. An intensive study has been reported in HMs removal from wastewater by the batch method. Rigorous studies have been carried out to implement biosorption processes for wastewater treatment. Several studies have reported good biosorption capacities using effluents from highly polluted sources, instead of laboratory-level heavy metal aqueous solutions. Due to their abundant existence and cost-efficient expense in industrial manufacturing, diverse fungal strains like *Trametes versicolor, Phanerochaete chrysosporium, Pleurotus ostreatus, Aspergillus niger, Rhizopus arrhizus* and *Fusarium* sp. have been studied and implemented for the elimination of toxic HMs from wastewater (Lu et al., 2016).

Industrial toxic waste containingCu^{+2} along with Zn^{+2}, Cr^{+6}, Na^+, Ca^{+2} and K^+ was detoxified by agricultural waste biomass by batch method at a pH of 6, and the biosorption efficiency of all present metals was between 77 to 95% (Singha and Das, 2013). Simultaneous and individual potential removal of Cd and Cr ions by fungi *Phanerochaete chrysosporium* has been optimized by Rudakiya and colleagues, who reported its biosorption capacity for Cd (18 mg/L), Cr (26 mg/L) and As (30 mg/L) at optimum pH. The obtained results help to conclude that the decrease that occurs in removal efficiency in simultaneous metal sorption compared to individual sorption is due to interference of other metals in biosorption cell mechanisms (Rudakiya et al., 2018).

The detoxification competence of genuine effluents can be hampered by the existence of contaminants such as metals, organic debris, anions and other substances that compete for binding sites. *Pleurotus ostreatus*, a fungus biosorbent, was employed to clean wastewater collected from electroplating industrial drainage. When biomass was employed for commercial treatment of wastewater, there was a modest decrease in biosorption capacity. Metal sorption capacities for Ni, Cu, Zn and Cr in effluents were 59.22, 46.01, 9.1, and 9.40%, respectively, but with a single synthetic metal solution, they were 52, 63.52, 10.9, and 11.8%, respectively. As documented in several other researches, this moderate to modest decrease in biosorption capacity could be due to the competition of diverse pollutants for binding sites. Another complicating element is a high COD level, which reduces biosorption (Javaid et al., 2011).

Pb^{+2} was captured from the waste released by battery industries using *Aspergillus versicolor* biomass. The absorption efficiency of Pb (II) ions was found to be more than 85%, which was nearly identical to the value achieved by the individual metallic solution. As a result, the report concluded that the existence of extra metal ions and anions in industrial discharge has no effect on the biomass employed in the experimental biosorption if optimum conditions are maintained (Himadri et al., 2011). To attain removal rates equivalent to those reported with standard single solutions, changes to the process parameters with real industrial effluents may be recommended.

A packed column of *Spirogyra* granule was implemented to treat diverse industrial discharges(1L at a rate 0.6 ml/min) from several industrial activities including the carpet, paper mill and electroplating sectors. Many heavy metals were removed from the industrial wastes, with a removal efficiency of more than 90% using packed columns. This entailed a lowering of the pH range from 7.5 to 4.5 (in the case of metals, excluding chromium) and pH 2 (for chromium), despite the fact that metal solutions should have a pH of 5.0 (Singh et al., 2012).

Dey and colleagues have reported a conclusive study of various fungi including *A. terreus* AML02, *P. fumosoroseus* 4099, *B. bassiana* 4580, *A.s terreus* PD-17 and *A. fumigatus* PD-18 for removal of multiple metals simultaneously. *A. fumigatus* has exhibited an utmost metal tolerance index value for each metal, followed by *B. bassiana*. Furthermore, these fungal strains were screened for multiple metal combinations in varied concentrations. In comparison to the rest of the fungi, *B. bassiana* and *A. fumigates* are found to show superior cube root growth constants (k) representing their superior versatility towards multiple metal exposures (Dey P. et al., 2016).

CONCLUSION

Different methods such as chemical, physical and biological methods are frequently used for the decontamination of water, particularly heavy metals, dyes and other pollutants. However, in recent times, genetically engineered microorganisms (GEMs) are being used for the removal of HMs from water, soil and the environment. The method is gaining attention of a large number of researchers. Considering its future importance, improvements in GEMs in terms of their survival and other properties towards the removal of HMs are still needed. The applications of these microorganisms for HMs remediation are still limited, although they have a lot of advantages and may be considered as future materials for the decontamination of heavy metals.

REFERENCES

Abbas, Salman H., Ibrahim M. Ismail, Tarek M. Mostafa, and Abbas H. Sulaymon. "Biosorption of heavy metals: A review." *Journal of Chemical Science Technology*, 3, no. 4 (2014): 74–102.

Abdi, Omran, and Mosstafa Kazemi. "A review study of biosorption of heavy metals and comparison between different biosorbents." *Journal of Materials Environmental Science*, 6, no. 5 (2015): 1386–1399.

Adler, R. A., and J. Rascher. 2007. "A strategy for the management of acid mine drainage from gold mines in Gauteng." Contract Report for Thutuka (Pty) Ltd. Submitted by the Water Resource Governance Systems Research Group, CSIR: Pretoria. Report No. CSIR/NRE/PW/ER/2007/0053/C.2007.

Afraz, Vahideh, Habibollah Younesi, Marzieh Bolandi, and Mohammad Rasoul Hadiani. "Optimization of lead and cadmium biosorption by Lactobacillus acidophilus using response surface methodology." *Biocatalysis and Agricultural Biotechnology*, 29 (2020): 101828.

Aftab, Kiran, Kalsoom Akhtar, Muzammil Hussain, and Kinza Aslam. "Synthesis, characterization and application of bio-composites based on aspergillus flavus NA9 for extraction of zinc ions from synthetic and real waste water effluents." *Journal of Polymers & the Environment*, 28, no. 5 (2020).

Ahalya, N., T. V. Ramachandra, and R. D. Kanamadi. "Biosorption of heavy metals." *Research Journal of Chemistry and Environment*, 7, no. 4 (2003): 71–79.

Alvarez, G. S., Foglia, M. L., Camporotondi, D. E., Tuttolomondo, M. V., Desimone, M. F., and Díaz, L. E. (2011). "A functional material that combines the Cr(vi) reduction activity of Burkholderia sp. with the adsorbent capacity of sol-gel materials." *Journal of Materials Chemistry*, 21:6359–6364. doi:10.1039/c0jm04112b.

Arıca, M. Y., Ç. İ. Ğ. D. E. M. Arpa, B. Kaya, S. Bektaş, A. Denizli, and Ö. Genç. "Comparative biosorption of mercuric ions from aquatic systems by immobilized live and heat-inactivated Trametes versicolor and Pleurotus sajur-caju." *Bioresource Technology*, 89, no. 2 (2003): 145–154.

Arunakumara, K. K. I. U., and X. Zhang. "Heavy metal bioaccumulation and toxicity with special reference to microalgae." *Journal of Ocean University of China*, 7, no. 1 (2008):60–64. https://doi.org/10.1007/s11802-008-0060-y.

Azimi, A., A. Azari, M. Rezakazemi, and M. Ansarpour. "Removal of heavy metals from industrial wastewaters: A review." *ChemBioEng Reviews*, 4, no. 1 (2017): 37–59.

Baddiley, James. "Bacterial cell wall biosynthesis." *Polymerization in Biological Systems*, 7 (1972): 87.

Bairagi, Himadri, Md Motiar R. Khan, Lalitagauri Ray, and Arun K. Guha. "Adsorption profile of lead on Aspergillus versicolor: A mechanistic probing." *Journal of Hazardous Materials*, 186, no. 1 (2011): 756–764.

Bakirova, S. F., L. V. Shestoperova, A. V. Kotova, O. S. Turkov, V. G. Benkovskii, and G. N. Aleshin. "New data on trace elements composition of the ash in western Kazahstanpetroleums. Izv. Akad. NaukKaz." *SSR, Ser. Khim*, no. 4 (1980): 63–67.

Baran, M. F., and M. Z. Duz. "Removal of cadmium (II) in the aqueous solutions by biosorption of Bacillus licheniformis isolated from soil in the area of Tigris River." *International Journal of Environmental Analytical Chemistry*, 101(2021): 533–548. doi:10.1080/03067319.2019.1669583.

Bashir, Arshid, Lateef Ahmad Malik, Sozia Ahad, Taniya Manzoor, Mudasir Ahmad Bhat, G. N. Dar, and Altaf Hussain Pandith. "Removal of heavy metal ions from aqueous system by ion-exchange and biosorption methods." *Environmental Chemistry Letters*, 17, no. 2 (2019): 729–754.

Baumann, Hans A., Liam Morrison, and Dagmar B. Stengel. "Metal accumulation and toxicity measured by PAM—chlorophyll fluorescence in seven species of marine macroalgae." *Ecotoxicology and Environmental Safety*, 72, no. 4 (2009): 1063–1075.

Bazaka, Kateryna, Russell J. Crawford, Evgeny L. Nazarenko, and Elena P. Ivanova. "Bacterial extracellular polysaccharides." *Bacterial Adhesion* (2011): 213–226.

Beveridge, T. J. "Ultrastructure, chemistry, and function of the bacterial wall." *International Review of Cytology*, 72 (1981): 229–317.

Bibak, Mehdi, Masoud Sattari, Saeid Tahmasebi, Ali Agharokh, and Javid Imanpour Namin. "Marine macro-algae as a bio-indicator of heavy metal pollution in the marine environments, Persian Gulf." *Indian Journal of Geo Marine Sciences*, 49 (2020): 357–363.

Birge, Wesley J., Jeffrey A. Black, and Barbara A. Ramey. "The reproductive toxicology of aquatic contaminants." *Hazard Assessment of Chemicals: Current Developments*, 1 (1981): 59–67.

Braconi, Daniela, Giulia Bernardini, and Annalisa Santucci. "Linking protein oxidation to environmental pollutants: Redox proteomic approaches." *Journal of Proteomics*, 74, no. 11 (2011): 2324–2337.

Brodin, M., M. Vallejos, M. T. Opedal, M. C. Area, and G. Chinga-Carrasco. "Lignocellulosics as sustainable resources for production of bioplastics–A review." *Journal of Cleaner Production*, 162 (2017):646–664.

Carrier, Trent A., and J. D. Keasling. "Controlling messenger RNA stability in bacteria: Strategies for engineering gene expression." *Biotechnology Progress*, 13, no. 6 (1997): 699–708.

Chen, Can, and Jianlong Wang. "Removal of Pb^{2+}, $Ag+$, $Cs+$ and Sr^{2+} from aqueous solution by brewery's waste biomass." *Journal of Hazardous Materials*, 151, no. 1 (2008): 65–70.

Chen, L., Q. S. Zheng, Z. P. Liu, and L. Jin. "Effects of different concentrations of copper ion on the growth and chlorophyll fluorescence characteristics of Scendesmus obliquus L." *Ecological Environmental Science*, 18 (2009): 1231–1235.

Chen, Z., X. Pan, H. Chen, Z. Lin, and X. Guan. "Investigation of lead(II) uptake by Bacillus thuringiensis 016." *World Journal of Microbiology Biotechnology*, 31 (2015):1729–1736. doi:10.1007/s11274-015-1923-1.

Chouke, P. B., A. K. Potbhare, N. P. Meshram, M. M. Rai, K. M. Dadure, K. Chaudhary, A. R. Rai, M. F. Desimone, R. G. Chaudhary, and D. T. Masram. "Bioinspired NiO nanospheres: Exploring in vitro toxicity using Bm-17 and L. rohita liver cells, DNA degradation, docking, and proposed vacuolization mechanism." *ACS Omega*, 7 (2022a): 6869–6884.

Chouke, P. B., T. Shrirame, A. K. Potbhare, A. Mondal, A. R. Chaudhary, S. Mondal, S. R. Thakare, E. Nepovimova, M. Valis, K. Kuca, R. Sharma, and R. G. Chaudhary. "Bioinspired metal/metal oxide nanoparticles: A road map to potential applications." *Materials Today Advances*, 16 (2022b): 100314.

Coelho, Ednei, Tatiana Alves Reis, Marycel Cotrim, Marcia Rizzutto, and Benedito Corrêa. "Bioremediation of water contaminated with uranium using *Penicillium piscarium*." *Biotechnology Progress*, 36, no. 5 (2020): e30322.

DalCorso, Giovanni, Anna Manara, and Antonella Furini. "An overview of heavy metal challenge in plants: From roots to shoots." *Metallomics*, 5, no. 9 (2013): 1117–1132.

Dey, Priyadarshini, Deepak Gola, Abhishek Mishra, Anushree Malik, Peeyush Kumar, Dileep Kumar Singh, Neelam Patel, Martin von Bergen, and Nico Jehmlich. "Comparative performance evaluation of multimetal resistant fungal strains for simultaneous removal of multiple hazardous metals." *Journal of Hazardous Materials*, 318 (2016): 679–685.

Dey, Uttiya, Soumendranath Chatterjee, and Naba Kumar Mondal. "Isolation and characterization of arsenic-resistant bacteria and possible application in bioremediation." *Biotechnology Reports*, 10 (2016): 1–7.

Dhal, Biswaranjan, and Banshi Dhar Pandey. "Mechanism elucidation and adsorbent characterization for removal of Cr (VI) by native fungal adsorbent." *Sustainable Environment Research*, 28, no. 6 (2018): 289–297.

Dijkstra, Arnoud J., and Wolfgang Keck. "Peptidoglycan as a barrier to transenvelope transport." *Journal of Bacteriology*, 178, no. 19 (1996): 5555–5562.

Dixit, Ruchita, Deepti Malaviya, Kuppusamy Pandiyan, Udai B. Singh, Asha Sahu, Renu Shukla, Bhanu P. Singh et al. "Bioremediation of heavy metals from soil and aquatic environment: An overview of principles and criteria of fundamental processes." *Sustainability*, 7, no. 2 (2015): 2189–2212.

Doyle, Ronald J., Timothy H. Matthews, and Uldis N. Streips. "Chemical basis for selectivity of metal ions by the Bacillus subtilis cell wall." *Journal of Bacteriology*, 143, no. 1 (1980): 471–480.

Droual, R., C. U. Meteyer, and F. D. Galey. "Zinc toxicosis due to ingestion of a penny in a gray-headed chachalaca (Ortalis cinereiceps)." *Avian Diseases* (1991): 1007–1011.

Drozdova, Jarmila, Helena Raclavska, Konstantin Raclavsky, and Hana Skrobankova. "Heavy metals in domestic wastewater with respect to urban population in Ostrava, Czech Republic." *Water and Environment Journal*, 33, no. 1 (2019): 77–85.

Fashola, Muibat Omotola, Veronica Mpode Ngole-Jeme, and Olubukola Oluranti Babalola. "Heavy metal pollution from gold mines: Environmental effects and bacterial strategies for resistance." *International Journal of Environmental Research and Public Health*, 13, no. 11 (2016): 1047.

Ferrey, M. L., M. C. Hamilton, W. J. Backe, and K. E. Anderson. "Pharmaceuticals and other anthropogenic chemicals in atmospheric particulates and precipitation." *Science of the Total Environment*, 612 (2018):1488–1497.

Filote, Cătălina, Mihaela Roşca, Raluca Maria Hlihor, Petronela Cozma, Isabela Maria Simion, Maria Apostol, and Maria Gavrilescu. "Sustainable application of biosorption and bioaccumulation of persistent pollutants in wastewater treatment: Current practice." *Processes*, 9, no. 10 (2021): 1696.

Garai, P., P. Banerjee, P. Mondal, and N. C. Saha. "Effect of heavy metals on fishes: Toxicity and bioaccumulation." *Journal of Clinical and Toxicology S*, 18 (2021).

García-Mendieta, Alfredo, M. Teresa Olguín, and Marcos Solache-Ríos. "Biosorption properties of green tomato husk (Physalis philadelphica Lam) for iron, manganese and iron–manganese from aqueous systems." *Desalination*, 284 (2012): 167–174.

Gavrilescu, Maria. "Removal of heavy metals from the environment by biosorption." *Engineering in Life Sciences*, 4, no. 3 (2004): 219–232.

Gimeno-García, Eugenia, Vicente Andreu, and Rafael Boluda. "Heavy metals incidence in the application of inorganic fertilizers and pesticides to rice farming soils." *Environmental Pollution*, 92, no. 1 (1996): 19–25.

Gluskoter, Harold J. 1975. "Mineral matter and trace elements in coal." *Advances in Chemistry*, Vol. 141, pp. 1–22, ACS Publication, DOI: 10.1021/ba-1975-0141.ch001

Golab, Z., M. Breitenbach, and A. Jezierski. "Sites of copper binding in Streptomyces pilosus." *Water, Air, and Soil Pollution*, 82, no. 3 (1995): 713–721.

González-Dávila, Melchor. "The role of phytoplankton cells on the control of heavy metal concentration in seawater." *Marine Chemistry*, 48, no. 3–4 (1995): 215–236.

Gu, Y.-G., J.-J. Ning, C.-L. Ke, and H.-H. Huang."Bioaccessibility and human health implications of heavy metals in different trophic level marine organisms: A case study of the South China Sea." *Ecotoxicology and Environmental Safety*, 163(2018):551–557.

Gupta, Dharmendra K., Hiroshi Tohoyama, Masanori Joho, and Masahiro Inouhe. "Changes in the levels of phytochelatins and related metal-binding peptides in chickpea seedlings exposed to arsenic and different heavy metal ions." *Journal of Plant Research*, 117, no. 3 (2004): 253–256.

Hayat, K., S. Menhas, J. Bundschuh, and H. J. Chaudhary. "Microbial biotechnology as an emerging industrial wastewater treatment process for arsenic mitigation: A critical review." *Journal of Cleaner Production*, 151 (2017): 427–438. https://doi.org/10.1016/j.jclepro.2017.03.084.

Hitch, B. F., and R. E. Mesmer. "The ionization of aqueous ammonia to 300 C in KCl media." *Journal of Solution Chemistry*, 5, no. 10 (1976): 667–680.

Hossain, Mohammad Anwar, Pukclai Piyatida, Jaime A. Teixeira da Silva, and Masayuki Fujita. "Molecular mechanism of heavy metal toxicity and tolerance in plants: Central role of glutathione in detoxification of reactive oxygen species and methylglyoxal and in heavy metal chelation." *Journal of Botany*, 2012 (2012): 1–37.

Igiri, B. E., S. I. R. Okoduwa, G. O. Idoko, E. P. Akabuogu, A. O. Adeyi, and I. K. Ejiogu. "Toxicity and bioremediation of heavy metals contaminated ecosystem from tannery wastewater: A review." *Journal of Toxicology*, 2018 (2018):2568038. doi:10.1155/2018/2568038.

Ismail, Melor, Siew-Moi Phang, Soo-Loong Tong, and Murray T. Brown. "A modified toxicity testing method using tropical marine microalgae." *Environmental Monitoring and Assessment*, 75, no. 2 (2002): 145–154.

Javaid, Amna, Rukhsana Bajwa, Umer Shafique, and Jamil Anwar. "Removal of heavy metals by adsorption on Pleurotus ostreatus." *Biomass and Bioenergy*, 35, no. 5 (2011): 1675–1682.

Jeon, Choong, and Wolfgang H. Höll. "Chemical modification of chitosan and equilibrium study for mercury ion removal." *Water Research*, 37, no. 19 (2003): 4770–4780.

Jungers, Robert H., Robert E. Lee Jr, and Darryl J. Von Lehmden. "The EPA national fuels surveillance network. I. Trace constituents in gasoline and commercial gasoline fuel additives." *Environmental Health Perspectives*, 10 (1975): 143–150.

Karn, S. K., X. Pan, and I. R. Jenkinson. "Bio-transformation and stabilization of arsenic (As) in contaminated soil using arsenic oxidizing bacteria and FeCl3 amendment."*3Biotech*, 7 (2017):50. doi:10.1007/s13205-017-0681-1.

Keasling, Jay D., and Sang-Weon Bang. "Recombinant DNA techniques for bioremediation and environmentally-friendly synthesis." *Current Opinion in Biotechnology*, 9, no. 2 (1998): 135–140.

Kiran, M. Gopi, Kannan Pakshirajan, and Gopal Das. "Heavy metal removal from multicomponent system by sulfate reducing bacteria: Mechanism and cell surface characterization." *Journal of Hazardous Materials*, 324 (2017): 62–70.

Kirova, Gergana, Zdravka Velkova, Margarita Stoytcheva, and Velizar Gochev. "Tetracycline removal from model aqueous solutions by pretreated waste Streptomyces fradiae biomass." *Biotechnology & Biotechnological Equipment*, 35, no. 1 (2021): 953–963.

Kolenbrander, P. E., and J. C. Ensign. "Isolation and chemical structure of the peptidoglycan of Spirillum serpens cell walls." *Journal of Bacteriology*, 95, no. 1 (1968): 201–210.

Krishna, A. Keshav, and K. Rama Mohan. "Distribution, correlation, ecological and health risk assessment of heavy metal contamination in surface soils around an industrial area, Hyderabad, India." *Environmental Earth Sciences*, 75, no. 5 (2016): 411.

Kumar, M., A. Seth, A. K. Singh, M. S. Rajput, and M. Sikandar. "Remediation strategies for heavy metals contaminated ecosystem: A review." *Environmental and Sustainability Indicators*, 12 (2021): 100155. https://doi.org/10.1016/j.indic.2021.100155.

Kuroda, Kouichi, and Mitsuyoshi Ueda. "Yeast biosorption and recycling of metal ions by cell surface engineering." In *Microbial Biosorption of Metals*, pp. 235–247. Springer, Dordrecht, 2011.

Kweon, Chol-Bum, David E. Foster, James J. Schauer, and Shusuke Okada. "Detailed chemical composition and particle size assessment of diesel engine exhaust. No. 2002–01–2670." (2002) SAE Technical paper.

Lata, S., P. K. Singh, and S. R. Samadder. "Regeneration of adsorbents and recovery of heavy metals: A review." *International Journal of Environmental Science and Technology*, 12, no. 4 (2015): 1461–1478.

Laurén, Darrel Jon, and D. G. McDonald. "Influence of water hardness, pH, and alkalinity on the mechanisms of copper toxicity in juvenile rainbow trout, Salmo gairdneri." *Canadian Journal of Fisheries and Aquatic Sciences*, 43, no. 8 (1986): 1488–1496.

Lee, Seon-Woo, Eric Glickmann, and Donald A. Cooksey. "Chromosomal locus for cadmium resistance in Pseudomonas putida consisting of a cadmium-transporting ATPase and a MerR family response regulator." *Applied and Environmental Microbiology*, 67, no. 4 (2001): 1437–1444.

Li, Fang, Xinju Li, Le Hou, and Anran Shao. "Impact of the coal mining on the spatial distribution of potentially toxic metals in farmland tillage soil." *Scientific Reports*, 8, no. 1 (2018): 1–10.

Liao, Qi, Guangyuan Tu, Zhihui Yang, Haiying Wang, Lixu He, Jiaqi Tang, and Weichun Yang. "Simultaneous adsorption of As (III), Cd (II) and Pb (II) by hybrid bio-nanocomposites of nano hydroxy ferric phosphate and hydroxy ferric sulfate particles coating on Aspergillus niger." *Chemosphere*, 223 (2019): 551–559.

Liu, J., X. Luo, Y. Sun, D. C. W. Tsang, J. Qi, W. Zhang et al. "Thallium pollution in China and removal technologies for waters: A review." *Environmental International*, 126(2019):771–790. https://doi.org/10.1016/j.envint.2019.01.076.

Lu, Tao, Qi-Lei Zhang, and Shan-Jing Yao. "Application of biosorption and biodegradation functions of fungi in wastewater and sludge treatment." In *Fungal Applications in Sustainable Environmental Biotechnology*, pp. 65–90. Springer, Cham, 2016.

Maguffin, S. C., M. F. Kirk, A. R. Daigle, S. R. Hinkle, and Q. Jin. "Substantial contribution of biomethylation to aquifer arsenic cycling." *Natural Geosciences*, 8 (2015):290–293. doi:10.1038/ngeo2383.

Mahbub, K. R., K. Krishnan, R. Naidu, and M. Megharaj. "Mercury remediation potential of a mercury resistant strain Sphingopyxis sp. SE2 isolated from contaminated soil." *Journal of Environmental Sciences*, 51 (2017):128–137. https://doi.org/10.1016/j.jes.2016.06.032.

Malandrino, Mery, Ornella Abollino, Agnese Giacomino, Maurizio Aceto, and Edoardo Mentasti. "Adsorption of heavy metals on vermiculite: Influence of pH and organic ligands." *Journal of Colloid and Interface Science*, 299, no. 2 (2006): 537–546.

Malik, Anushree. "Metal bioremediation through growing cells." *Environment International*, 30, no. 2 (2004): 261–278.

Mallatt, Jon. "Fish gill structural changes induced by toxicants and other irritants: A statistical review." *Canadian Journal of Fisheries and Aquatic Sciences*, 42, no. 4 (1985): 630–648.

Mathema, V. B., B. C. Thakuri, and M. Sillanpää. "Bacterial mer operon-mediated detoxification of mercurial compounds: A short review." *Archives of Microbiology*, 193 (2011):837–844. doi:10.1007/s00203-011-0751-4.

McDonald, D. G., and C. M. Wood. 1993. "Branchial mechanisms of acclimation to metals in freshwater fish." In *Fish Ecophysiology*, pp. 297–321. Springer, Dordrecht.

Michalak, Izabela, Katarzyna Chojnacka, and Anna Witek-Krowiak. "State of the art for the biosorption process—a review." *Applied Biochemistry and Biotechnology*, 170, no. 6 (2013): 1389–1416.

Morgan, I. J., F. Galvez, R. S. Munger, C. Wood, and R. Henry. "The physiological effects of acute silver exposure in rainbow trout (Oncorhynchus mykiss)." In *Proceedings of the 3rd International Conference of Transport, Fate and Effects of Silver in the Environment*, pp. 303–306, 1995.

Mount, Donald I. "Chronic toxicity of copper to fathead minnows (Pimephales promelas, Rafinesque)." *Water Research*, 2, no. 3 (1968): 215–223.

Muñoz, Antonio J., Francisco Espínola, Encarnación Ruiz, Aneli M. Barbosa-Dekker, Robert F. H. Dekker, and Eulogio Castro. "Biosorption mechanisms of Ag (I) and the synthesis of nanoparticles by the biomass from Botryosphaeria rhodina MAMB-05." *Journal of Hazardous Materials*, 420 (2021): 126598.

Muraleedharan, T. R., and L. Iyengarand Venkobachar. "Further insight into the mechanism of biosorption of heavy metals by Ganoderma lucidum." *Environmental Technology*, 15, no. 11 (1994): 1015–1027.

Ndiokwere, C. L.. "A study of heavy metal pollution from motor vehicle emissions and its effect on roadside soil, vegetation and crops in Nigeria." *Environmental Pollution Series B, Chemical and Physical*, 7, no. 1 (1984): 35–42.

Nishitani, Takashi, Mariko Shimada, Kouichi Kuroda, and Mitsuyoshi Ueda. "Molecular design of yeast cell surface for adsorption and recovery of molybdenum, one of rare metals." *Applied Microbiology and Biotechnology*, 86, no. 2 (2010): 641–648.

Njoku, K. L., O. R. Akinyede, and O. F. Obidi. "Microbial remediation of heavy metals contaminated media by Bacillus megaterium and Rhizopus stolonifer." *Scientific African* 10 (2020): e00545.

Nkwunonwo, Ugonna C., Precious O. Odika, and Nneka I. Onyia. "A review of the health implications of heavy metals in food chain in Nigeria." *The Scientific World Journal*, 2020 (2020).

Ogden, Richard, and D. A. Adams. "Recombinant DNA technology: Applications." *Carolina Tips*, 52 (1989): 18–19.

Opeolu, Beatrice O., O. Bamgbose, T. A. Arowolo, and M. T. Adetunji. "Utilization of biomaterials as adsorbents for heavy metals removal from aqueous matrices." *Scientific Research and Essays*, 5, no. 14 (2010): 1780–1787.

Pagnanelli, Francesca, Carolina Cruz Viggi, and Luigi Toro. "Isolation and quantification of cadmium removal mechanisms in batch reactors inoculated by sulphate reducing bacteria: Biosorption versus bioprecipitation." *Bioresource Technology*, 101, no. 9 (2010): 2981–2987.

Pan, X., Z. Chen, L. Li, W. Rao, Z. Xu, and X. Guan. "Microbial strategy for potential lead remediation: A review study." *World Journal of Microbiology and Biotechnology*, 33 (2017):35. doi:10.1007/s11274-017-2211-z.

Paria, Kishalay, Smritikana Pyne, and Susanta Kumar Chakraborty. "Optimization of heavy metal (lead) remedial activities of fungi Aspergillus penicillioides (F12) through extra cellular polymeric substances." *Chemosphere*, 286 (2022): 131874.

Perpetuo, Elen Aquino, Cleide Barbieri Souza, and Claudio Augusto Oller Nascimento. "Engineering bacteria for bioremediation." In *Progress in Molecular and Environmental Bioengineering-From Analysis and Modeling to Technology Applications*. Intechopen, 2011. London, UK.

Plumlee, Konnie H. *Clinical Veterinary Toxicology*. Mosby, St. Louis, MO, 2004.

Potbhare, A. K., P. B. Chouke, A. Mondal, R. B. Thakare, S. K. Mondal, R. G. Chaudhary, and A. R. Rai. "*Rhizoctonia solani* assisted biosynthesis of silver nanoparticles for antibacterial assay." *Material Today: Proceedings*, 29 (2020): 939–945.

Puschner, Birgit, and Robert H. Poppenga. "Lead and zinc intoxication in companion birds." *Compendium (Yardley, PA)*, 31, no. 1 (2009): E1–12.

Ragnarsdottir, K. V., S. R. Gislason, T. Thorvaldsson, A. J. Kemp, and A. Andresdottir. "Ejection of trace metals from volcanoes." *Mineralogical Magazine*, 58, no. Goldschmidt 1994 Conference Abstracts (1994): 752–753.

Ramya, Sri Lakshmi. "Application of biosorption for removal of heavy metals from wastewater." IntechOpen, 2018.

Rangsayatorn, N., E. S. Upatham, M. Kruatrachue, P. Pokethitiyook, and G. R. Lanza. "Phytoremediation potential of Spirulina (Arthrospira) platensis: Biosorption and toxicity studies of cadmium." *Environmental Pollution* 119, no. 1 (2002): 45–53.

Rao, C. Prasada, and Harold J. Gluskoter. "Occurrence and distribution of minerals in Illinois coals." *Circular*, no. 476 (1973).

Rascio, Nicoletta, and Flavia Navari-Izzo. "Heavy metal hyperaccumulating plants: How and why do they do it? And what makes them so interesting?" *Plant Science*, 180, no. 2 (2011): 169–181.

Redha, Ali Ali. "Removal of heavy metals from aqueous media by biosorption." *Arabian Journal of basic and Applied Sciences*, 27, no. 1 (2020): 183–193.

Reece, R. L., D. B. Dickson, and P. J. Burrowes. "Zinc toxicity (new wire disease) in aviary birds." *Australian Veterinary Journal*, 63, no. 6 (1986): 199–199.

Ren, Deyi, Fenhua Zhao, Junying Zhang, and D. Xu. "A preliminary study on genetic type of enrichment for hazardous minor and trace elements in coal." *Earth Science Frontier*, 6, no. Suppl (1999): 17–22.

Romero-Puertas, M. C., J. M. Palma, M. Gómez, L. A. Del Rio, and L. M. Sandalio. "Cadmium causes the oxidative modification of proteins in pea plants." *Plant, Cell & Environment*, 25, no. 5 (2002): 677–686.

Ruch, Rodney R., Harold J. Gluskoter, and Neil F. Shimp. "Occurrence and distribution of potentially volatile trace elements in coal: A final report." *Environmental Geology*, no. 072 (1974).

Rudakiya, Darshan M., Vignesh Iyer, Darsh Shah, Akshaya Gupte, and Kaushik Nath. "Biosorption potential of phanerochaete chrysosporium for arsenic, cadmium, and chromium removal from aqueous solutions." *Global Challenges*, 2, no. 12 (2018): 1800064.

Sa Costa, Heloisa Pereira de, da Silva Meuris Gurgel Carlos, Vieira Melissa Gurgel Adeodato. "Biosorption of aluminum ions from aqueous solutions using non-conventional low-cost materials: A review." *Journal of Water Process Engineering*, 40 (2021):101925.

Salton, Milton R. J. *The Bacterial Cell Wall*. No. 04; QR75, S3, 1964.

Sanders, Alison P., Sloane K. Miller, Viet Nguyen, Jonathan B. Kotch, and Rebecca C. Fry. "Toxic metal levels in children residing in a smelting craft village in Vietnam: A pilot biomonitoring study." *BMC Public Health*, 14, no. 1 (2014): 1–8.

Sar, P., S. K. Kazy, R. K. Asthana, and S. P. Singh. "Metal adsorption and desorption by lyophilized Pseudomonas aeruginosa." *International Biodeterioration & Biodegradation*, 44, no. 2–3 (1999): 101–110.

Scheuhammer, A. M. "The chronic toxicity of aluminium, cadmium, mercury, and lead in birds: A review." *Environmental Pollution*, 46, no. 4 (1987): 263–295.

Schutzendubel, Andres, and Andrea Polle. "Plant responses to abiotic stresses: Heavy metal-induced oxidative stress and protection by mycorrhization." *Journal of Experimental Botany*, 53, no. 372 (2002): 1351–1365.

Sedlakova-Kadukova, J., A. Kopcakova, L. Gresakova, A. Godany, and P. Pristas. "Bioaccumulation and biosorption of zinc by a novel Streptomyces K11 strain isolated from highly alkaline aluminium brown mud disposal site." *Ecotoxicology and Environmental Safety*, 167 (2019): 204–211.

Sekhar, K. Chandra, S. Subramanian, J. M. Modak, and K. A. Natarajan. "Removal of metal ions using an industrial biomass with reference to environmental control." *International Journal of Mineral Processing*, 53, no. 1–2 (1998): 107–120.

Shah, K., R. Filby, and W. Haller. "Determination of trace elements in petroleum by neutron activation analysis: II. Determination of Sc, Cr, Fe, Co, Ni, Zn, As, Se, Sb, Eu, Au, Hg and U." *Journal of Radioanalytical and Nuclear Chemistry*, 6, no. 2 (1970): 413–422.

Shamim, Saba. "Biosorption of heavy metals." *Biosorption*, 2 (2018): 21–49.

Sharma, Pooja, Sirohi Ranjna, Tong Yen Wah, Kim Sang Hyoun, and Pandey Ashok. "Metal and metal(loids) removal efficiency using genetically engineered microbes: Applications and challenges." *Journal of Hazardous Materials*, 416 (2021): 125855.

Sharma, Shanti S., and Karl-Josef Dietz. "The relationship between metal toxicity and cellular redox imbalance." *Trends in Plant Science*, 14, no. 1 (2009): 43–50.

Stoddart, R. W. "Biophysical Characterisation of the Cell Surface." *British Journal of Cancer*, 40, no. 2 (1979): 326–327.

Singh, Alpana, Dhananjay Kumar, and J. Gaur. "Continuous metal removal from solution and industrial effluents using Spirogyra biomass-packed column reactor." *Water Research*, 46, no. 3 (2012): 779–788.

Singh, Biswajit, and Sudip Kumar Das. "Adsorptive removal of Cu (II) from aqueous solution and industrial effluent using natural/agricultural wastes." *Colloids and Surfaces B: Biointerfaces*, 107 (2013): 97–106.

Singh, Raghvendra Pratap, Geetanjali Manchanda, Zhi-Feng Li, and Alok R. Rai. "Insight of proteomics and genomics in environmental bioremediation." In *Handbook of Research on Inventive Bioremediation Techniques*, pp. 46–69. IGI Global, 2017. Hershey, PA 17033, USA.

Singh, N. B., M. F. Desimone, R. G. Chaudhary, and W. B. Gurnule. "Management of nanomaterial wastes." In *Nanomaterials Recycling*, pp. 125–144. Elsevier, 2022. Amsterdam, Netherlands.

Sivashankar, R., A. B. Sathya, J. Kanimozhi, and B. Deepanraj. "Characterization of the biosorption process." *Biosorption for Wastewater Contaminants* (2021): 102–116.

Smily, John Rose Mercy Benila, and Pasumalai Arasu Sumithra. "Optimization of chromium biosorption by fungal adsorbent, Trichoderma sp. BSCR02 and its desorption studies." *Hayati Journal of Biosciences*, 24, no. 2 (2017): 65–71.

Smith, A. "Zinc toxicosis in a flock of Hispaniolan Amazons." In *Proceedings of the Annual Conference Association Avian Veterinarian*, pp. 447–453, 1995.

Sonnenfeld, E. M., T. J. Beveridge, A. L. Koch, and R. J. Doyle. "Asymmetric distribution of charge on the cell wall of Bacillus subtilis." *Journal of Bacteriology*, 163, no. 3 (1985): 1167–1171.

Spry, D. J., and C. M. Wood. "A kinetic method for the measurement of zinc influx in vivo in the rainbow trout, and the effects of waterborne calcium on flux rates." *Journal of Experimental Biology*, 142, no. 1 (1989): 425–446.

Stagg, R. M., and T. J. Shuttleworth. "The effects of copper on ionic regulation by the gills of the seawater-adapted flounder (Platichthys flesus L.)." *Journal of Comparative Physiology*, 149, no. 1 (1982): 83–90.

Starr, M., A-J. Lindroos, L. Ukonmaanaho, T. Tarvainen, and H. Tanskanen. "Weathering release of heavy metals from soil in comparison to deposition, litterfall and leaching fluxes in a remote, boreal coniferous forest." *Applied Geochemistry*, 18 (2003): 607–613.

Sundararaju, Sathyavathi, Arumugam Manjula, Vignesh Kumaravel, Thillaichidambaram Muneeswaran, and Thirumalaisamy Vennila. "Biosorption of nickel ions using fungal biomass Penicillium sp. MRF1 for the treatment of nickel electroplating industrial effluent." *Biomass Conversion and Biorefinery* (2020): 1–10.

Tewari, Neetu, P. Vasudevan, and B. K. Guha. "Study on biosorption of Cr (VI) by Mucor hiemalis." *Biochemical Engineering Journal*, 23, no. 2 (2005): 185–192.

Thomas, W. H., J. T. Hollibaugh, D. L. R. Seibert, and G. T. Wallace Jr. "Toxicity of a mixture of ten metals to phytoplankton." *Marine Ecology Progress Series*, 2, no. 3 (1980): 212–220.

Tian, C., B. Zhang, A. G. L. Borthwick, Y. Li, and W. Liu. "Electrochemical oxidation of thallium (I) in groundwater by employing single-chamber microbial fuel cells as renewable power sources." *International Journal Hydrogen Energy*, 42 (2017):29454–29462. https://doi.org/10.1016/j.ijhydene.2017.10.026.

Todorova, Kostadinka, Zdravka Velkova, Margarita Stoytcheva, Gergana Kirova, Sonia Kostadinova, and Velizar Gochev. "Novel composite biosorbent from Bacillus cereus for heavy metals removal from aqueous solutions." *Biotechnology & Biotechnological Equipment*, 33, no. 1 (2019): 730–738.

Tuzen, Mustafa, Kadriye Ozlem Saygi, Canan Usta, and Mustafa Soylak. "Pseudomonas aeruginosa immobilized multiwalled carbon nanotubes as biosorbent for heavy metal ions." *Bioresource Technology*, 99, no. 6 (2008): 1563–1570.

Upadhyay, Kinjal H., Avni M. Vaishnav, Devayani R. Tipre, Bhargav C. Patel, and Shailesh R. Dave. "Kinetics and mechanisms of mercury biosorption by an exopolysaccharide producing marine isolate Bacillus licheniformis."*3Biotech*, 7, no. 5 (2017): 1–10.

Valls, Marc, Sílvia Atrian, Víctor de Lorenzo, and Luis A. Fernández. "Engineering a mouse metallothionein on the cell surface of *Ralstonia eutropha* CH34 for immobilization of heavy metals in soil." *Nature Biotechnology*, 18, no. 6 (2000): 661–665.

Van der Zee, J., P. Zwart, and A. J. H. Schotman. "Zinc poisoning in a nicobar pigeon." *Journal of Zoo Animal Medicine*, 16, no. 2 (1985): 68–69. (Van der Zee, Zwart, and Schotman 1985, 68).

Veglio, Francesco, and F. Beolchini. "Removal of metals by biosorption: A review." *Hydrometallurgy*, 44, no. 3 (1997): 301–316.

Verbost, P. M., G. E. R. T. Flik, R. A. C. Lock, and S. E. Wendelaar Bonga. "Cadmium inhibits plasma membrane calcium transport." *The Journal of Membrane Biology*, 102, no. 2 (1988): 97–104.

Verbost, P. M., G. E. R. T. Flik, R. A. C. Lock, and S. E. Wendelaar Bonga. "Cadmium inhibition of Ca2+ uptake in rainbow trout gills." *American Journal of Physiology-Regulatory, Integrative and Comparative Physiology*, 253, no. 2 (1987): R216–R221.

Verma, Samakshi, and Kuila Arindam. "Bioremediation of heavy metals by microbial process." *Environmental Technology & Innovation*, 14 (2019): 100369.

Vigneri, R., P. Malandrino, F. Giani, M. Russo, and P. Vigneri. "Heavy metals in the volcanic environment and thyroid cancer." *Molecular and Cellular Endocrinology*, 457 (2017): 73–80.

Vijayaraghavan, K., and Yeoung-Sang Yun. "Bacterial biosorbents and biosorption." *Biotechnology Advances*, 26, no. 3 (2008): 266–291.

Vijayaraghavan, K., and Yeoung-Sang Yun. "Chemical modification and immobilization of Corynebacterium glutamicum for biosorption of reactive black 5 from aqueous solution." *Industrial & Engineering Chemistry Research*, 46, no. 2 (2007): 608–617.

Volesky, Bohumil. "Biosorption process simulation tools." *Hydrometallurgy*, 71, no. 1–2 (2003): 179–190.

Wang, Jianlong, and Can Chen. "Biosorption of heavy metals by Saccharomyces cerevisiae: A review." *Biotechnology Advances*, 24, no. 5 (2006): 427–451.

Wang, L., D. Hou, Y. Cao, Y. S.Ok, F. M. G. Tack, J. Rinklebe et al. "Remediation of mercury contaminated soil, water, and air: A review of emerging materials and innovative technologies." *Environmental International*, 134 (2020): 105281. https://doi.org/10.1016/j.envint.2019.105281.

Wang, N., Y. Qiu, K. Hu, C. Huang, J. Xiang, H. Li, J. Tang, J. Wang, and T. Xiao. "One-step synthesis of cake-like biosorbents from plant biomass for the effective removal and recovery heavy metals: Effect of plant species and roles of xanthation." *Chemosphere* (2020):129129.

Wang, Z., B. Zhang, Y. Jiang, Y. Li, and C. He. "Spontaneous thallium (I) oxidation with electricity generation in single-chamber microbial fuel cells." *Applied Energy*, 209 (2018):33–42. https://doi.org/10.1016/j.apenergy.2017.10.075.

Wedekind, K. J., and D. H. Baker. "Zinc bioavailability in feed-grade sources of zinc." *Journal of Animal Science*, 68, no. 3 (1990): 684–689.

Wen, Xiaofeng, Chunyan Du, Guangming Zeng, Danlian Huang, Jinfan Zhang, Lingshi Yin, Shiyang Tan et al. "A novel biosorbent prepared by immobilized Bacillus licheniformis for lead removal from wastewater." *Chemosphere*, 200 (2018): 173–179.

Wood, Chris M., C. Hogstrand, F. Galvez, and R. S. Munger. "The physiology of waterborne silver toxicity in freshwater rainbow trout (Oncorhynchus mykiss) 1. The effects of ionic Ag+." *Aquatic Toxicology*, 35, no. 2 (1996): 93–109.

Xu, Shaozu, Yonghui Xing, Song Liu, Xiuli Hao, Wenli Chen, and Qiaoyun Huang. "Characterization of Cd2+ biosorption by Pseudomonas sp. strain 375, a novel biosorbent isolated from soil polluted with heavy metals in Southern China." *Chemosphere*, 240 (2020): 124893.

Xu, Yiyuan, et al. "Distribution and dispersion of heavy metals in the rock -soil -moss system of the black shale areas in the southeast of Guizhou Province, China." *Environmental Science and Pollution Research* 29.1 (2022): 854–867.

Yahya, Bayda A. "The use of the fungi penicillium and rhizopus to remove some heavy metals from the wastewater in hospital in Mosul city." *Annals of the Romanian Society for Cell Biology* (2021): 5096–5103.

Yin, Kun, Qiaoning Wang, Min Lv, and Lingxin Chen. "Microorganism remediation strategies towards heavy metals." *Chemical Engineering Journal*, 360 (2019): 1553–1563.

Yu, J., M. Tong, X. Sun, and B. Li. "A simple method to prepare poly(amic acid)- modified biomass for enhancement of lead and cadmium adsorption." *Biochemical England Journal*, 33 (2007): 126–133.

Zafar, Muhammad Nadeem, Azra Parveen, and Raziya Nadeem. "A pretreated green biosorbent based on Neem leaves biomass for the removal of lead from wastewater." *Desalination and Water Treatment*, 51, no. 22–24 (2013): 4459–4466.

Zaynab, Madiha, Al-Yahyai Rashid, Ameen Ayesha, Sharif Yasir, Ali Liaqat, Fatima Mahpara, Khan Khalid Ali, and Li Shuangfei. "Health and environmental effects of heavy metals." *Journal of King Saud University Science* (2021): 101653.https://doi.org/10.1016/j.jksus.2021.101653.

Zhang, B., R. Fan, Z. Bai, S. Wang, L. Wang, and J. Shi. "Biosorption characteristics of Bacillus gibsonii S-2 waste biomass for removal of lead (II) from aqueous solution." *Environmental Science and Pollution Research*, 20(2013): 1367–1373. doi:10.1007/s11356-012-1146-z.

Zhao, Lili, Jue Wang, Pengpeng Zhang, Qiaoqiao Gu, and Chuancai Gao. "Absorption of heavy metal ions by alginate." *Bioactive seaweeds for Food Applications* (2018): 255–268.

Zhou, W., D. Liu, H. Zhang, W. Kong, and Y. Zhang. "Bioremoval and recovery of Cd(II) by Pseudoalteromonas sp. SCSE709–6: Comparative study on growing and grown cells." *Bioresources Technology*, 165 (2014):145–151. https://doi.org/10.1016/j.biortech.2014.01.119.

12 Genetically Modified Microorganisms for Remediation of Pollutants Containing Heavy Metals

Sahidul Islam and Ujjwal Mandal

INTRODUCTION

The earth's crust contains natural mineral deposits in the form of various compounds (Balaram 2019). Various metals and their compounds are extracted from ores for the purpose of use in various fields such as electronics, clean energy sources, and other uses for the betterment of human life. Presently, there is an international level priority to control carbon emissions in order to prevent climate change (Fawzy et al. 2020). Thus, there is a great demand for the supply of metals; however, despite increased amounts of funding, primary-grade metal reserves are discovered less and less frequently (Jowitt et al. 2020). Cost-cutting in mining and metal refineries was a common practice during the 20th century for economic profit, which caused a great decrease in public confidence. Besides the lack of availability of primary-grade minerals, the other difficulty in mining is opposition by local inhabitants and governments that are safeguarded by larger organisations. In mining industries, significant improvement is now made by building trust and giving assurance about the management of effluents. Treatment of metal-containing effluents is quite challenging from techno-economic, environmental and social viewpoints (Ali et al. 2019). A single technology is not capable to do that, so several technologies are necessary for treatment of effluents and pollution control in the mining, industrial and neighbouring aquatic sites. The presently available technologies for removing ions of heavy metals (HMs) include precipitation, coagulation, flocculation, membrane filtration, photocatalysis and adsorption using inorganic materials (Fu and wang 2011). The main advantages of traditional methods include compatibility with high concentrations of HMs, rapidness, ease of operation and clear understanding of the molecular basis. These points need to be considered while designing for heavy metal removal technologies, which ultimately determine the capital and operational costs. Green engineering principles are the contemporary basis of heavy metal removal. Although in many conventional technologies, HMs are removed efficiently, in many cases contaminant byproducts are produced that are difficult to dispose of and involve too much energy cost. Moreover, difficulties arise because of the utility of several substances derived from ion-exchanging resins, non-renewable sources like coal, oil and activated carbons (Diep et al. 2018).

The biomass-driven removal of HMs is considered to be cost-effective, techno-economically feasible, environmentally friendly and simple in terms of operating processes. However, such assessments are not yet performed widely. Biologically driven HM removal—whether obeying green principles or not—is still a subject matter of research and development (Ayangbenro and Babalola 2017).

Among numerous biological phenomena, bioaccumulation and biosorption are in the interest of study as methods of removing HMs. Being a natural process, biosorption and bioaccumulation can be assessed for microorganisms to study their potential to remove HMs. With clear knowledge

DOI: 10.1201/9781003188568-12

about the metabolism of HMs and identification of the genes in microorganisms, genetically modified microorganisms can be created for efficient HM removal activities (Diep et al. 2018).

TOXICITY OF HMS

Elements with atomic numbers greater than 22 and densities greater than 5g/ml are considered HMs. According to this definition, 69 elements, including 16 artificially made, are considered heavy metals (Aquino et al. 2011). The toxicity of some of the heavy metals is prominent even in low concentrations. The toxic HMs cause potential danger to the environment and ecosystem. HM ions can be classified as (i) essential HM ions: Na^+, K^+, Mg^{2+}, Ca^{2+}, V^{n+}, Mn^{n+}, Fe^{n+}, Co^{n+}, Ni^{n+}, Cu^{2+}, Zn^{2+}, Mo^{n+} and W^{n+} (ii) toxic HM ions: $Hg^{2+,}$ Cr^{3+}, Pb^{n+}, Cd^{n+}, As^{n+}, Sr^{n+}, Ag^+, Si^{4+}, Al^{3+}, Tl^{n+} (the higher valent forms are extremely toxic) (iii) radioactive toxic HM ions: U^{n+}, Rn^{n+}, Th^{n+}, Ra^{n+}, Am^{n+}, Tc^{n+} (iv) ions of semi-metals: B^{4+}, Si^{4+}, Ge^{4+}, As^{n+}, Sn^{n+}, Te^{n+}, Po^{n+}, At^{n+}, Se^{n+} (these ions of metalloids exhibit distinct kinds of biological effects) (Aquino et al. 2011; Lunch 2012). Based on the impacts on the environment, metal ions can be categorised as (i) bio-available and (ii) bio-non-available (Lunch 2012). Bio-available metal ions are water soluble, non-absorbed and mobile. Bio-non-available metal ions undergo complex formation, precipitation and sorption (Olaniran et al. 2013). The cationic form and oxidation state of metals determine their bio-availability and fate (Olaniran et al. 2013). The cationic form of a metal ion is sorbed (absorbed/adsorbed) in the negatively charged (generally in the pH range 4–8) cell surface. The cell surface can be negative due to the presence of excess SH- or OH-, phosphate, sulphate moiety in the branch or humic acids and colloidal clay minerals (Chianese et al. 2020). The HMs bind to the negatively charged surface of living cells owing to coulombic attraction (Bhattacharjee et al. 2020). Due to this HM binding, the structure of the assembly with nucleic acids, proteins and cell wall causes destabilization; consequently, there may be mutagenesis and genetic disorder (O'Brien et al. 2018). With the help of microarray technology, the gene expression patterns induced by six metal ions ($Cr^{n+,}$ $Ni^{4+,}$ As^{n+}, Sb^{n+}, Cd^{n+}, Hg^{n+}) were found similar to that which occurs in the presence of reactive singlet oxygen species, chemical generating agents and DMNQ (2,3-dimethoxy-1, 4-naphthoquinone) (Mello and Hess 2005). The HMs cause damage in a cell by generating reactive oxygen and destroying antioxidants present in the cell (Jan et al. 2015).

HMS AND THE ENVIRONMENT

HMs cause significant contamination in the environment, like other pollutants. The metal ions can be accumulated in the ecosystem in various forms and move to the various components of the ecosystem through the food chain (Yan et al. 2020). HM toxicity in living creatures is diverse, therefore appropriate treatment of industrial and mining waste is required using the combination of various modern technologies (Briffa et al. 2020). The conventional methods are not proper for pollutant management and are very costly.

CONVENTIONAL METHODS OF MANAGEMENT OF POLLUTANTS CONTAINING HMS

Removal of environment-contaminating HMs by the traditional method is based on the physico-chemical method, which is inefficient and cost-prohibitive (Tangahu et al. 2011). For example, one of the methods of HM ion removal from wastewater is to increase the pH, which renders the hydroxide formations that are precipitated. But in this method, a large amount of mudwater is produced, containing a high concentration of metal ions in Mg/ml level; hence it is difficult to dispose of appropriately (Barakat 2011). The other methods are comparatively complex and often involve multiple steps. A few such methods are (i) the precipitation of metal ions as oxides, hydroxides, carbonates,

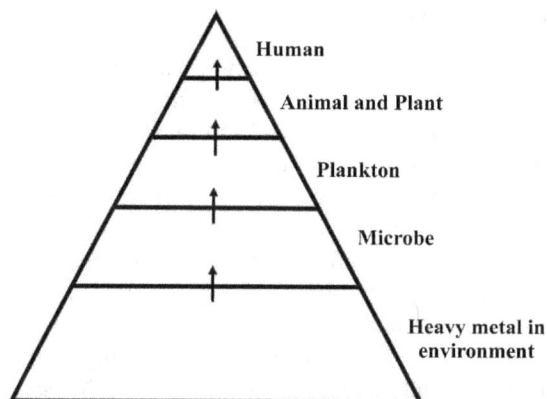

FIGURE 12.1   Destiny of heavy metals throughout the food chain.

sulphides and so on (Pohl 2020), (ii) the redox chemical method (Qasem et al. 2021), (iii) the electrochemical method (Vidu et al. 2020), (iv) adsorption/absorption (sorption using activated charcoal) (Karnib et al. 2014), (v) membrane filtration method (e.g., reverse osmosis, electrodialysis, membrane filtration) (Khulbe and Matsuura 2018), (vi) evaporation (Wu et al. 2015), (vii) solvent extraction (Marta 1995), (viii) electrolytic recovery (Maarof et al. 2016), and (ix) electrodeposition (Tonini and Ruotolo 2017). Researchers across the world have given attention to bioremediation due to the advantage of easy operation, simplicity, efficiency and cost-friendliness (Azubuike et al. 2016).

BIOREMEDIATION PROCESS

Remediation of the contaminated components and sub-components of the environment, such as soil, water, deposits and sediments, can be done by exploiting biologically assisted oxidation state changes in the pollutants. In the bioremediation process, microorganisms change toxic substances like hydrocarbons, agrochemicals and other toxic organic materials into non-toxic forms (Wuana and Okieimen 2011). For inorganic origin pollutants such as HM ions, microbes cannot convert them into non-toxic forms easily, and the task is microbe-specific (Yin et al. 2018). Bioremediation for HM ions involves metabolization in microbes (Jin et al. 2018). It was found that several microorganisms consume HM ions as micronutrients. One such example is the requirement of Fe^{3+} by all bacteria; anaerobic bacteria need Fe^{2+} (Zhang et al. 2009). The consumption of HM ions depends on the biomass of the microorganism, the geochemistry in their colony and redox catalysed conversion into insoluble forms (Rehm and Reed 2001). This redox-accompanied conversion is enzyme-catalysed (Sharma et al. 2014). Microbes improve soil quality by consuming pollutants; hence crops are produced in good yields (Alori and Babalola 2018). Appropriate selection of microbes can be done by the proper understanding of the adsorption and immobilisation of HM ions in restoring soil quality (Rana et al. 2021). Microorganisms inherently can act as metal ion accumulators; therefore, it is necessary to identify their genes that may be transferred through the microarray development to the microorganism in which this trait is absent (Arber 2014). Restoration of soil health by detoxification using microorganisms is a safe and effective method in areas contaminated by ore mining, oil plants, insecticides, pesticides, pigments, plastic, organic solvents and refineries where toxic pollutants are released (Sales da Silva et al. 2020). The major limitations in the successful exploitation of microorganisms are the lack of information on their cellular response to trace elements and HM ions. It is a matter of interest to study the increase of sensitivity of microorganisms towards HM pollutants by genetic modification and promotion of their widespread applications (Samuel et al. 2021).

BIOREMEDIATION MECHANISM

The development of a microorganism colony in metal-contaminated sites facilitates the restoration of soil quality, as microorganisms convert toxic metal ions into their non-toxic form (Dixit et al. 2015). Microorganisms convert organic pollutants into small molecules such as carbon dioxide and water as end products or some intermediate products that are required for cell growth (Gougoulias et al. 2014). Microorganisms can act as a two-way defence system: (i) they produce pollutant-degradative enzymes (Karigar et al. 2011) and (ii) resist the toxicity of heavy metal ions (Mark et al. 2000). Many explanations of the mechanism of bioremediation have been proposed. These include biosorption (Oyewole et al. 2019), metal microorganism interactions (Tsezos 2009), bioaccumulation (Emenike et al. 2018), biotransformation (Emenike et al. 2018), biomineralization (Li et al. 2013) and bioleaching (Yun et al. 2008). Microorganisms consume HMs as a part of their nutrients, and the redox process occurs leading to their conversion into a final state which is less toxic or nontoxic. The design of microbes, their growth, activity, metabolism and resilience towards environmental changes determine the success of bioremediation (Azubuike et al. 2016).

BIOREMEDIATION BY SORPTION

HMs bind with extracellular polymeric substances in the microorganism (Costa et al. 2018). It is revealed that various mechanisms such as micro-precipitation of metals and exchange of H^+ ions are involved in the binding process (Hussain 2018). The bioremediation process is still unpopular due to a lack of understanding of the role of genetics and the genome, metabolic pathways and the relevant kinetics in the absorption process (Chandran et al. 2020).

PHYSIO-BIO-CHEMICAL MECHANISM OF BIOREMEDIATION

In the biosorption process, the biosorbent shows an affinity for heavy metal ions (sorbate) and the sorption process continues until the establishment of equilibrium. Adsorption of Zn^{2+} and Cd^{2+} by *Saccharomyces cerevisiae* involves the mechanism of ion exchange (Stanislav et al. 2007). *Cunninghamella elegans* has the potential to act as a sorbent of HM ions present in the

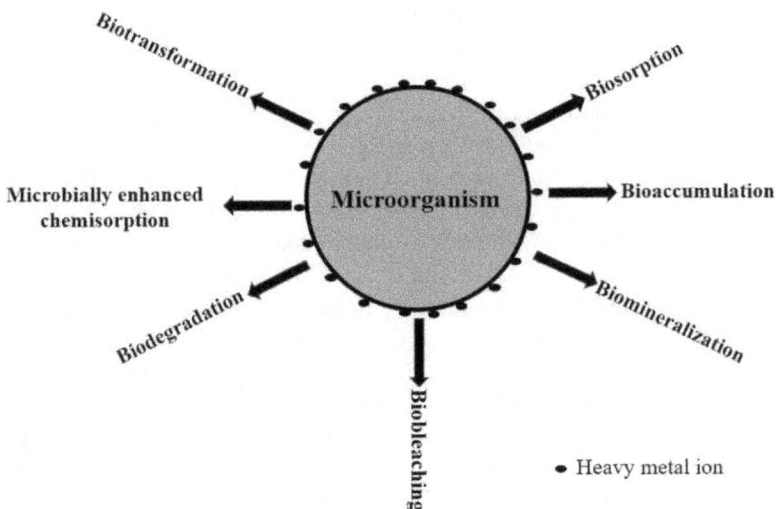

FIGURE 12.2   Heavy metal ion-microorganism interaction and bioremediation.

wastewater released from the textile industry (Valeria et al. 2011). The degradation of HM ions by metabolism involves a process where energy is required. The combination of active and passive modes of bioremediation of toxic metals is called bioaccumulation. The biocatalytic activity of several fungi, which act in accessing HMs and converting them into less toxic form, were found (Deshmukh et al. 2016). Fungi like *Allescheriella sp., Phlebia sp., Klebsiella oxytoca, Stachybotrys sp., Botryosphaeria rhodian* and *Pleurotus pulmonarius* can bind with HMs. Fungi like *Cephalosporium aphidicola* and *Aspergillus parasitica* can biodegrade Pb^{2+}-contaminated soil. *Neocosmospora vasinfecta* and *Hymenoscyphus ericae* convert the toxic Hg^{2+} form of mercury into its nontoxic state (Dixit et al. 2015). The microbes secrete biosurfactants which form a complex with the metal ions by strong ionic interaction and get discharged to the water from the soil matrix due to less interfacial tension (Khan et al. 2011). Bioremediation mechanisms by aerobic and anaerobic microbes are different (Eltarahony et al. 2020). In aerobic degradation, the introduction of oxygen during the reaction occurs due to nascent oxygen atoms generated by enzymes such as peroxidases, ligninases or mediated by enzymes such as oxidative dehalogenases, hydroxylases, monooxygenases and dioxygenases. In case of anaerobic degradation, the reaction proceeds through initial activation, and then anoxic electron acceptor mediated oxidative catabolism occurs. The technique of decreasing mobilization of heavy metals from the polluted sites by changing their physical or chemical state is called immobilization (Dixit et al. 2015). Destruction of HMs is impossible, but their state can be changed due to redox reactions or organic complex formation. In resistant bacteria, the main two mechanisms are detoxification and pumping of the toxic metals from the cell (Tang et al. 2021). The required energy for growth is achieved as microorganisms oxidise organic pollutants where metal ions such as Fe (III) and Mn (IV) are reduced (Weber et al. 2006). Due to the change of state of the metals, their solubility changes. As an example, *Geobaccter species* reduce soluble U(VI) to insoluble U(IV) (Fletcher et al. 2010). After binding the HMs, the system always tries to neutralise the developed stress. Studying the expression of metal-binding proteins, the HM ion accumulation in the microorganism can be studied. *Synechococcus* sp. is known to produce a metal-binding protein with gene expression of smt A (Morita et al. 2012). Genetic modification of *Ralstonia eutropha* gives the protein expression of mouse metallotheionein and reduces the toxicity of Cd (II) (Valls et al. 2000). *Escherichia coli* produces the expression of proteins that regulate the range of Cd accumulation (Deng et al. 2007). Co-expression of Phytochelatins (PC) and glutathione (GSH) precursor shows a two-fold increase in Cd accumulation (Zhang et al. 2018). Metalloregulatory proteins present in microorganisms regulate the natural resistance to heavy metals like Hg.

BIOREMEDIATION PROCESS AND MOLECULAR MECHANISM

Several mechanisms for the removal of metal ions by microorganisms are proposed and established from factual support. Hg reduction at high temperatures was found using genetically modified bacteria like *Deinococcus geothemalis*. In this case, mer operon from *E. coli* is present which reduces Hg (II) (Brim et al. 2003). Genetic modification of *Cupriavidus metallidurans* (a mercury-resistant bacteria) by introducing pTP6, gives genes (MerB and MerG) that regulate the biodegradation of Hg; at the same time, the synthesis of MerA (mercuric reductase) and MerB (organomercurial lyase protein) occur (Rojas et al. 2011). Mercury-resistant property is also found when the *Pseudomonas* strain is modified with pMR68 plasmid (Sone et al. 2013). *Klebsiella pneumonia* M426 bacteria degrade mercury in two different pathways, one is the precipitation of mercury and another one is the conversion of Hg (II) to volatile Hg (0) (Ashraf et al. 2005). Radiation-resisting bacteria *Deinococcus radiodurans* that convert Cr (VI) to Cr (III) when genetically modified by introducing *xyl* operon of *Pseudomonas putida* and cloned gene of tod, can completely degrade toluene (Brim 2006). In the pathway of degradation, metal-bound coenzymes and siderophores-type microbial metabolites are involved (Khan et al. 2017).

GENETIC MODIFICATION OF MICROORGANISMS AND HM REMOVAL

Indigenous microorganisms are not enough to remove HMs from highly contaminated sites. Natural microbes cannot remove Hg efficiently. However, bioremediation efficiency can be improved by recombinant DNA technology, where a foreign gene is inserted into the genome or plasmid (extra-chromosomal region) from an organism of another species (same or different); thus, genetically modified microorganisms are created. In earlier days of genetic modification technologies, genetically modified *Pseudomonas putida* and *E. coli* strain M109 containing mer A gene was used for the removal of Hg (Deckwer et al. 2004). It is necessary to identify the appropriate gene which plays the role in the removal of HMs and detoxification; it is the primary requisite for the genetic modification of the microorganism. With the advancement of genetic technologies, researchers are able to study the catabolism of organic pollutants in microorganisms (Chakraborty and Das 2016). The appropriate tailoring of microbial genes by recombinant nucleic acid technologies improves or creates new metabolic pathways in modified microorganisms that enhance the bioremediation process. Genetically modified microorganisms with unique and specific capabilities and improved microbial metabolisms are cost-effective in removing and detoxifying HM ions from contaminated sites. Genetically modified organisms are eco-friendly and are now used in removing various metal ions such as Fe, Cu, Ni, As, Cd and Hg. In the new metabolic pathways, toxic forms of heavy metals are more effectively converted into less/non-toxic forms, thus the bioremediation process is improved (Borchert et al. 2021).

BIOREMEDIATION OF HMS FROM CONTAMINATED SOIL BY SYMBIOTIC PLANTS AND BACTERIA

The growth and development of microbes can be improved by adjusting pH as well as the levels of nutrients and oxygen. Rhizosphere bacteria grow and thrive by obtaining nutrients from substances like organic acids, amino acids, enzymes and carbohydrates excreted by plant roots; in reciprocation, these bacteria increase the availability of nutrients for the plants.

TABLE 12.1
HMs and their Bioremediation by Genetically Modified Bacteria

Heavy Metals	Genetically modified bacteria	References
Cr	*Deinococcus radiodurans*	Brim et al. 2003
Cr	*Methylococcus capsulatus*	Al Hasin et al. 2010
Cr	*Pseudomonas putida*	Ackerley et al. 2004
As	*Bacillus Idriensis*	Liu et al. 2011
As	*Bacillus subtilis*	Huang et al. 2015
As	*Escherichia coli* strain	Yuan et al. 2008
Cd	*Bacillus subtilis* BR151 (pT0024)	Ivask et al. 2011
Cd	*Escherichia coli* strain	Freeman et al. 2005
Cd	*Mesorhizobium huakuii*	Porter et al. 2017
Hg	*Achromobacter* sp. AO22	Ng et al. 2009
Hg	*Acidithiobacillus ferrooxidans*	Valdés et al. 2008
Hg	*Deinococcus geothemalis*	Dixit et al. 2015
Hg	*Escherichia coli* JM109	Zhao et al. 2005
Hg	*Escherichia coli* MC1061	Bondarenko et al. 2008

Rhizobium is a popularly known bacterium that grows in the roots of leguminous plants and builds symbiotic relationships. In the root of legumes, Rhizobia form nitrogen-fixing nodules that contain up to 108 bacteria progenies (Datta et al. 2015). This tendency can be exploited in biotechnology for gene expression, such as metallothioneins (MTs) that sequester HMs from contaminated sites (Sarma and Prasad 2019). After the uptake, HMs are then accumulated in plant roots, transported through the xylem and converted into less toxic forms, or get accumulated in the rhizospheres and nodules. Thus, this becomes a less expensive method of HM removal from the soil. One example of such a microbe is genetically modified *Mesorhizobium* sp. and *M. huakuii* subsp. rengei, by AtPCS gene coding PC synthase shows increased Cd accumulation in bacterial cells and inoculation of this modified *Mesorhizobium* with *Astragalus sinicus* increases Cd accumulation in root nodules. With this symbiotic entity, phytoremediation of Cd-polluted paddy soil is achieved. The increased Cd accumulation was found in the root, which contributes 10% to the removal of Cd from the ground after two months of plant cultivation. Gupta et al. (2002) developed phosphate-solubilizing *Pseudomonas* sp. NBRI 4014 that has a very high resistance to Cr, Ni and Cd in Glycine max plants grown in metal-contaminated soil (Gupta et al. 2002).

EXPLOITATION OF GENETICALLY MODIFIED MICROORGANISMS AND SAFETY ASPECTS

Genetically modified microorganisms designed for bioremediation may have adverse consequences on human health and the environment. The wide-scale application of modified microorganisms is not prevalent now due to low psychological acceptance by the public. One of the most important issues of their application is containment. After the control of pollution using the modified microorganism, there may be huge ecological impact. Before releasing the modified microorganism, it is necessary to study the risk assessment, containment procedure and design of the required shield. The survival ability of the modified microorganism is also very important to deplete the HM contamination up to the desired level (Dietmar and Walter 2000).

CONCLUSION

One of the major challenges for the application of modified microorganisms in HM removal is metal tolerance. The improvement of HM tolerance capability in plants can be done by using appropriate microorganisms which have symbiotic relationships. For legume plants, this is very

TABLE 12.2
Enhancement of Phytoremediation by Genetically Modified Bacteria

Heavy Metals	GM microbe	Plant	Gene expression of the GM Microbe	References
Cu	*Ensifer medicae*	*Medicago truncatula* with mt4a gene	copAB	Pérez-Palacios et al. 2017
As	*Meshorhizobium huakuii subsp.* rengei strain B3	*Astragalus sinicus*	Iron regulated transporter 1 gene from *Arabidopsis thaliana* (ATIRT1)	IKe et al. 2008
Cd	*Meshorhizobium huakuii subsp.* rengei strain B3	*Astragalus sinicus*	MTL4 and ATPCS	Ike et al. 2007

useful, as there is a very strong symbiotic relationship between the host plant and microorganism. It was found that the addition of nodule bacteria, PGPR or mycorrhiza has considerable impact on plant growth in highly HM-contaminated soil. After the appropriate genetic modification to the microbes, this effect became more prominent. These practical aspects support the exploitation of modified microorganisms. In conclusion, a number of sophisticated biotechnological tools have been developed for the genetic modification and bioengineering of microbes for the bioremediation of environmental pollutants including HMs. The construction and genetic modification of microbes is now very easy and quick with advancements in biotechnological tools. However, it is necessary to assess the direct effects and repercussions on the biodiversity of existing beneficial microbes in the releasing sites, as well as symbiotic relations in the presence of HMs and overall risks and benefits.

REFERENCES

Ackerley, D., Gonzalez, C., Keyhan, M., Blake, R. and Matin, A. 2004. Mechanism of chromate reduction by the Escherichia coli protein, NfsA, and the role of different chromate reductases in minimizing oxidative stress during chromate reduction. *Environ. Microbiol.* 6(8): 851–860.

Al Hasin, A., Gurman, S.J., Murphy, L.M., Perry, A., Smith, T.J. and Gardiner, P.H. 2010. Remediation of chromium (VI) by a methane-oxidizing bacterium. *Environ. Sci. Technol.* 44(1): 400–405.

Ali, H., Khan, E. and Ilahi, I. 2019. Environmental chemistry and ecotoxicology of hazardous heavy metals: Environmental persistence, toxicity, and bioaccumulation. *J. Chem.* 2019: 1–14.

Alori, E.T. and Babalola, O.O. 2018. Microbial inoculants for improving crop quality and human health in Africa. *Front. Microbiol.* 9: 2213–2248.

Aquino, E., Barbieri, C. and Oller Nascimento, C.A. 2011. *Engineering bacteria for bioremediation.* Progress in Molecular and Environmental Bioengineering—From Analysis and Modeling to Technology Applications.

Arber, W. 2014. Horizontal gene transfer among bacteria and its role in biological evolution. *Life.* 4(2): 217–224.

Ashraf, M.M.E., Lynne, E.M. and Nigel, L. B. 2005. A new method for mercury removal. *Biotechnol Lett.* 27(21): 1649–1655.

Ayangbenro, A. and Babalola, O. 2017. A new strategy for heavy metal polluted environments: A review of microbial biosorbents. *Int. J. Environ. Res.* 14(1): 94–129.

Azubuike, C.C., Chikere, C.B. and Okpokwasili, G.C. 2016. Bioremediation techniques– classification based on site of application: Principles, advantages, limitations and prospects. *World J. Microbiol. Biotechnol.* 32(11): 180–207.

Balaram, V. 2019. Rare earth elements: A review of applications, occurrence, exploration, analysis, recycling, and environmental impact. *Geosci. Front.* 10(4): 1285–1303.

Barakat, M.A. 2011. New trends in removing heavy metals from industrial wastewater. *Arab. J. Chem.* 4(4): 361–377.

Bhattacharjee, C., Dutta, S. and Saxena, V.K. 2020. A review on biosorptive removal of dyes and heavy metals from wastewater using watermelon rind as biosorbent. *Environ. Adv.* 2: 100007.

Bondarenko, O., Rõlova, T., Kahru, A. and Ivask, A. 2008. Bioavailability of Cd, Zn and Hg in soil to nine recombinant luminescent metal sensor bacteria. *Sensors* 8(11): 6899–6923.

Borchert, E., Hammerschmidt, K., Hentschel, U. and Deines, P. 2021. Enhancing microbial pollutant degradation by integrating eco-evolutionary principles with environmental biotechnology. *Trends Microbiol.* 29(10): 908–918.

Briffa, J., Sinagra, E. and Blundell, R. 2020. Heavy metal pollution in the environment and their toxicological effects on humans. *Heliyon.* 6(9): 4691–4709.

Brim, H. 2006. Deinococcus radiodurans engineered for complete toluene degradation facilitates Cr (VI) reduction. *Microbiology.* 152(8): 2469–2477.

Brim, H., Venkateswaran, A., Kostandarithes, H.M., Fredrickson, J.K. and Daly, M.J. 2003. Engineering deinococcus geothermalis for bioremediation of high-temperature radioactive waste environments. *Appl. Environ. Microbiol.* 69(8): 4575–4582.

Chakraborty, J. and Das, S. 2016. Molecular perspectives and recent advances in microbial remediation of persistent organic pollutants. *Environ. Sci. Pollut. Res.* 23(17): 16883–16903.

Chandran, H., Meena, M. and Sharma, K. 2020. Microbial biodiversity and bioremediation assessment through omics approaches. *Front. Environ. Chem.* 1: 570326–570353.

Chianese, S., Fenti, A., Iovino, P., Musmarra, D. and Salvestrini, S. 2020. Sorption of organic pollutants by humic acids: A review. *Molecules*. 25(4): 918–940.

Costa, O.Y.A., Raaijmakers, J.M. and Kuramae, E.E. 2018. Microbial extracellular polymeric substances: Ecological function and impact on soil aggregation. *Front. Microbiol*. 9: 1636–1658.

Datta, A., Singh, R.K., Kumar, S. and Kumar, S. 2015. An effective and beneficial plant growth promoting soil bacterium "Rhizobium": A review. *Ann. Plant Sci*. 4(1): 933–942.

Deckwer, W.D., Becker, F.U., Ledakowicz, S. and Wagner-Döbler, I. 2004. Microbial removal of ionic mercury in a three-phase fluidized bed reactor. *Environ. Sci. Technol*. 38(6): 1858–1865.

Deng, X., Yi, X.E. and Liu, G. 2007. Cadmium removal from aqueous solution by gene- modified Escherichia coli JM109. *J. Hazard. Mater*. 139(2): 340–344.

Deshmukh, R., Khardenavis, A.A. and Purohit, H.J. 2016. Diverse metabolic capacities of fungi for bioremediation. *Indian J. Microbiol*. 56(3): 247–264.

Diep, P., Mahadevan, R. and Yakunin, A.F. 2018. Heavy metal removal by bioaccumulation using genetically engineered microorganisms. *Front. Bioeng. Biotechnol*. 6: 429–456.

Dietmar, H.P. and Walter, R. 2000. Engineering bacteria for bioremediation. *Curr. Opin. Biotechnol*. 11(3): 262–270.

Dixit, R., Wasiullah, M.D., Pandiyan, K., Singh, U., Sahu, A., Shukla, R., Singh, B., Rai, J., Sharma, P., Lade, H. and Paul, D. 2015. Bioremediation of heavy metals from soil and aquatic environment: An overview of principles and criteria of fundamental processes. *Sustainability*. 7(2): 2189–2212.

Eltarahony, M., Zaki, S. and Abd-El-Haleem, D. 2020. Aerobic and anaerobic removal of lead and mercury via calcium carbonate precipitation mediated by statistically optimized nitrate reductases. *Sci. Rep*. 10(1): 4029–4060.

Emenike, C.U., Barasarathi, J., Pariatamby, A., Shahul, H. and Fauziah. 2018. Biotransformation and removal of heavy metals: A review of phyto and microbial remediation assessment on contaminated soil. *Environ. Rev*. 26(2): 156–158.

Fawzy, S., Osman, A.I., Doran, J. and Rooney, D.W. 2020. Strategies for mitigation of climate change: A review. *Environ. Chem. Lett*. 18(6): 2069–2094.

Fletcher, K.E., Boyanov, M.I., Thomas, S.H., Wu, Q., Kemner, K.M. and Löffler, F.E. 2010. U(VI) reduction to mononuclear U(IV) by desulfitobacterium species. *Environ. Sci. Technol*. 44(12): 4705–4709.

Freeman, J.L., Persans, M.W., Nieman, K. and Salt, D.E. 2005. Nickel and cobalt resistance engineered in Escherichia coli by overexpression of serine acetyltransferase from the nickel hyperaccumulator plant Thlaspi goesingense. *Appl. Environ. Microbiol*. 71(12): 8627–8633.

Fu, F. and Wang, Q. 2011. Removal of heavy metal ions from wastewaters: A review. *J. Environ. Manage*. 92(3): 407–425.

Gadd, G.M. 2001. Accumulation and transformation of metals by microorganisms. *Biotechnology Set*. 1: 225–264.

Gougoulias, C., Clark, J.M. and Shaw, L.J. 2014. The role of soil microbes in the global carbon cycle: Tracking the below-ground microbial processing of plant-derived carbon for manipulating carbon dynamics in agricultural systems. *J. Sci. Food Agric*. 94(12): 2362–2371.

Gupta, A., Meyer, J.M. and Goel, R. 2002. Development of heavy metal resistant mutants of phosphate solubilizing Pseudomonas sp. NBRI 4014 and their characterization. *Curr. Microbiol*. 45(5): 323–327.

Huang, K., Chen, C., Shen, Q., Rosen, B.P. and Zhao, F. J. 2015. Genetically engineering Bacillus subtilis with a heat-resistant Arsenite methyltransferase for bioremediation of arsenic-contaminated organic waste. *Appl. Environ. Microbiol*. 81(19): 6718–6724.

Hussain, C.M. 2018. Handbook of environmental materials management ‖ micro- remediation of metals: A new frontier in bioremediation. 10.1007/978–3–319- 58538–3(Chapter 10–1): 1–36.

Ike, A., Sriprang, R., Ono, H., Murooka, Y. and Yamashita, M. 2007. Bioremediation of cadmium contaminated soil using symbiosis between leguminous plant and recombinant rhizobia with the MTL4 and the PCS genes. *Chemosphere*. 66(9): 1670–1676.

Ike, A., Sriprang, R., Ono, H., Murooka, Y. and Yamashita, M. 2008. Promotion of metal accumulation in nodule of *Astragalus sinicus* by the expression of the iron-regulated transporter gene in *Mesorhizobium huakuii* subsp. rengei B3. *J. Biosci. Bioeng*. 105(6): 642–648.

Ivask, A., Dubourguier, H.C., Põllumaa, L. and Kahru, A. 2011. Bioavailability of Cd in 110 polluted top soils to recombinant bioluminescent sensor bacteria: Effect of soil particulate matter. *J. Soils Sediments*. 11(2): 231–237.

Jan, A., Azam, M., Siddiqui, K., Ali, A., Choi, I. and Haq, Q., 2015. Heavy metals and human health: Mechanistic insight into toxicity and counter defense system of antioxidants. *Int. J. Mol. Sci*. 16(12): 29592–29630.

Jin, Y., Luan, Y., Ning, Y. and Wang, L. 2018. Effects and mechanisms of microbial remediation of heavy metals in soil: A critical review. *Appl. Sci.* 8(8), 1336–1159.

Jowitt, S.M., Mudd, G.M. and Thompson, J.F.H. 2020. Future availability of non- renewable metal resources and the influence of environmental, social, and governance conflicts on metal production. *Commun. Earth Environ.* 1(1): 1–29.

Karigar, C.S. and Rao, S.S. 2011. Role of microbial enzymes in the bioremediation of pollutants: A review. *Enzyme Res.* 1–11.

Karnib, M., Kabbani, A., Holail, H. and Olama, Z. 2014. Heavy metals removal using activated carbon, silica and silica activated carbon composite. *Energy Procedia.* 50: 113–120.

Khan, A., Singh, P. and Srivastava, A. 2017. Synthesis, nature and utility of universal iron chelator—Siderophore: A review. *Microbiol. Res.* 212–213: 103–111.

Khan, M.S., Zaidi, A., Goel, R. and Musarrat, J. 2011. [Environmental Pollution] biomanagement of metal-contaminated soils volume 20 ‖ use of biosurfactants in the removal of heavy metal ions from soils. 10.1 007/978-94-007-1914-9(Chapter 8): 183–223.

Khulbe, K.C. and Matsuura, T. 2018. Removal of heavy metals and pollutants by membrane adsorption techniques. *Appl. Water Sci.* 8(1): 19–48.

Li, M., Cheng, X. and Guo, H. 2013. Heavy metal removal by biomineralization of urease producing bacteria isolated from soil. *Int. Biodeterior. Biodegr.* 76: 81–85.

Liu, S., Zhang, F., Chen, J. and Sun, G. 2011. Arsenic removal from contaminated soil via biovolatilization by genetically engineered bacteria under laboratory conditions. *Res. J. Environ. Sci.* 23(9): 1544–1550.

Lunch, A. 2012. [Experientia supplementum] molecular, clinical and environmental toxicology volume 101 ‖ heavy metal toxicity and the environment. 10.1007/978- 3-7643-8340-4(Chapter 6): 133–164.

Maarof, H.I., Daud, W., Mohd, A.W. and Aroua, M.K. 2016. Recent trends in removal and recovery of heavy metals from wastewater by electrochemical technologies. *Rev. Chem. Eng.* 23: 228–255.

Mark, R.B., Sanjay, K. and Frederick, W.O. 2000. Microbial resistance to metals in the environment. *Ecotoxicol. Environ. Saf.* 45(3): 0–207.

Marta, Č. 1995. Use of solvent extraction for the removal of heavy metals from liquid wastes. *Environ. Monit. Assess.* 34(2): 151–162.

Mello, V.D. and Hess, K.L. 2005. A conceptual and practical overview of cDNA microarray technology: Implications for basic and clinical sciences. *Braz. J. Med. Biol. Res.* 38(10): 1543–1552.

Morita, E.H., Kawamoto, S., Abe, S., Nishiyama, Y., Ikegami, T. and Hayashi, H. 2012. Comparative study of the different mechanisms for zinc ion stress sensing in two cyanobacterial strains, Synechococcus sp. PCC 7942 and Synechocystis sp. PCC 6803. *Biophys.* 8: 103–109.

Ng, S.P., Davis, B., Palombo, E.A. and Bhave, M. 2009. A Tn 5051-like mer-containing transposon identified in a heavy metal tolerant strain Achromobacter sp. AO22. *BMC Res. Notes* 2(1): 1–7.

O'Brien, J., Hayder, H., Zayed, Y. and Peng, C. 2018. Overview of MicroRNA biogenesis, mechanisms of actions, and circulation. *Front. Endocrinol.* 9: 402–424.

Olaniran, A., Balgobind, A. and Pillay, B. 2013. Bioavailability of heavy metals in soil: Impact on microbial biodegradation of organic compounds and possible improvement strategies. *Int. J. Mol. Sci.* 14(5): 10197–10228.

Oyewole, O.A., Zobeashia, S., Suanu, L.T., Oladoja, E.O., Raji, R.O., Odiniya, E.E. and Musa, A.M. 2019. Biosorption of heavy metal polluted soil using bacteria and fungi isolated from soil. *SN Appl. Sci.* 1(8): 857–870.

Pérez-Palacios, P., Romero-Aguilar, A., Delgadillo, J., Doukkali, B., Caviedes, M.A., Rodríguez-Llorente, I.D. and Pajuelo, E. 2017. Double genetically modified symbiotic system for improved Cu phytostabilization in legume roots. *Environ. Sci. Pollut. Res.* 24(17): 14910–14923.

Pohl, A. 2020. Removal of heavy metal ions from water and wastewaters by sulfur- containing precipitation agents. *Water Air Soil Pollut.* 231(10): 503–525.

Porter, S.S., Chang, P.L., Conow, C.A., Dunham, J.P. and Friesen, M.L. 2017. Association mapping reveals novel serpentine adaptation gene clusters in a population of symbiotic Mesorhizobium. *ISME J.* 11(1): 248–262.

Qasem, N.A.A., Mohammed, R.H. and Lawal, D.U. 2021. Removal of heavy metal ions from wastewater: A comprehensive and critical review. *NPJ Clean Water* 4(1): 450–472.

Rana, A., Sindhu, M., Kumar, A., Dhaka, R.K., Chahar, M., Singh, S. and Nain, L. 2021. Restoration of heavy metal-contaminated soil and water through biosorbents: A review of current understanding and future challenges. *Physiol. Plant.* 173(1): 398–417.

Rojas, L.A., Yáñez, C., González, M., Lobos, S., Smalla, K. and Seeger, M. 2011. Characterization of the metabolically modified heavy metal-resistant cupriavidus metallidurans strain MSR33 generated for mercury bioremediation. *PLoS One.* 6(3): e17555.

Sales da Silva, I., Gomes, A., FabÃola C., Padilha, R.S., Nath, Maria., Casazza, A.A., Converti, A. and Asfora S.L. 2020. Soil bioremediation: Overview of technologies and trends. *Energies.* 13(18): 4664–4685.

Samuel, M.S., Datta, S., Khandge, R.S. and Selvarajan, E. 2021. A state of the art review on characterization of heavy metal binding metallothioneins proteins and their widespread applications. *Sci. Total Environ.* 775: 145829.

Sarma, H. and Prasad, M.N.V. 2019. Metabolic engineering of rhizobacteria associated with plants for remediation of toxic metals and metalloids. In *Transgenic Plant Technology for Remediation of Toxic Metals and Metalloids.* Elsevier, pp. 299–318.

Sharma, B., Singh, S. and Siddiqi, N.J. 2014. Biomedical implications of heavy metals induced imbalances in redox systems. *Biomed Res. Int*: 1–26.

Sone, Y., Mochizuki, Y., Koizawa, K., Nakamura, R., Pan-Hou, H., Itoh, T. and Kiyono, M. 2013. Mercurial-resistance determinants in Pseudomonasstrain K-62 plasmid pMR68. *AMB Express.* 3(1): 41–65.

Stanislav, V., Tomas, R. and Pavel, K. 2007. Biosorption of Cd^{2+} and Zn^{2+} by cell surface- engineered Saccharomyces cerevisiae. *Int. Biodeterior. Biodegradation.* 60(2): 96–102.

Tang, X., Huang, Yi., Li, Ying., Wang, Li., Pei, X., Zhou, D., He, P. and Hughes, S.S. 2021. Study on detoxification and removal mechanisms of hexavalent chromium by microorganisms. *Ecotoxicol. Environ. Saf.* 208: 111699–111720.

Tangahu, B.V., Sheikh, A., Siti, R., Basri, H., Idris, M., Anuar, N. and Mukhlisin, M. 2011. A Review on heavy metals (As, Pb, and Hg) Uptake by plants through phytoremediation. *Int. J. Chem. Eng.* 1–31.

Tonini, G.A. and Ruotolo, L.A.M. 2017. Heavy metal removal from simulated wastewater using electrochemical technology: Optimization of copper electrodeposition in a membraneless fluidized bed electrode. *Clean. Technol. Environ. Policy.* 19(2): 403–415.

Tsezos, M. 2009. Metal—microbes interactions: Beyond environmental protection. *Adv. Mat. Res.* 71: 527–532.

Valdés, J., Pedroso, I., Quatrini, R., Dodson, R.J., Tettelin, H., Blake, R., Eisen, J.A. and Holmes, D.S. 2008. Acidithiobacillus ferrooxidans metabolism: From genome sequence to industrial applications. *BMC Genom.* 9(1): 597–608.

Valeria, T., Valeria, P., Ilaria, D., Antonella, A., Giuliano, F., Pietro, G., Antonella, M. and Giovanna, C.V. 2011. Cunninghamella elegans biomass optimisation for textile wastewater biosorption treatment: An analytical and ecotoxicological approach. *Appl Microbiol Biotechnol.* 90(1): 343–352.

Valls, M., Atrian, S., Lorenzo, V. and Fernández, L.A. 2000. Engineering a mouse metallothionein on the cell surface of Ralstonia eutropha CH34 for immobilization of heavy metals in soil. *Nat. Biotechnol.* 18(6): 661–666.

Vidu, R., Matei, E., Predescu, A.M., Alhalaili, B., Pantilimon, C., Tarcea, C. and Predescu, C. 2020. Removal of heavy metals from wastewaters: A challenge from current treatment methods to nanotechnology applications. *Toxics* 8(4): 101–129.

Weber, K.A., Achenbach, L.A. and Coates, J.D. 2006. Microorganisms pumping iron: Anaerobic microbial iron oxidation and reduction. *Nat Rev Microbiol.* 4(10): 752–764.

Wu, S., Xu, Y., Sun, J., Cao, Z., Zhou, J., Pan, Y. and Qian, G. 2015. Inhibiting evaporation of heavy metal by controlling its chemical speciation in MSWI fly ash. *Fuel.* 158: 764–769.

Wuana, R.A., Okieimen, F.E. 2011. Heavy metals in contaminated soils: A review of sources, chemistry, risks and best available strategies for remediation. *ISRN Ecol*: 1–20.

Yan, A., Wang, Y., Tan, S.N., Mohd, Y., Mohamed, L., Ghosh, S. and Chen, Z. 2020. Phytoremediation: A promising approach for revegetation of heavy metal-polluted land. *Front. Plant Sci.* 11: 359–378.

Yin, K., Wang, Q. and Chen, L. 2018. Microorganism remediation strategies towards heavy metals. *Chem. Eng. J.* S138589471832190.

Yuan, C., Lu, X., Qin, J., Rosen, B.P. and Le, X.C. 2008. Volatile arsenic species released from Escherichia coli expressing the AsIII S-adenosylmethionine methyltransferase gene. *Environ. Sci. Technol.* 42(9): 3201–3206.

Yun, G.L., Ming, Z., Guang, M.Z., X.W., Xin, L., Ting, F. and W.X. 2008. Bioleaching of heavy metals from mine tailings by indigenous sulfur-oxidizing bacteria: Effects of substrate concentration. *Bioresour. Technol.* 99(10): 4124–4129.

Zhang, G., Dong, H., Jiang, H., Kukkadapu, R.K., Kim, J., Eberl, D. and Xu, Z. 2009. Biomineralization associated with microbial reduction of Fe^{3+} and oxidation of Fe^{2+} in solid minerals. *Am. Mineral.* 94(7): 1049–1058.

Zhang, X., Rui, H., Zhang, F., Hu, Z., Xia, Y. and Shen, Z. 2018. Overexpression of a functional vicia sativa pcs1 homolog increases cadmium tolerance and phytochelatins synthesis in arabidopsis. *Front. Plant Sci.* 9: 107–129.

Zhao, X., Zhou, M.H., Li, Q.B., Lu, Y.H., He, N., Sun, D.H. and Deng, X. 2005. Simultaneous mercury bioaccumulation and cell propagation by genetically engineered Escherichia coli. *Process Biochem.* 40(5): 1611–1616.

13 Application of Genetically Modified Microorganisms in the Remediation of Industrial Waste

El Asri Ouahid and Inamuddin

INTRODUCTION

The industrial sector is the fundamental pillar for the socio-economic development of each nation. It is generally positioned after the agricultural sector, hence its name, the secondary industry. Currently, the industrial sector has considerably broadened its spectrum of action and diversified its application fields. We have witnessed in the last decades the increase of three critical parameters: the permanent demands of raw bioresources, natural energy resources, and the consumption of exhaustible non-renewable sources. So we are facing a global expansion of the industry sector (Altawell, 2021; Simandan, 2009).

We are facing increasing industrialization processes to satisfy the needs of the growing and urbanizing world population. The industrial sector has the most extensive multiplier action of all fields of the economy. Lenchuk (2016) states that one US dollar invested in the industrial area produces 1.5 US dollar growth in the gross domestic product. So, this industrialization growth will decrease natural resources, living space, and the accumulation of unwanted and toxic substances. Lebanon's industrial activities produce 346 730 tons of solid waste, 20 million m^3 of wastewater, and 15 000 tons of hazardous waste (El-Fadel et al., 2001). The annual world production of solid residues has exceeded 2 billion tons; this quantity can increase exponentially with the development of industrialization Kolekar et al., 2016). Some researchers have described that the yearly industrial waste generated worldwide is roughly 9.2 billion tons; 1.74 tons of industrial waste per capita (Vignesh et al. 2021). So, we are faced with a considerable amount of industrial waste generated every day.

Physicochemical and microbiological characteristics of industrial waste vary depending on the industry origin, the raw materials used, and its manufacturing processes. Generally, we see an enormous spectrum of industrial wastes (putrescible residue, carton, sludge, etc.) (Arockiam JeyaSundar et al., 2020). High toxicity, hazardous, and non-biodegradable are distinct industrial wastes compared to other wastes (Ameta et al., 2018). Some researchers have qualified this waste into two categories: the first type is called non-hazardous industrial waste, such as organic waste, cardboard, steel and tin cans, glass, and pebbles; this waste does not present a menace to public health or natural ecosystems (Millati et al., 2019). The second type is unsafe manufacturing residues like flammable, corrosive, active, and poisonous materials; this waste harms citizens' health and natural ecosystems. So, the rising industrial waste amounts and its diversity of compositions are associated with the expansion of industrialization. This result will inevitably amplify the appearance sooner or later of different types of pollution and diseases.

There exist various treatment techniques for industrial waste such as anaerobic digestion, composting, landfill, and incineration (El Asri et al., 2021). These treatments, except incineration, require the intervention of microorganisms to ensure the conversion of the organic fraction of industrial

DOI: 10.1201/9781003188568-13

waste into products for several uses. In the beginning, the researchers pursued the isolation and purification of microbial strains to ensure an effective conversion. However, these wild strains have proven insufficient for a significant recovery of these wastes. The appearance of technological progress and the development of genetic engineering allows modification of the genetic composition of these strains to make them very efficient in biological conversion. This natural treatment based on genetically modified microorganisms (GMMs) has become the new sustainable industrial waste management vision.

This chapter is an excellent opportunity for studying the role of genetically modified microorganisms in treatment technologies of industrial waste. Therefore, we have divided this chapter into big axes. The first axis answers two crucial questions: what are GMMs, and why are they chosen in the treatment instead of wild strains? In the second axis, we will address specific industries that use GMMs to treat their waste, such as management, treatment, and recovery.

GENETICALLY MODIFIED MICROORGANISMS: PRINCIPLE, TYPES, AND BENEFITS

WHAT ARE GENETICALLY MODIFIED MICROORGANISMS?

In present times, microorganisms such as fungi, bacteria, and yeast have undergone modifications in their genetic makeup by recombinant DNA technology. These microbes are named genetically modified or engineered microorganisms (GMM). These GMM products, absent in natural ecosystems, can generate new or enhanced functionalities. These engineered microorganisms are characterized by various properties such as being robust and safe microbes.

Several methods can be used for GMM generation. (i) Non-targeted mutagenesis: the genetic modification of microorganisms is made by applying selection pressure and random mutagenesis. (ii) Genetic engineering: *in vitro*, the genes (one or more DNA sequences) are inserted or deleted into the selected microorganisms' genome, which gives enhanced or new functionality from one species to another. (iii) Genome editing: a recent biotechnology that cuts and pastes nucleic acid sequences by specialized proteins called engineered endonucleases such as zinc finger nucleases, transcription activator-like endonucleases, and clustered regularly interspaced palindromic repeats (Gaj et al., 2013; Lee et al., 2018). GMMs have been modified by humans on a small scale (laboratory) by these different recombinant DNA technologies. Thes genetic modification produces microbes to improve the biodegradation of industrial wastes (hazardous and non-hazardous) in lab conditions. The processes may then be scaled up for industrial applications.

WHY DO WE USE GENETICALLY MODIFIED MICROORGANISMS?

We have several traditional ways to treat and remediate environmental pollution due to industrial activity. But unfortunately, these ways are based on physical and chemical techniques, which eventually would create more menaces to the ecosystems of the planet. The current trend is towards microbiological treatment because it is a natural process.

The microbial treatment uses diverse microbes like bacteria, algae, yeast, and fungi to transform hazardous and toxic compounds of industrial waste into simple carbon dioxide and water. The wild microbial strains are characterized by slow degradation, less efficiency, and incomplete conversion of industrial waste. In addition, in several cases, the naturally occurring microbes cannot degrade some hazardous compounds, particularly xenobiotic compounds. Drzymała and Kalka (2020) have demonstrated that the mixture of two pharmaceuticals residues (diclofenac and sulfamethoxazole) presents high toxicity for *Aliivibrio fischeri* bacteria. So, the natural microbial strains have high toxicity to some industrial waste; this toxicity increased with some waste mixtures.

Genetically modified microorganisms can overcome these problems; the GMMs can enhance the xenobiotic degradation process and rapid conversion. The GMM communities also have adequate and edited catabolic pathways for high treatment compared to wild microbes. GMMs have a remarkable ability to change the environment and adapt to extreme conditions. To highlight the vital role of these strains in the treatment of industrial waste, we will discuss in the following axis how GMMs ensure rapid conversion and efficient treatment in several industrial fields.

GMMs Improve the Quantity and Quality of Biodegradable Enzymes

Currently, there is an emphasis on the production of enzymes. An enzyme is a protein sequence of amino acids characterized by highly efficient biocatalysts. The global market of enzymes produced by industrial units approached USD 5,01 billion in 2016, increasing to USD 6.32 billion in 2022 (Chapman et al., 2018). Enzymes have long been extracted from animals, plants, and microbial sources. But in recent decades, microbes are the primary enzyme-producing sources; they have reached 80% of the production of the total enzyme market (Verma et al., 2019). Due to the drawbacks of using wild microbes, as described previously, GMMs have become a new production source.

The optimization of enzyme secretion and the limitation of the output of unwanted secondary metabolites are two big reasons for the production of GMMs. Genetic modification techniques can combine the output of desired enzymes with the appropriate production amounts. Finally, we will be discussing the employment of GMMs in the treatment of waste from the most promising industries.

INDUSTRIES USING GMMS IN WASTE TREATMENT

Keratinous Wastes of Poultry Industry

The poultry industry generates vast amounts of keratinous wastes, principally from feathers. These wastes are estimated to amount to more than 8 million tons annually (Yang et al., 2016). The feathers represent about 8–10% of chicken weight; this organic matter is distinctive due to its recalcitrant character, making it foremost among global waste pollutants (Brandelli et al., 2015; Grazziotin et al., 2006). The feathers are made principally of keratin matter; it represents over 90% of the chemical composition (Li, 2019), it is a recalcitrant protein that is resistant to biodegradation, and it is insoluble in organic and aqueous solutions (Brandelli et al., 2015; Li, 2019). Keratinolytic and proteolytic enzymes-keratinases produced by GMMs can biodegrade this unmanageable waste.

We found several studies showing that the genes of production of keratinases are present in different microbes such as *Stenotrophomonas maltophilia*, *Bacillus subtilis*, *Bacillus licheniformis*, *Chryseobacterium* sp., *Streptomyces fradiae*, and *Serratia* sp. (Cao et al., 2009; Grazziotin et al., 2006; Khardenavis et al., 2009; Li et al., 2007; Riffel et al., 2007). Therefore, the genes of keratinase production are isolated, purified, transferred, and inserted into the genome of other microbes like *Bacillus subtilis*, *Bacillus amyloliquefaciens*, *Bacillus licheniformis*, and *E. coli* (Figure 13.1).

We next discuss the non-pathogenic microbial potential for the safe, healthy, and environmental-friendly valorization of feather wastes. Yong et al. (2020) have used the recombinant plasmid called pSUGV4-KER71 to transfer keratinase genes (BsKER71) into *Bacillus subtilis* WB600 to produce *Bacillus subtilis* S1-4. This last GMM produces keratinase that exhibited thermal adaptation and more decisive enzymatic action at alkaline pH. Yang et al. (2016) have successfully purified the keratinase-encoding gene called kerK of *B. subtilis* and inserted them into *Bacillus amyloliquefaciens* K11 using recombinant plasmids. *Bacillus amyloliquefaciens* K11 acquired an excellent characteristic—it can degrade chicken feathers in a short time (12 hours maximum) entirely. Fang et al. (2017) have isolated the keratinase genes from *S. maltophilia* BBE 11-1 and transformed them into *E. coli* BL21 by the pET22b-FDD and pET22b-DDF plasmids. This transformation process can triple the extracellular and thermophilic properties of keratinase activity. Finally, these studies proved the GMMs to be a powerful and quick method for feather-degrading and recommended it for applications in waste feather disposal (Figure 13.1). So, choosing the gene of interest (the keratinase enzyme) and the host-microbe strain is necessary, referring to its conversion potential, quick growth, condition adaptation, and nutrition preferences to ensure good treatment.

THE WASTE OF THE PACKAGING AND PAPER INDUSTRY

Packaging and paper manufacturing is considered one of the most dangerous industrial sectors on the planet. It is the fifth-largest energy consumer and the third chief industrial wastewater producer globally (Gopal et al., 2019). It releases more than 100 thousand tons of toxic pollutants per year (Cheremisinoff and Rosenfeld, 2010). The laborers of this industry present an elevated risk of lung and bladder cancer (Gopal et al., 2019). It can produce as high as 60 m³ of wastewater per ton of paper produced (Ince et al., 2011). The paper and pulp industry uses several enzymes such as cellulase, xylanase, laccase, and lipase. In 2011, the xylanase enzyme alone had bleached 10 million tonnes of paper (Demuner et al., 2011). So, this industry produces a significant amount of chemical components in wastewater, solid waste, and air emissions, which leads to the destruction of ecosystems and human life.

Cellulase and lipase are mainly used to stimulate the modification of fibers, while xylanase and laccase have extensive use in delignification and bleaching technologies (Demuner et al., 2011). Xylanase is an essential application in mills worldwide of paper and pulp bleaching technology. It decreases bleach chemical agents like chlorine dioxide and hydrogen peroxide, and produces good brightness. So, reducing the chemical process and using enzymes of microbial origin can decrease the cost and discharge of chlorinated organic compounds in industry effluents (Bajoub et al., 2016). Thus, different techniques of genetic engineering (encoding, recombinant DNA, mutagenesis) have intervened in this industry to offer it recombinant enzymes.

We found several fungi species such as *Thermomyces lanuginosus*, *Aspergillus niger*, and *Aureobasidium pullulans* have xylanase genes (xynA). This gene encoded the xylanases with 221 to

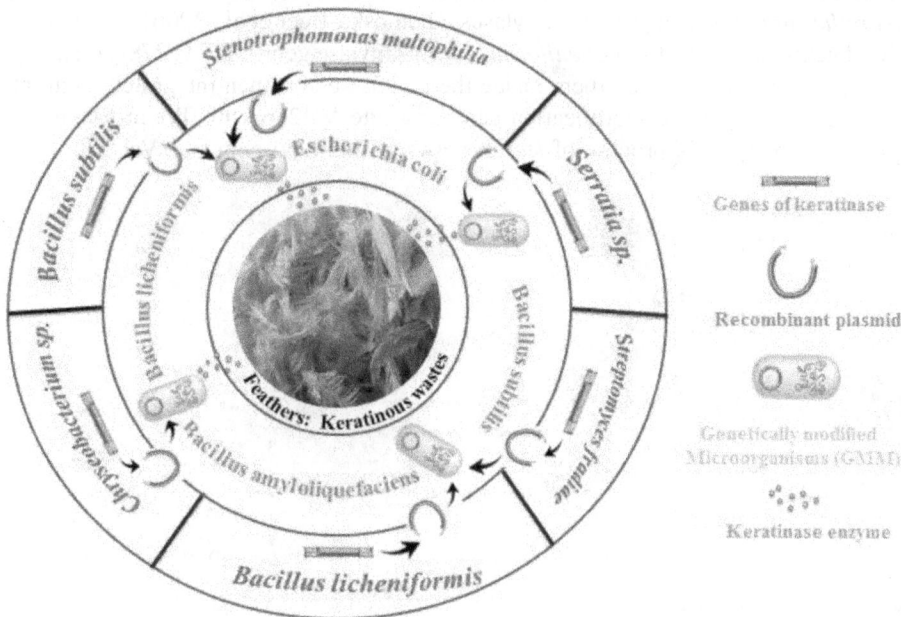

FIGURE 13.1 Application of some microbes for the production of GMMs used in the recovery of keratinous wastes.

225 amino acids (Krisana et al., 2005; Schlacher et al., 1996). *Penicillium purpurogenum* has a gene (xynB) that encodes 208 amino acids, *Cochliobolus sativus* has a gene (xyl2), and *Aspergillus usamii* has a gene (xynII) that encodes 184 amino acids (Díaz et al., 1997; Zhou et al., 2008). *E. coli*, *Saccharomyces cerevisiae*, *Aspergillus oryzae*, *Trichoderma reseei*, and *Pichia pastoris* are cloning hosts for these fungal genes (Ahmed et al., 2009). The recombinant xylanase produced by different types of GMMs has superior properties: it has high secretion levels, and it is an alkali-stable and thermostable enzyme with high expression under its promoters.

Other work has studied bacteria as sources of enzymes of interest. Rathod et al. (2017) transferred the azoA gene from *Enterococcus sp.*, which encodes the enzyme azoreductase in *E. coli* and *Pseudomonas fluorescens* using the expression vector PBBRMCS2. This transformation has led to the production of a GMM capable of degrading the recalcitrant azo dye in wastewater from the paper industry. Saleem et al. (2014) have used *Bacillus cereus* to decolorize pulp and paper industrial effluents. Hence, some researchers, such as Vélez-Lee et al. (2016) have recommended the genetic modification of this strain by using pUB110 plasmid with the vgb gene in *Vitrocilla stercoraria*. This genetic modification increases the microbial conversion of phenol and p-nitrophenol in wastewater from the paper industry compared to wild-type strains (Vélez-Lee et al., 2016). Chauhan and Thakur (2002) have confirmed that *Pseudomonas fluorescens* reduces color, phenol, and lignin amounts in pulp and paper factory effluent. Pant et al., 2021 have recommended exploring *Pseudomonas fluorescens* genetically modified for pollution reduction in industrial discharge. These recently described microbes can produce genetically modified organisms to treat packaging and paper manufacturing wastewater. Therefore, these GMMs comprise an eco-friendly and sustainable technology (Figure 13.2).

Bakery Industry Waste

The most abundant bakery industry waste is the residue of bread, cakes, and pastries. Starch is a significant component of these wastes. To ensure good starch hydrolysis, we recommend using an α-amylase enzyme that increases the release of glucose, maltose, and various oligosaccharides from the starch structure. Some *Bacillus* strains such as *B. licheniformis*, *B. amyloliquefaciens*, and *B. stearothermophilus* can produce amylases (Olempska-Beer et al., 2006). Rivera et al. (2003) have released genomic DNA of *B. licheniformis*, the α-amylase gene (ATCC-27811). They produced mutations at position 286, inserted them inside the pET3a vector, then integrated them into the *E. coli* strain BL21. This genetic modification can substitute Val286 with Tyr in α-amylase. When we compared the hydrolysis process of starch between this new enzyme (Val 286 Tyr) and the

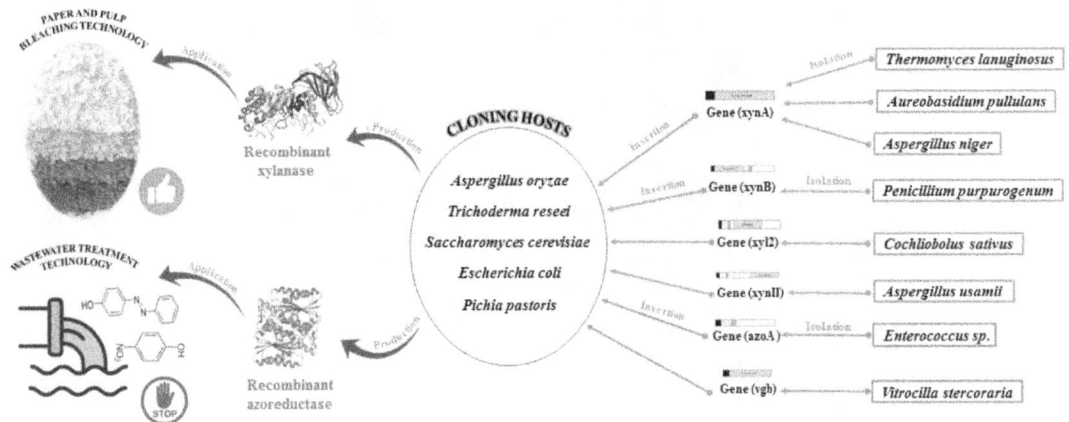

FIGURE 13.2   Microbes used to produce GMMs in the packaging and paper industry.

wild-type, we see that this mutant enzyme is efficient because it increases hydrolysis by five times its normal rate. So, the amylase gene was over-expressed in this GMM.

The high costs, consumption of chemicals, and energy requirements are three significant parameters to the treatment of the bakery industry waste. Therefore, the resolution of these problems needs increased capital investment. Using recombined enzymes produced by GMMs is a good investment because it has many sustainability benefits: it can reduce operating costs, land usage, energy requirements, and waste generation, as well as simplify production routes.

AGRO-FOOD INDUSTRY RESIDUES

The global production of sugarcane is about 1.6 billion tons, and this agro-industry produces 279 million tons of organic waste (Chandel et al., 2012). During two months of olive oil production in the Mediterranean area, one ton of olive fruits produces more than one ton of wastewater and 800 kg of solid waste (Medouni-Haroune et al., 2018; Paraskeva and Diamadopoulos, 2006). The yearly production of olive pomace in Spain is between 2 and 4 million tons, but Turkey fluctuates between 100 and 120 thousand tons (Borja et al., 2005; Tekin and Dalgıç, 2000). Kim and Dale (2004) estimated that the residue of corn, rice, sugarcane, and wheat is more than 70 million tons globally. In addition, Hadar (2013) has described the agro-industrial waste generated by cereal and tomatoes from three European countries as abundant lignocellulosic wastes (Hadar, 2013). These various waste types have the most carbon content, principally in lignocellulosic form. They contain 35 to 50% cellulose, 20 to 35% hemicellulose, and 10 to 25% lignin (Arora et al., 2020). So, we have an enormous amount of lignocellulosic waste produced daily by the agro-industry field.

The conversion of agro-industry residues requires a large quantity and numerous glycoside hydrolase enzymes such as endoglucanases, exoglucanases, and β-glucosidases. It is necessary to use these enzymes to convert lignocellulose into fermentable sugars, then ethanol. Faraco and Hadar (2011) have estimated the conversion of lignocellulosic waste in the Mediterranean basin to 13 million tons of oil equivalent of ethanol. Kim and Dale (2004) calculated that the global conversion of lignocellulosic wastes generates 491 GL year^{-1} bioethanol. This portion of ethanol could substitute more than 30% of the worldwide gasoline utilization and 3.6% of worldwide electricity production. So, genetic modification has become a key point for improving agro-industrial waste conversion into bioethanol (Table 13.1).

Several researchers have genetically modified and improved fungi to convert lignocellulosic residues from agricultural industries into bioethanol. Den Haan et al. (2007) encoded two gene types. Firstly, the cellulase from *Trichoderma reesei* is an endoglucanase type, and another from *Saccharomycopsis fibuligera* is a beta-glucosidase type. They have inserted these genes on the episomal plasmid YEp352 into *Saccharomyces cerevisiae*. This GMM can transform the lignocellulosic residues into ethanol via one step. Kotaka et al. (2008) successfully inserted three genes of *Aspergillus oryzae* (beta-glucosidase and two endoglucanases genes) into *Saccharomyces cerevisiae*. The incorporation of the endo-β-1,4-xylanase 2 genes of *Aspergillus nidulan* into *Fusarium oxysporum* through *Agrobacterium tumefaciens* can convert more than 60% of corn cob to ethanol (Anasontzis et al., 2011). This transforming microbe presents the highest activity of conversion of cellobiose to ethanol.

Some researchers have modified the metabolic pathways of *Clostridium thermocellumin* for redirecting the carbon flux (acetate and lactate) to ethanol. Biswas et al. (2014) deleted two genes (hypoxanthine phosphoribosyltransferase and lactate dehydrogenase) of *C. thermocellumin*. This modification can increase ethanol production by 30% compared to the wild strain. We can delete the genes controlling some enzymes like formate lyase, malic enzyme, and hydrogenase maturase (Deng et al., 2013; Rydzak et al., 2015). Some researchers have used the engineering of *Geobacillus thermoglucosidasius* for the conversion of palm residues (palm kernel cake) to ethanol (90% yield) (Raita et al., 2016). Zhou et al. (2016) have increased the performance of this bacterium by suppressing the expression of lactate and formate genes. Liu et al. (2018) succeeded in producing a catalog

TABLE 13.1

GMMs Used for the Production of Bioethanol from Agro-food Residues

Microbes	Genes	Vector	New character	References
Trichoderma reesei	Endoglucanase gene	Plasmid YEp352	Conversion of lignocellulosic residues to ethanol	(Den Haan et al., 2007)
Saccharomycopsis fibuligera	bglxA gene	Plasmid YEp352		
Aspergillus oryzae	endo-β-1,4-xylanase 2 genes	*Agrobacterium tumefaciens*	Conversion corn cob to ethanol	(Kotaka et al., 2008)
Clostridium thermocellumin	HPRT and LDH genes	–	Production of ethanol	(Biswas et al., 2014)
Geobacillus thermoglucosidasius	Lactate and formate genes	–	Conversion of palm residues to ethanol	(Raita et al., 2016)
Thermus thermophilus	xylA gene	Plasmid pUC19-XI	Ethanolic production by xylose isomerase	(Walfridsson et al., 1996)
Caldicellulosiruptor bescii	NADH-dependent AdhE	Plasmid pDCW180 Plasmids pDCW183	Conversion of switchgrass to acetate, lactate, and hydrogen	(Chung et al., 2015)
Fusarium oxysporum	endo-b-1,4-xylanase 2 gene	*Agrobacterium tumefaciens*	Conversion of corn cob and wheat bran to ethanol	(Anasontzis et al., 2011)

of several genetically modified microbes that allows cellulose conversion to bioethanol. So, all these studies confirmed that the use of GMMs can improve the production of ethanol more than natural species.

CONCLUSION

We have shown in this work the importance of using and adapting genetically modified microorganisms in the management and recovery of poultry, packaging and pulp, bakery residue, and agro-food industries. These modified microbes present powerful and quick degradation of various industrial wastes. We have also described the experimental protocols to create them. The integration of GMMs in the industrialization process can provide several advantages: (a) reduce treatment costs and land usage, (b) limit contamination of citizens' health and pollution processes, (c) improve renewable energy resources, and (d) simplify production routes. In addition, it can increase the industry's revenues by the direct sale of its byproducts for reuse. So, it is a good investment because it has many socioeconomic and sustainability benefits.

REFERENCES

Ahmed, S., Riaz, S., Jamil, A., 2009. Molecular cloning of fungal xylanases: An overview. Appl. Microbiol. Biotechnol. 84, 19–35. https://doi.org/10.1007/s00253-009-2079-4.

Altawell, N. (Ed.), 2021. 12—Energy technologies and energy storage systems for sustainable development. In: Rural Electrification. Academic Press, pp. 231–248. https://doi.org/10.1016/B978-0-12-822403-8.00012-6.

Ameta, R., Solanki, M.S., Benjamin, S., Ameta, S.C., 2018. Photocatalysis. In: Advanced Oxidation Processes for Waste Water Treatment. Elsevier, pp. 135–175. https://doi.org/10.1016/B978-0-12-810499-6.00006-1.

Anasontzis, G.E., Zerva, A., Stathopoulou, P.M., Haralampidis, K., Diallinas, G., Karagouni, A.D., Hatzinikolaou, D.G., 2011. Homologous overexpression of xylanase in Fusarium oxysporum increases ethanol productivity during consolidated bioprocessing (CBP) of lignocellulosics. J. Biotechnol. 152, 16–23. https://doi.org/10.1016/j.jbiotec.2011.01.002.

Arockiam JeyaSundar, P.G.S., Ali, A., Guo, di, Zhang, Z., 2020. Waste treatment approaches for environmental sustainability. In: Microorganisms for Sustainable Environment and Health. Elsevier, pp. 119–135. https://doi.org/10.1016/B978-0-12-819001-2.00006-1.

Arora, A., Nandal, P., Singh, J., Verma, M.L., 2020. Nanobiotechnological advancements in lignocellulosic biomass pretreatment. Mater. Sci. Energy Technol. 3, 308–318. https://doi.org/10.1016/j.mset.2019.12.003.

Bajoub, A., Ajal, E.A., Fernández-Gutiérrez, A., Carrasco-Pancorbo, A., 2016. Evaluating the potential of phenolic profiles as discriminant features among extra virgin olive oils from Moroccan controlled designations of origin. Food Res. Int. 84, 41–51. https://doi.org/10.1016/j.foodres.2016.03.010.

Biswas, R., Prabhu, S., Lynd, L.R., Guss, A.M., 2014. Increase in ethanol yield via elimination of lactate production in an ethanol-tolerant mutant of clostridium thermocellum. PLoS One 9, e86389. https://doi.org/10.1371/journal.pone.0086389.

Borja, R., Martín, A., Sánchez, E., Rincón, B., Raposo, F., 2005. Kinetic modelling of the hydrolysis, acidogenic and methanogenic steps in the anaerobic digestion of two-phase olive pomace (TPOP). Process Biochem. 40, 1841–1847. https://doi.org/10.1016/j.procbio.2004.06.026.

Brandelli, A., Sala, L., Kalil, S.J., 2015. Microbial enzymes for bioconversion of poultry waste into added-value products. Food Res. Int. 73, 3–12. https://doi.org/10.1016/j.foodres.2015.01.015.

Cao, Z.-J., Zhang, Q., Wei, D.-K., Chen, L., Wang, J., Zhang, X.-Q., Zhou, M.-H., 2009. Characterization of a novel Stenotrophomonas isolate with high keratinase activity and purification of the enzyme. J. Ind. Microbiol. Biotechnol. 36, 181–188. https://doi.org/10.1007/s10295-008-0469-8.

Chandel, A.K., da Silva, S.S., Carvalho, W., Singh, O.V., 2012. Sugarcane bagasse and leaves: Foreseeable biomass of biofuel and bio-products. J. Chem. Technol. Biotechnol. 87, 11–20. https://doi.org/10.1002/jctb.2742.

Chapman, J., Ismail, A., Dinu, C., 2018. Industrial applications of enzymes: Recent advances, techniques, and outlooks. Catalysts 8, 238. https://doi.org/10.3390/catal8060238.

Chauhan, N., Thakur, I., 2002. Treatment of pulp and paper mill effluent by Pseudomonas fluorescens in fixed film bioreactor. Pollut. Res. 21, 429–434.

Chung, D., Cha, M., Snyder, E.N., Elkins, J.G., Guss, A.M., Westpheling, J., 2015. Cellulosic ethanol production via consolidated bioprocessing at 75 °C by engineered Caldicellulosiruptor bescii. Biotechnol. Biofuels 8, 163. https://doi.org/10.1186/s13068-015-0346-4.

Demuner, B.J., Pereira Junior, N., Antunes, A.M.S., 2011. Technology prospecting on enzymes for the pulp and paper industry. J. Technol. Manag. Innov. 6, 148–158. https://doi.org/10.4067/S0718-27242011000300011.

Den Haan, R., Rose, S.H., Lynd, L.R., van Zyl, W.H., 2007. Hydrolysis and fermentation of amorphous cellulose by recombinant Saccharomyces cerevisiae. Metab. Eng. 9, 87–94. https://doi.org/10.1016/j.ymben.2006.08.005.

Deng, Y., Olson, D.G., Zhou, J., Herring, C.D., Joe Shaw, A., Lynd, L.R., 2013. Redirecting carbon flux through exogenous pyruvate kinase to achieve high ethanol yields in Clostridium thermocellum. Metab. Eng. 15, 151–158. https://doi.org/10.1016/j.ymben.2012.11.006.

Díaz, R., Sapag, A., Peirano, A., Steiner, J., Eyzaguirre, J., 1997. Cloning, sequencing and expression of the cDNA of endoxylanase B from Penicillium purpurogenum. Gene 187, 247–251. https://doi.org/10.1016/S0378-1119(96)00762-7.

Drzymała, J., Kalka, J., 2020. Ecotoxic interactions between pharmaceuticals in mixtures: Diclofenac and sulfamethoxazole. Chemosphere 259, 127407. https://doi.org/10.1016/j.chemosphere.2020.127407.

El Asri, O., Fadlaoui, S., Ramdani, M., Errochdi, S., 2021. Microbial degradation of biowaste for hydrogen production In: Inamuddin, Ahamed M.I., Prasad, R. (Eds.), Recent Advances in Microbial Degradation, Environmental and Microbial Biotechnology. Springer Singapore, Singapore, pp. 431–447. https://doi.org/10.1007/978-981-16-0518-5_17.

El-Fadel, M., Zeinati, M., El-Jisr, K., Jamali, D., 2001. Industrial-waste management in developing countries: The case of lebanon. J. Environ. Manage. 61, 281–300. https://doi.org/10.1006/jema.2000.0413.

Fang, Z., Zhang, J., Du, G., Chen, J., 2017. Rational protein engineering approaches to further improve the keratinolytic activity and thermostability of engineered keratinase KerSMD. Biochem. Eng. J. 127, 147–153. https://doi.org/10.1016/j.bej.2017.08.010.

Faraco, V., Hadar, Y., 2011. The potential of lignocellulosic ethanol production in the mediterranean basin. Renew. Sustain. Energy Rev. 15, 252–266. https://doi.org/10.1016/j.rser.2010.09.050.

Gaj, T., Gersbach, C.A., Barbas, C.F., 2013. ZFN, TALEN, and CRISPR/Cas-based methods for genome engineering. Trends Biotechnol. 31, 397–405. https://doi.org/10.1016/j.tibtech.2013.04.004.

Gopal, P.M., Sivaram, N.M., Barik, D., 2019. Paper industry wastes and energy generation from wastes. In: Energy from Toxic Organic Waste for Heat and Power Generation. Elsevier, pp. 83–97. https://doi.org/10.1016/B978-0-08-102528-4.00007-9.

Grazziotin, A., Pimentel, F.A., de Jong, E.V., Brandelli, A., 2006. Nutritional improvement of feather protein by treatment with microbial keratinase. Anim. Feed Sci. Technol. 126, 135–144. https://doi.org/10.1016/j.anifeedsci.2005.06.002.

Hadar, Y., 2013. Sources for lignocellulosic raw materials for the production of Ethanol. In: Faraco, V. (Ed.), Lignocellulose Conversion. Springer, Berlin, Heidelberg, pp. 21–38. https://doi.org/10.1007/978-3-642-37861-4_2.

Ince, B. K., Cetecioglu, Z., Ince, O., 2011. Pollution prevention in the pulp and paper industries. In: Broniewicz, E. (Ed.), Environmental Management in Practice. InTech. https://doi.org/10.5772/23709.

Khardenavis, A.A., Kapley, A., Purohit, H.J., 2009. Processing of poultry feathers by alkaline keratin hydrolyzing enzyme from Serratia sp. HPC 1383. Waste Manag. 29, 1409–1415. https://doi.org/10.1016/j.wasman.2008.10.009.

Kim, S., Dale, B.E., 2004. Global potential bioethanol production from wasted crops and crop residues. Biomass Bioenergy 26, 361–375. https://doi.org/10.1016/j.biombioe.2003.08.002.

Kolekar, K.A., Hazra, T., Chakrabarty, S.N., 2016. A review on prediction of municipal solid waste generation models. Procedia Environ. Sci. 35, 238–244. https://doi.org/10.1016/j.proenv.2016.07.087.

Kotaka, A., Bando, H., Kaya, M., Kato-Murai, M., Kuroda, K., Sahara, H., Hata, Y., Kondo, A., Ueda, M., 2008. Direct ethanol production from barley β-glucan by sake yeast displaying Aspergillus oryzae β-glucosidase and endoglucanase. J. Biosci. Bioeng. 105, 622–627. https://doi.org/10.1263/jbb.105.622.

Krisana, A., Rutchadaporng, S., Jarupan, G., Lily, E., Sutipa, T., Kanyawim, K., 2005. Endo-1,4-β-xylanase B from Aspergillus cf. niger BCC14405 Isolated in Thailand: Purification, characterization and gene isolation. BMB Rep. 38, 17–23. https://doi.org/10.5483/BMBRep.2005.38.1.017.

Lee, J.G., Sung, Y.H., Baek, I.-J., 2018. Generation of genetically-engineered animals using engineered endonucleases. Arch. Pharm. Res. 41, 885–897. https://doi.org/10.1007/s12272-018-1037-z.

Lenchuk, E.B., 2016. Course on new industrialization: A global trend of economic development. Stud. Russ. Econ. Dev. 27, 332–340. https://doi.org/10.1134/S1075700716030102.

Li, J., Shi, P.-J., Han, X.-Y., Meng, K., Yang, P.-L., Wang, Y.-R., Luo, H.-Y., Wu, N.-F., Yao, B., Fan, Y.-L., 2007. Functional expression of the keratinolytic serine protease gene sfp2 from Streptomyces fradiae var. k11 in Pichia pastoris. Protein Expr. Purif. 54, 79–86. https://doi.org/10.1016/j.pep.2007.02.012.

Li, Q., 2019. Progress in microbial degradation of feather waste. Front. Microbiol. 10, 2717. https://doi.org/10.3389/fmicb.2019.02717.

Liu, H., Sun, J., Chang, J.-S., Shukla, P., 2018. Engineering microbes for direct fermentation of cellulose to bioethanol. Crit. Rev. Biotechnol. 38, 1089–1105. https://doi.org/10.1080/07388551.2018.1452891.

Medouni-Haroune, L., Zaidi, F., Medouni-Adrar, S., Kecha, M., 2018. Olive pomace: From an olive mill waste to a resource, an overview of the new treatments. J. Crit. Rev. 1–6. https://doi.org/10.22159/jcr.2018v5i5.28840.

Millati, R., Cahyono, R.B., Ariyanto, T., Azzahrani, I.N., Putri, R.U., Taherzadeh, M.J., 2019. Agricultural, industrial, municipal, and forest wastes. In: Sustainable Resource Recovery and Zero Waste Approaches. Elsevier, pp. 1–22. https://doi.org/10.1016/B978-0-444-64200-4.00001-3.

Olempska-Beer, Z.S., Merker, R.I., Ditto, M.D., DiNovi, M.J., 2006. Food-processing enzymes from recombinant microorganisms—a review. Regul. Toxicol. Pharmacol. 45, 144–158. https://doi.org/10.1016/j.yrtph.2006.05.001.

Pant, G., Garlapati, D., Agrawal, U., Prasuna, R.G., Mathimani, T., Pugazhendhi, A., 2021. Biological approaches practised using genetically engineered microbes for a sustainable environment: A review. J. Hazard. Mater. 405, 124631. https://doi.org/10.1016/j.jhazmat.2020.124631.

Paraskeva, P., Diamadopoulos, E., 2006. Technologies for olive mill wastewater (OMW) treatment: A review. J. Chem. Technol. Biotechnol. 81, 1475–1485. https://doi.org/10.1002/jctb.1553.

Raita, M., Ibenegbu, C., Champreda, V., Leak, D.J., 2016. Production of ethanol by thermophilic oligosaccharide utilising Geobacillus thermoglucosidasius TM242 using palm kernel cake as a renewable feedstock. Biomass Bioenergy 95, 45–54. https://doi.org/10.1016/j.biombioe.2016.08.015.

Rathod, J., Dhebar, S., Archana, G., 2017. Efficient approach to enhance whole cell azo dye decolorization by heterologous overexpression of Enterococcus sp. L2 azoreductase (azoA) and Mycobacterium vaccae formate dehydrogenase (fdh) in different bacterial systems. Int. Biodeterior. Biodegrad. 124, 91–100. https://doi.org/10.1016/j.ibiod.2017.04.023.

Riffel, A., Brandelli, A., Bellato, C. de M., Souza, G.H.M.F., Eberlin, M.N., Tavares, F.C.A., 2007. Purification and characterization of a keratinolytic metalloprotease from Chryseobacterium sp. kr6. J. Biotechnol. 128, 693–703. https://doi.org/10.1016/j.jbiotec.2006.11.007.

Rivera, M.H., López-Munguía, A., Soberón, X., Saab-Rincón, G., 2003. α-Amylase from Bacillus licheniformis mutants near to the catalytic site: Effects on hydrolytic and transglycosylation activity. Protein Eng. Des. Sel. 16, 505–514. https://doi.org/10.1093/protein/gzg060.

Rydzak, T., Lynd, L.R., Guss, A.M., 2015. Elimination of formate production in Clostridium thermocellum. J. Ind. Microbiol. Biotechnol. 42, 1263–1272. https://doi.org/10.1007/s10295-015-1644-3.

Saleem, M., Ahmad, S., Ahmad, M., 2014. Potential of Bacillus cereus for bioremediation of pulp and paper industrial waste. Ann. Microbiol. 64, 823–829. https://doi.org/10.1007/s13213-013-0721-y.

Schlacher, A., Holzmann, K., Hayn, M., Steiner, W., Schwab, H., 1996. Cloning and characterization of the gene for the thermostable xylanase XynA from Thermomyces lanuginosus. J. Biotechnol. 49, 211–218. https://doi.org/10.1016/0168-1656(96)01516-7.

Simandan, D., 2009. Industrialization. In: International Encyclopedia of Human Geography. Elsevier, pp. 419–425. https://doi.org/10.1016/B978-008044910-4.00178-4.

Tekin, A.R., Dalgıç, A.C., 2000. Biogas production from olive pomace. Resour. Conserv. Recycl. 30, 301–313. https://doi.org/10.1016/S0921-3449(00)00067-7.

Vélez-Lee, A.E., Cordova-Lozano, F., Bandala, E.R., Sanchez-Salas, J.L., 2016. Cloning and expression of vgb gene in Bacillus cereus, improve phenol and p-nitrophenol biodegradation. Phys. Chem. Earth Parts ABC 91, 38–45. https://doi.org/10.1016/j.pce.2015.10.017.

Vignesh, K.S., Rajadesingu, S., Arunachalam, K.D., 2021. Challenges, issues, and problems with zero-waste tools. In: Concepts of Advanced Zero Waste Tools. Elsevier, pp. 69–90. https://doi.org/10.1016/B978-0-12-822183-9.00004-0.

Walfridsson, M., Bao, X., Anderlund, M., Lilius, G., Bülow, L., Hahn-Hägerdal, B., 1996. Ethanolic fermentation of xylose with Saccharomyces cerevisiae harboring the Thermus thermophilus xylA gene, which expresses an active xylose (glucose) isomerase. Appl. Environ. Microbiol. 62, 4648–4651. https://doi.org/10.1128/aem.62.12.4648-4651.1996.

Yang, L., Wang, H., Lv, Y., Bai, Y., Luo, H., Shi, P., Huang, H., Yao, B., 2016. Construction of a rapid feather-degrading bacterium by overexpression of a highly efficient alkaline keratinase in its parent strain bacillus amyloliquefaciens K11. J. Agric. Food Chem. 64, 78–84. https://doi.org/10.1021/acs.jafc.5b04747.

Yong, B., Fei, X., Shao, H., Xu, P., Hu, Y., Ni, W., Xiao, Q., Tao, X., He, X., Feng, H., 2020. Recombinant expression and biochemical characterization of a novel keratinase BsKER71 from feather degrading bacterium Bacillus subtilis S1–4. AMB Express 10, 9. https://doi.org/10.1186/s13568-019-0939-6.

Zhou, C., Bai, J., Deng, S., Wang, J., Zhu, J., Wu, M., Wang, W., 2008. Cloning of a xylanase gene from Aspergillus usamii and its expression in Escherichia coli. Bioresour. Technol. 99, 831–838. https://doi.org/10.1016/j.biortech.2007.01.035.

Zhou, J., Wu, K., Rao, C.V., 2016. Evolutionary engineering of Geobacillus thermoglucosidasius for improved ethanol production. Biotechnol. Bioeng. 113, 2156–2167. https://doi.org/10.1002/bit.25983.

Index

Note: Page numbers in *italics* indicate a figure or photo and page numbers in **bold** indicate a table on the corresponding page.

For Product Safety Concerns and Information please contact our EU
representative GPSR@taylorandfrancis.com
Taylor & Francis Verlag GmbH, Kaufingerstraße 24, 80331 München, Germany